Praise for

CHERNOBYL

"A masterful account of how the USSR's bureaucratic dysfunction, censorship, and impossible economic targets produced the disaster and hindered the response." —*New York Review of Books*

"Gripping, meticulously researched.... [Mr. Plokhy] mercilessly chronicles the absurdities of the Soviet system and the arrogance of its apparatchiks. But the fact that he grew up fewer than 500 kilometers south of Chernobyl probably accounts for his vividly empathetic descriptions of the people on the ground—the plant managers and employees, the firefighters, soldiers, and others—who risked their lives to contain the damage." —*Wall Street Journal*

"The bare outline of the Chernobyl fire and the Soviet silence have been well covered.... Mr. Plokhy, who directs the Ukrainian Research Institute at Harvard, adds much detail to the... construction that caused the failure, and the false assignment of blame to operating engineers.... [His] most telling disclosures deal with how the Soviet subterfuges played a major role in Ukraine's decision to become an independent nation once the Soviet Union disintegrated." —*Washington Times*

"A lucid account of how the Soviet mania for nuclear power combined with endemic shoddiness in the industrial sector and near-paranoid habits of state secrecy led to the 1986 disaster.... The most comprehensive and convincing history of Chernobyl yet to appear in English." —*Financial Times*

"The first comprehensive history of the Chernobyl disaster.... Here at last is the monumental history the disaster deserves." —Julie McDowall, *Times*

"A work of deep scholarship and powerful storytelling. Plokhy is the master of the telling detail." —Victory Sebestyen, *Sunday Times*

"Compelling.... Plokhy's well-paced narrative plunges the reader into the sweaty, nervous tension of the Chernobyl control room on the fateful night when human frailty and design flaws combined to such devastating effect." —*Guardian*

"Plokhy's book...sustains a tone of thoughtful observation that is neither too detached nor heavily invested in a particular agenda.... [He] delves deeper into the political fallout of Chernobyl, which played a significant role in the break-up of the Soviet Union." —*New Statesman*

"Haunting.... [Plokhy's] voice is humane and inflected with nostalgia. His Chernobyl and Prypiat emerge vividly—as perhaps all disaster-afflicted cities must—as shattered idylls." —*Spectator*

"Plokhy recounts the circumstances of the accident and its aftermath in painstaking detail.... He tells the story with great assurance and style.... A fierce and at times personal indictment of the ideology, bureaucracy, and overconfidence of the Soviet system, as well as a strident condemnation of all modern states that continue to pursue military or economic objectives to the detriment of their populations and the environment." —*Literary Review*

"A history of the nuclear disaster that set precedents—and standards—for future mishaps of the kind.... [Plokhy] concludes that even in the wake of Chernobyl, we have not gotten much better at containing meltdowns.... A thoughtful study of catastrophe, unintended consequences, and, likely, nuclear calamities to come." —*Kirkus Reviews*

"The most comprehensive exploration of the events that led to the Chernobyl disaster.... Engrossing." —*Library Journal*

"Plokhy . . . is a brilliant interpreter not only of the events themselves but of their long-term historical significance. . . . As moving as it is painstakingly researched, this book is a tour de force and a cracking read." —*Observer*

"Historian Serhii Plokhy's deft, richly detailed account draws on newly opened archives and weaves in stories of players such as Chernobyl director Viktor Briukhanov." —*Nature*

"[Plokhy] casts his lyrical eye on a vast amount of detail, giving readers a sense of dramatic urgency that makes his account difficult to put down. . . . The further Chernobyl recedes in time, Plokhy writes, the more it fades into myth. His book, however, should help bring us back to reality." —Kristen Iversen, *American Scholar*

"Serhii Plokhy provides the definitive story of the Chernobyl crisis and its aftermath, skillfully covering all angles from the scientific story, the humanitarian and economic costs of the clean-up, the manner in which the explosion forced Gorbachev to jump-start his perestroika reforms, and the igniting of Ukrainian nationalism." —Andrew Wilson, Professor of Ukrainian Studies, UCL

"Serhii Plokhy has produced a highly readable account of the Chernobyl disaster and its political impact. It is destined to be the authoritative account for years to come." —John Herbst, director, Eurasia Center, Atlantic Council

"Serhii Plokhy is uniquely qualified to tell this tragic story: he writes not only as a major historian, but also as someone who was living with his family under the cloud of the Chernobyl disaster at the time. The result is as riveting as a novel." —Mary Elise Sarotte, author of *The Collapse: The Accidental Opening of the Berlin Wall*

"*Chernobyl: The History of a Nuclear Catastrophe* lays out in devastating detail how the Soviets were vastly unprepared, in ways small and large, for what became the worst disaster in the history of nuclear energy....A riveting account....Is it possible that the world might someday forget the horrors that unfolded there three decades ago? Books like Plokhy's should help ensure that that doesn't happen."

—Henry Fountain, *Undark*

"An insightful and important book that often reads like a good thriller, and that exposes the danger of mixing powerful technology with irresponsible politics."

—Yuval Noah Harari, author of *Sapiens* and *21 Lessons for the 21st Century*

CHERNOBYL

THE HISTORY OF A NUCLEAR
CATASTROPHE

SERHII PLOKHY

BASIC BOOKS
New York

To the Children of the Nuclear Age

Basic Books
Hachette Book Group
1290 Avenue of the Americas, New York, NY 10104
www.basicbooks.com

Printed in the United States of America

First Edition: May 2018

Published by Basic Books, an imprint of Perseus Books, LLC, a subsidiary of Hachette Book Group, Inc. The Basic Books name and logo is a trademark of the Hachette Book Group.

The Hachette Speakers Bureau provides a wide range of authors for speaking events. To find out more, go to www.hachettespeakersbureau.com or call (866) 376-6591.

The publisher is not responsible for websites (or their content) that are not owned by the publisher.

Print book interior design by Amy Quinn

The Library of Congress has cataloged the hardcover as follows:

Names: Plokhy, Serhii, 1957- author.
Title: Chernobyl : the history of a nuclear catastrophe / Serhii Plokhy.
Description: First edition. | New York : Basic Books, [2018] | Includes bibliographical references and index.
Identifiers: LCCN 2017056459 (print) | LCCN 2017059722 (ebook) | ISBN 9781541617087 (ebook) | ISBN 9781541617094 (hardcover)
Subjects: LCSH: Chernobyl Nuclear Accident, Chornobyl, Ukraine, 1986. | Nuclear power plants—Accidents—Ukraine—Chornobyl. | Nuclear energy—Political aspects.
Classification: LCC TK1362.U38 (ebook) | LCC TK1362.U38 P555 2018 (print) | DDC 363.17/99094777—dc23
LC record available at https://lccn.loc.gov/2017056459

ISBNs: 978-1-5416-1709-4 (hardcover), 978-1-5416-1708-7 (ebook),
        978-1-5416-1707-0 (paperback)

LSC-H

Printing 4, 2022

CONTENTS

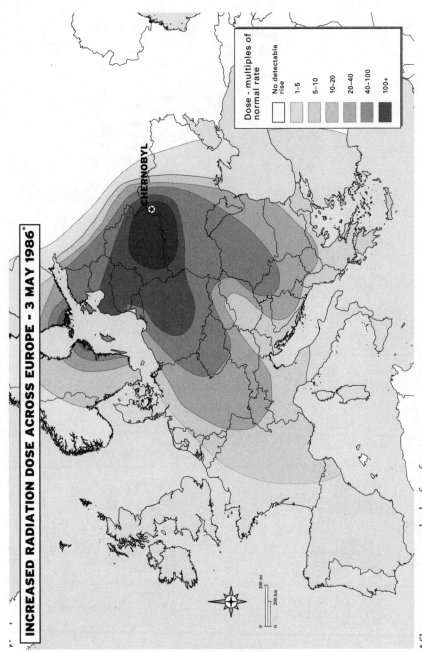

INCREASED RADIATION DOSE ACROSS EUROPE – 3 MAY 1986*

CHERNOBYL

Dose - multiples of normal rate

- No detectable rise
- 1–5
- 5–10
- 10–20
- 20–40
- 40–100
- 100+

200 mi
200 km
0

* Shown over contemporary borders for reference.

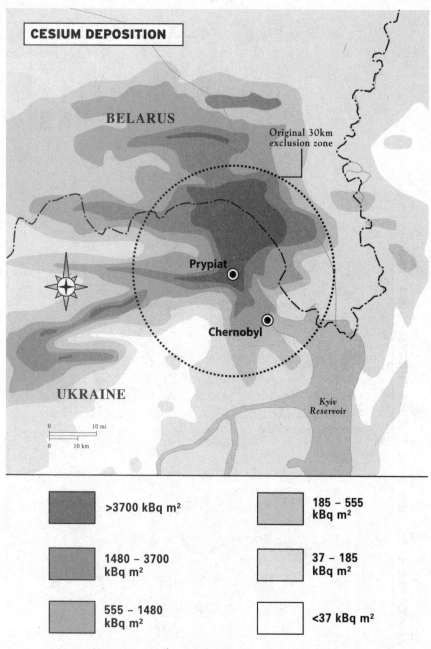

CESIUM DEPOSITION

BELARUS

Original 30km
exclusion zone

Prypiat

Chernobyl

UKRAINE

Kyiv
Reservoir

0   10 mi
0   10 km

>3700 kBq m²

1480 – 3700
kBq m²

555 – 1480
kBq m²

185 – 555
kBq m²

37 – 185
kBq m²

<37 kBq m²

1 microcurie (µCi) = 37 kilobecquerel (kBq)

PREFACE

THERE ARE eight of us on the trip to Chernobyl, marked on my Ukrainian map as "Chornobyl." Besides me, there are three science and engineering students from Hong Kong who are on a tour of Russia and Eastern Europe. Then, as far as I can tell from their accent, there are four Brits—three men and one woman, all in their twenties. I soon learn that the men are indeed British, while the woman, whose name is Amanda, is proudly Irish. They are getting along quite well.

A few weeks earlier, when Amanda asked her British husband, Stuart, what he wanted to do on their forthcoming vacation, he said he wanted to go to Chernobyl. So they came, accompanied by Stuart's brother and a family friend. Two computer games had provided the inspiration for the trip. In *S.T.A.L.K.E.R.: Shadow of Chernobyl*, a shooter-survival horror game, the action takes place in the Chernobyl Exclusion Zone after a fictional second nuclear explosion. In *Call of Duty: Modern Warfare*, the main action figure, Captain Jon Price, goes to the abandoned city of Prypiat to hunt down the leader of the Russian ultranationalists. Stuart and his team decided to see the place for themselves.

Vita, our animated young Ukrainian guide, first takes us to the 30-kilometer exclusion zone and then to the more restricted 10-kilometer one—two circles, one inside the other, with the former nuclear power plant at their center and a radius of 30 kilometers (18.6

miles) and 10 kilometers (6.2 miles), respectively. We get to see the Soviet radar called Duga, or Arch—a response to Ronald Reagan's "Star Wars" Strategic Defense Initiative—by today's standards a low-tech system. It was designed to detect a possible nuclear attack from the East Coast of the United States. From there we proceeded to the city of Chernobyl, its nuclear power station, and the neighboring city of Prypiat, a ghost town that once housed close to 50,000 construction workers and operators of the destroyed plant. Vita gives us radiation counters that beep when levels exceed the established norm. In some areas, including those close to the damaged reactor, they beep nonstop. Vita then takes the dosimeters away and shuts them off, just as Soviet workers sent to deal with the consequences of the disaster did back in 1986. They had to do their work, and the dosimeters showed unacceptable levels of radiation. Vita has her own job to do. She tells us that in our whole day in the zone, we will get the same amount of radiation as an airplane passenger absorbs in an hour. We trust her assurance that the radiation levels are not too crazy.

At least 50 million curies of radiation were released by the Chernobyl explosion, the equivalent of hundreds of Hiroshima bombs. All that was required for such catastrophic fallout was the escape of less than half of the reactor's nuclear fuel. Originally it had contained close to 400,000 pounds of enriched uranium—enough to pollute and devastate a good part of Europe. And if the other three reactors of the Chernobyl power plant had been damaged by the explosion of the first, then hardly any living and breathing organisms would be left unaffected on the planet. For weeks after the accident, scientists and engineers did not know whether the explosion of the radioactive Chernobyl volcano would be followed by even deadlier ones. It was not, but the damage done by the first explosion will last for centuries. The half-life of the plutonium-239 that was released by the Chernobyl explosion—and carried by winds all the way to Sweden—is 24,000 years.

Prypiat is sometimes referred to as the modern-day Pompeii. There are clear parallels between the two sites, but there are differences as well, if only because the Ukrainian city, its walls, ceilings, and even the occasional windowpane, are still basically intact. It was not the heat or magma of a volcano that claimed and stopped life there, but invisible particles of radiation, which drove out the inhabitants but spared most of the vegetation, allowing wild animals to come back and claim the space once built and inhabited by humans. There are numerous marks of the long-gone communist past on the streets of the city. Communist-era slogans are still there, and inside the abandoned movie theater, a portrait of a communist leader. Vita, our guide, says that no one can now tell who is depicted there, but I recognize a familiar face from my days as a young university professor in Ukraine at the time of the catastrophe—the painting is of Viktor Chebrikov, the head of the KGB from 1982 to 1988. It has miraculously survived the past thirty years, undamaged except for a tiny hole near Chebrikov's nose. Otherwise, the image is perfectly fine. We move on.

It is strange, I think to myself, that Vita, an excellent tour guide, cannot identify Chebrikov. She also seems at a loss to explain the signs saying "meat," "milk," and "cheese" hanging from what was once the ceiling of an abandoned Soviet-era supermarket. "How come," she asks, "they write that in the Soviet Union there were shortages of almost everything?" I explain that Prypiat was in many ways a privileged place because of the nuclear power plant, and that the workers were better supplied with agricultural produce and consumer goods than the general population. Besides, the fact that there were signs saying "meat" or "cheese" did not mean that those products were actually available. It was the Soviet Union, after all, where the gap between the image projected by government propaganda and reality was bridged only by jokes. I retell one of them: "If you want to fill your fridge with food, plug the fridge into the radio outlet." The radio was telling the story of ever-improving living standards; the empty fridge had its own story to tell.

It was on my trip to Prypiat that I decided I had to tell the saga of Chernobyl: it would be for the sake of those who were not around at the time but who wanted to know and understand what had happened on that fateful night of April 26, 1986, and in the days, months, and years that followed. Despite the Soviet government's initial efforts to conceal the Chernobyl disaster and downplay its consequences, it became well known in the Soviet Union and in the

West and received a great deal of public attention, starting with journalistic reports filed in the first days after the explosion and ending with documentary films, feature movies, nonfiction investigations, and novels. Although the key to understanding the causes, consequences, and lessons of the disaster is historical contextualization and interpretation, few historians have addressed the subject to date.

This book is a work of history—in fact, it is the first comprehensive history of the Chernobyl disaster from the explosion of the nuclear reactor to the closing of the plant in December 2000 and the final stages in the completion of the new shelter over the damaged reactor in May 2018. As I embarked on my research of the history of Chernobyl, I was helped enormously by the recent opening of previously closed archival collections dealing with the disaster. Many government archives opened their doors more widely than before, making it easier to consult documents issued by the Communist Party and government agencies at the time and in the aftermath of the disaster. The Maidan uprising and the Revolution of Dignity of 2014 in Ukraine also produced an archival revolution that allowed unprecedented access to previously closed KGB files.

I was writing this book both as a historian and as a contemporary of the events being discussed. At the time of the explosion I lived in Ukraine less than 500 kilometers downstream Dnieper of the damaged reactor. My family and I were not directly affected by the ordeal. But a few years later, doctors in Canada, where I was a visiting professor at the time, told me that at some point my thyroid had been inflamed—a worrisome sign of radiation exposure. Fortunately, my wife and children were fine. But radiation acts in unpredictable ways: One of my former university classmates was sent to Chernobyl as a policeman a few days after the accident; he still spends at least a month in the hospital every year. Another university colleague who spent time near the station after the explosion seems to be fine—he now teaches Soviet history in the United States. Talking with them and with other participants in the events and recollecting my own memories of the disaster helped me re-create the thoughts and motives of those who had sacrificed their health, or even their lives, to minimize the consequences of the Chernobyl meltdown.

The further we move in time from the disaster, the more it seems like a myth—and the more difficult it becomes to grasp its real-life roots and consequences. By putting the disaster in historical context, I attempt to provide better understanding of the world's worst nuclear accident. My use of newly available archival materials and recently published government documents, as well as interviews with eyewitnesses and accounts of other writers, such as Svetlana Alexievich and Yurii Shcherbak, has allowed me to present a long-term perspective on the disaster and its political, social, and cultural effects. In my narrative, I move from the control room of the damaged reactor to the abandoned villages of the exclusion zone and to the offices of those in power in Kyiv (Kiev), Moscow, and Washington. Placing the Chernobyl accident in the context of international history makes it possible to draw lessons of global significance.

Chernobyl as history is the story of a technological disaster that helped bring down not only the Soviet nuclear industry but the Soviet system as a whole. The accident marked the beginning of the end of the Soviet Union: a little more than five years later, the world superpower would fall apart, doomed not only by the albatross of its communist ideology but also by its dysfunctional managerial and economic systems.

The explosion at the Chernobyl nuclear plant had challenged and changed the old Soviet order. The policy of *glasnost*, or openness, which gave the media and citizens the right to discuss political and social problems and criticize the authorities, had its origins in the post-Chernobyl days. As the population demanded more and more information from the government, the official culture of secrecy slowly yielded. The Chernobyl disaster made the government recognize ecological concerns as a legitimate reason for Soviet citizens to create their own organizations, which broke the monopoly of the Communist Party on political activity. The first Soviet-era mass organizations and political parties began in the ecological movement, which engulfed the heavily polluted industrial centers of the Soviet Union.

Because radiation affected everyone, from party leaders to ordinary citizens, the Chernobyl accident also sharply increased discontent with Moscow and its policies across ethnic and social lines.

Nowhere was the political impact more profound than in Ukraine, the republic that was home to the failed reactor. Two conflicting political actors in Ukraine—the Ukrainian communist establishment and the nascent democratic opposition—discovered a common interest in opposing Moscow, and especially Soviet leader Mikhail Gorbachev. In December 1991, when Ukrainians voted for their country's independence, they also consigned the mighty Soviet Union to the dustbin of history—it was officially dissolved a few weeks after the Ukrainian referendum. While it would be wrong to attribute the development of glasnost in the Soviet Union, or the rise of the national movement in Ukraine and other republics, to the Chernobyl accident alone, the disaster's impact on those interrelated processes can hardly be overstated.

It would be easy to blame the Chernobyl accident on the failed communist system and the design flaws of Chernobyl-type reactors, implying that those problems belong to the past. But this confidence would be misplaced. The causes of the Chernobyl meltdown are very much in evidence today. Authoritarian rulers pursuing enhanced or great-power status—and eager to accelerate economic development and overcome energy and demographic crises, while paying lip service to ecological concerns—are more in evidence now than they were in 1986. Could the nuclear Armageddon called Chernobyl repeat itself? No one knows the answer to this question. But there is no doubt that a new Chernobyl-type disaster is more likely to happen if we do not learn the lessons of the one that has already occurred.

PROLOGUE

Around 7:00 a.m. on April 28, 1986, Cliff Robinson, a twenty-nine-year-old chemist working at the Forsmark Nuclear Power Plant two hours' drive from Stockholm, went to brush his teeth after breakfast. In order to get from the washroom to the locker room, he had to pass through a radiation detector, just as he had done thousands of times before. This time was different, though—the alarm went on. It made no sense, thought Robinson, as he had never even entered the control area, where he might have absorbed some radiation. He went through the detector a second time, and again it went on. Only on the third try did the alarm fall silent. Finally, an explanation—the damned thing had simply malfunctioned.

Robinson's job at the plant was monitoring radiation levels: how ironic, he thought, that the detector had chosen him to show how vigilant the system was. Good thing it had come back to its senses. Robinson went on with his duties, all but forgetting the unexpected alarm. But when he returned to the area later that morning, he saw a line of workers who also could not pass the detector without setting it off. Instead of checking the alarm, Robinson took a shoe from one of those waiting near the detector and took it to the lab for examination. What he discovered sent shivers up his spine. "I saw a sight that I will never forget," he recalled. "The shoe was highly contaminated. I could see the spectrum rising very quickly."

1

Robinson's first thought was that someone had detonated a nuclear bomb: the shoe emanated radioactive elements that they did not normally detect at the plant. He reported the findings to his boss, and from there they were passed on to the Swedish Radiation Safety Authority in Stockholm. The authorities in the capital thought the problem was probably in the power plant itself, and the Forsmark workers were promptly evacuated. Radioactive testing of the plant began, but turned up nothing, and after a few hours it was clear that the plant was not the cause of the contamination. The bomb hypothesis was ruled out as well—the radioactive elements did not fit a bomb profile. With radioactivity levels also high at other nuclear power stations, it was apparent that the radioactive particles were coming from abroad.

The calculations and wind direction pointed southeast to one of the world's two nuclear superpowers, the Union of Soviet Socialist Republics. Could something terrible have happened there? But the Soviets were silent. The Swedish Radiation Safety Authority contacted Soviet officials, who denied that anything taking place on their territory might have caused nuclear contamination. But safety services in the Scandinavian countries continued to register abnormally high levels of radiation: in Sweden, the level of gamma radiation was 30 to 40 percent higher than normal; in Norway it had doubled; and in Finland it was six times the norm.

Two radioactive gases, xenon and krypton, byproducts of the nuclear fission of uranium, were moving across Scandinavia, a region covering not only Finland, Sweden, and Norway, but also Denmark. Tests indicated that the source of the radioactive pollution, wherever it might be, was continuing to emit dangerous substances. The Swedes repeatedly called three Soviet agencies in charge of nuclear power management and generation, but they denied knowledge of any accident or explosion. The Swedish minister for the environment, Birgitta Dahl, declared that the country responsible for the spread of radioactivity was violating international agreements by withholding vital information from the world community. There was no response. Swedish diplomats reached out to their former foreign minister, Hans Blix, now in Vienna heading the International Atomic Energy Agency. The agency was also in the dark.

It was not clear what to expect. Although radiation levels were high, they did not yet pose a direct threat to human life and vegetation. But what if the contamination continued or even increased? And what had happened there, behind the Iron Curtain along the Soviet border? Was it the start of a new world war or a nuclear accident of enormous proportions? One way or another, the world would be involved. It was involved already. But the Soviets remained silent.[1]

I

WORMWOOD

CONGRESS

IT WAS a big day—many in Moscow and throughout the Soviet Union believed that it signaled the dawn of a new era. On the cold winter morning of February 25, 1986—the temperature during the previous night had fallen to minus two degrees Fahrenheit—close to 5,000 warmly dressed men and women, including senior Communist Party and state officials, military officers, scientists, directors of the large state companies, and representatives of workers and collective farmers (the "toiling masses"), descended on Red Square in downtown Moscow, which was decorated with a huge portrait of Vladimir Lenin. They were delegates to the Communist Party Congress, the twenty-seventh since the founding of the party by a handful of idealistic social democrats in the late nineteenth century. Their mission was to chart a new course for the country for the next five years.[1]

Once they reached the Kremlin, the crowds moved toward the Palace of Congresses, a modern glass-and-concrete building decorated with white marble plates. It had been erected in 1961 on the site of buildings belonging to the sixteenth-century tsar Boris Godunov. The Soviet premier at the time, Nikita Khrushchev, wanted to rival the Great Hall of the People that Mao Zedong had opened in Beijing in 1959. The Chinese palace could seat 10,000 people. The envious Soviets increased the seating capacity of their palace from 4,000 to 6,000 by putting almost half the building underground, where most of the seats of the meeting hall are located—only the balcony seats with boxes are above ground level. When it came to party congresses, which convened every five years, the Soviet leaders imposed a limit of 5,000 participants no matter how large the membership of the Communist Party became—and it was growing quickly—since filling the hall to capacity would have meant sacrificing the comfort of those in attendance. There was no venue in the Soviet Union, short of sports arenas, that could have seated more.[2]

Khrushchev inaugurated the new Palace of Congresses in October 1961, in time for the Twenty-second Party Congress. The congress decided to remove the corpse of Joseph Stalin from the mausoleum it then shared with that of Lenin, and it adopted a new program for building a communist society, with its foundations to be in place by the early 1980s. Now, in 1986, the delegates to the Twenty-seventh Congress had to take stock of what had been accomplished. The record was dismal, to say the least. As the population had increased, the economy had slowed, and the possibility of a complete breakdown was becoming ever more likely. The growth of national income, which Soviet economists had estimated at 10 percent in the 1950s, had fallen to barely 4 percent in 1985. The Central Intelligence Agency in the United States had made even grimmer estimates, putting the growth rate at 2 to 3 percent, and later reducing even that estimate to approximately 1 percent.[3]

With its goals for communism nowhere in sight, the economy in a tailspin, the Chinese launching their economic reforms by introducing market mechanisms, and the Americans rushing ahead not only in economic development but also in the arms race, under the leadership of the unfailingly optimistic Ronald Reagan, the Soviet leadership had lost its way. The people, ever more disillusioned with the communist experiment, had become despondent. And yet, with the communist religion in crisis, it suddenly appeared to have found a new messiah in a relatively young, energetic, and charismatic leader: Mikhail Gorbachev.

This was to be the fifty-four-year-old Gorbachev's first congress as general secretary of the party, and he was well aware that the eyes of the party leadership, of Soviet citizens—and indeed, of the entire world—were on him. The previous three years had become known as the era of Kremlin funerals. Leonid Brezhnev, who had ruled the Soviet Union since 1964, died a sick man in November 1982; the former head of the KGB, Yurii Andropov, who had inherited his position, spent half his brief tenure in a hospital bed and died in February 1984; his sickly successor, Konstantin Chernenko, followed suit in March 1985. It looked as if the leaders were about to take the country to the grave with them. Economic difficulties aside, they kept sending young boys to Afghanistan, where the Soviet Army had been bogged down since 1979, and preparing for nuclear confrontation with the West. KGB stations abroad were instructed to drop everything and look for signs of imminent nuclear attack.

Now hopes ran high, in both party and society, that Gorbachev, who was full of ideas, would be able to reverse the deadly trend. Hopes of rapprochement were rising in the West as well. In the United States, Reagan, tired of Soviet leaders dying on him, was looking for someone with whom he could do business. His close ally Prime Minister Margaret Thatcher of Great Britain told him that Gorbachev was such a man. Reagan's first meeting with Gorbachev, in Geneva in December 1985, was not without tension, but it opened the door to more productive subsequent dialogue, which was conducted not only by personal meetings and diplomatic channels but also by public pronouncements. In January 1986, Gorbachev surprised Reagan by putting forward a Soviet program for nuclear disarmament. It was expected that he would further challenge the American president on disarmament in his forthcoming speech to the party congress.[4]

Gorbachev, preoccupied with finding solutions to the multiple Soviet crises, put considerable thought and effort into his report to the congress. In the late fall of 1985, he summoned his two closest advisers—his chief assistant, Valerii Boldin, and Aleksandr Yakovlev, the former Soviet ambassador to Canada—to the state resort near Sochi on the Black Sea Coast. *Perestroika*—the radical restructuring of the Soviet political and economic system—still lay ahead; eventually, Yakovlev would become known as the grandfather of the movement. The key concept at the time was *uskorenie*, or acceleration. It was believed that the system was basically sound and simply needed a boost by means of "scientific and technical progress," the Soviet term for technological innovation.

In the days leading up to the congress, Gorbachev shut himself up at home, reading his long speech aloud and timing it. Read without a break or interruptions, it would be more than six hours in length. As Gorbachev practiced his oratorical skills, the delegates to the congress kept themselves busy visiting the stores of Moscow rather than galleries and museums. "Having come from all over the country, they were preoccupied with their own affairs," wrote Gorbachev's aide, Boldin, who had coauthored the speech. "They had to buy many things for themselves, their family members, and acquaintances, who had ordered so much that it would be hard to transport even by train."[5]

Most of the delegates came from the provinces, which were dogged by the shortages of agricultural products and consumer goods that had become a constant feature of Soviet life in the 1980s. The party leadership, unable to alleviate the shortages for the general population, did its best to supply the party elite. In hotels designated for congress delegates, party officials opened special branches of grocery and department stores, to which hard-to-get products were brought from all parts of the Soviet Union. There were stylish suits and dresses, shoes, caviar, cured meat, sausages, and, last but not least, bananas—all items desired by average Soviet citizens not only in the provinces but also in the much better supplied metropolitan centers such as Moscow, Leningrad, and Kyiv. The post office administration opened a special branch to handle all the merchandise that the delegates shipped back from Moscow.

For high-ranking party officials from the provinces and directors of large enterprises who had access to scarce goods at home because of their political power and connections, participation in the congress offered a different kind of opportunity. They used the time to lobby Moscow's potentates and ministers, asking for money and resources for their regions and firms. They also worked hard to maintain old networks of friends and acquaintances and make useful new connections. Networking meant drinking, often to excess—a hallmark and curse of the Soviet managerial style. The previous year, Gorbachev, alarmed by the level of alcoholism among the general population, had launched an anti-alcohol campaign. Party and state officials, in particular, were liable to prosecution for drunkenness.

Vitalii Vrublevsky, a close aide to the all-powerful party boss of Ukraine, Volodymyr Shcherbytsky, head of the Ukrainian delegation, recalled an episode in which KGB guards charged with checking passes to the congress smelled alcohol on one of the delegates and reported him to senior officials. The case, which involved a regional secretary in Ukraine's mining area of Luhansk, was reported all the way to the top of the party apparatus. "The secretary was expelled from the party on the spot," recalled Vrublevsky, who had barely avoided detection himself after spending a night drinking with some of the first cosmonauts—the Soviet equivalent of rock stars. "Volodymyr Shcherbytsky, sitting at the head table, kept glancing at his delegation," recalled Vrublevsky. "And, as ill luck would have it, my head kept drooping." He was saved by a friend who would squeeze his knee from time to time to wake him up in the middle of the speeches.[6]

VIKTOR BRIUKHANOV, the fifty-year-old director of the Chernobyl Nuclear Power Station in Ukraine, was a member of the 1986 Ukrainian delegation. It was the first congress that Briukhanov was attending after many years as a loyal party member and high-ranking manager. Three-quarters of the other delegates were also there for the first time, but managers such as Briukhanov accounted for slightly more than 350 of the party delegates, roughly 7 percent of the total. Below average in height, slim and erect, with curly black hair that he combed back and a somewhat awkward smile, Briukhanov made the impression of a kind and fair man. His subordinates valued him as a good engineer and effective manager. He was hardly a drinker. If anything, Briukhanov was a workaholic. He put in long hours, spoke little, and was known as one of a rare breed: a Soviet manager who got things done while showing consideration to his subordinates.[7]

The privilege of becoming a delegate was recognition for the work Briukhanov had done at the helm of the third most powerful nuclear power station in the world. He had built it from scratch, and now it had four nuclear reactors running, each producing 1,000 megawatts of electrical power (MWe). Two more reactors were under construction, and the station had overfulfilled its planned targets for 1985, producing 29 billion KWh. Briukhanov had received two high Soviet awards for his work, and many believed that he was poised to receive an even higher distinction, the Order of Lenin, as well as the gold star of a Hero of Socialist Labor. In late November 1985, the Ukrainian Supreme Soviet in Kyiv had marked his fiftieth birthday with a commendation. His selection as a delegate, with the corresponding lapel pin, was a distinction in its own right, equal if not superior to most government awards.

When on the eve of Briukhanov's birthday a reporter came from Kyiv to Prypiat, where the Chernobyl plant was located, to interview him about his accomplishments and plans for the future, Briukhanov, usually a man of few words, suddenly opened up to the visitor. He recalled a cold winter day in 1970 when he had come to Chernobyl and rented a room in the local hotel. Only thirty-five years old at the time, he had been appointed director of a power plant that was yet to be built. "To be frank, it was scary at first," Briukhanov told the reporter. That was then. Now Briukhanov was running an enterprise with thousands of highly qualified managers, engineers, and workers. He also bore de facto responsibility for running the company town of Prypiat, which housed close to 50,000 construction workers and plant personnel. He even complained to the reporter about the need to divert people and resources from the nuclear station to ensure the smooth running of the city's infrastructure. But there were also payoffs from the "father of the city" status that had been thrust upon him. Before and during the congress, photographs and profiles of Briukhanov were published in local and regional newspapers, including the one in Chernobyl.[8]

Photos of the Kyiv regional delegation taken in Red Square during the congress and then upon the group's return to Ukraine show Briukhanov dressed in a fancy muskrat fur hat and short sheepskin coat with a mohair scarf around his neck—all expensive and hard to get in the Soviet Union at the time, tokens of the prestige and power of their owner. Briukhanov did not need the shops set up for rank-and-file congress delegates, but time in Moscow gave him the opportunity to meet with colleagues in the industry and lobby the party's Central Committee and the Ministry of Energy and Electrification, which supervised his plant. The task was relatively easy, given that many officials in both places had once worked at the Chernobyl plant that Briukhanov ran.[9]

ON THE morning of February 25, 1986, Viktor Briukhanov and his fellow deputies took their seats in the Kremlin Palace of Congresses in the center of the hall before the podium. For those like Briukhanov who were attending their first party congress, the ritual opening presented an interesting spectacle whose main features went back to Stalin's times.

At ten in the morning, the party's Politburo members, led by Mikhail Gorbachev, marched to the podium. Like most people, Briukhanov knew them from their portraits, which were displayed on public buildings all over the Soviet Union. Among them was the head of the KGB, Viktor Chebrikov, whose portrait would survive for decades in the Prypiat palace of culture. Like everyone else, Briukhanov rose to his feet to welcome the leaders with applause. Once it subsided, Gorbachev made his way to the podium. "Comrade delegates," declared the general secretary, his voice betraying his excitement. "At Communist Party congresses of Union republics and territorial and oblast party conferences, 5,000 delegates were chosen to attend the 27th congress of the Communist Party of the Soviet Union [CPSU]. There are 4,993 delegates attending the congress. Seven people are absent for valid reasons. This gives us a basis to commence the work of the congress." There were no objections. The congress began its proceedings.[10]

Among the first items on the agenda was paying tribute to those who had passed away since the previous congress in 1981—six elderly members of the Politburo, including three general secretaries: Brezhnev, Andropov, and Chernenko. With the deceased duly honored, the way was clear for a new start—Gorbachev's political report to the congress. With lunch and coffee breaks, it lasted the rest of the day. A team of professional announcers took six hours to read it on Soviet radio afterward. Gorbachev almost equaled the new record for communist speeches established earlier that month by Fidel Castro, who had reported to the congress of the Communist

Party of Cuba for seven hours and ten minutes. Now seated in the guest row behind the general secretary, Castro listened carefully to the translation of Gorbachev's speech. It turned out to be the most critical address delivered by a Soviet leader since the end of the Stalin era.[11]

"For a number of years, not only because of objective factors but above all for subjective reasons, the activity of party and state bodies has fallen behind the demands of the time, of life itself," declared the general secretary. "Problems in the development of the country accumulated more quickly than they were solved. Inertia and stagnation in forms and methods of administration, decreasing dynamism in work, the growth of bureaucratism—all this made for considerable detriment to the cause. Elements of stagnation became apparent in the life of society." Such a critique of Soviet reality and the senior leadership had not been heard since Nikita Khrushchev's "secret" speech to the Twentieth Party Congress in February 1956. Gorbachev later noted that the 1956 congress and the current one had begun on the same calendar day, February 25. The word that he used for "stagnation," *zastoi*, would become the accepted term for the decline in Soviet economic development of the late 1970s and early 1980s.

Gorbachev wanted the party "to overcome negative aspects of socioeconomic development as quickly as possible, to impart the necessary dynamism and acceleration to that process, to learn the lessons of the past to the maximum." He set ambitious tasks for the Soviet economy and society—in fifteen years, before the end of the millennium, gross domestic product (GDP) was to be doubled by dramatically increasing the productivity of labor. He staked the achievement of that task on the scientific and technological revolution, including the introduction of new technologies, and a shift from fossil fuels, especially coal, oil, and gas, toward nuclear energy. "In the current five-year plan," declared Gorbachev, "atomic energy stations two and a half times more powerful than those in the previous plan will come online, and obsolete units at thermoelectric power stations will be replaced en masse."

Briukhanov knew the figures—they were part of the government energy program that had been prepared and announced before the congress. But now the party had given the program a ringing public endorsement. In five years, the next party congress would review the results, and the party leadership would punish those it held responsible for failing to achieve the projected result, should that become necessary. This meant that not only would the existing four units of the Chernobyl power plant have to fulfill and overfulfill their quotas, but the new fifth and sixth units would also have to be completed and connected to the electric grid. There were also plans to build two and then an additional four reactors on the other bank of the Prypiat River. The production capacity of these new units would significantly exceed that of the old ones, producing 1.5 thousand megawatts of electrical power per unit instead of 1 thousand. After fifteen years of juggling the tasks involved in building the plant and running it at the same time, Briukhanov felt exhausted, but the party demanded more nuclear energy, and he served at the party's pleasure.

In his report, Gorbachev paid much more attention to nuclear arms than to nuclear energy. He called on his comrades to think about new approaches to arms control, pointing out that the nuclear arms already accumulated by the opposing military blocs, the North Atlantic Treaty Organization (NATO) and the Moscow-led Warsaw Pact, threatened to destroy life on earth many times over. His proposed solution was a program that would eliminate all nuclear weapons before the end of the century. Now he reported to the congress that he had received President Reagan's response to his initiative. Gorbachev considered it largely negative. Reagan supported the destruction of nuclear arms in principle but insisted on maintaining his Strategic Defense Initiative (SDI), a project dubbed "Star Wars" because of its focus on the construction of a space-based antimissile system. "The reduction of strategic nuclear arsenals is predicated on our agreement to the 'Star Wars' programs and the reduction—unilateral, by the way—of Soviet conventional armaments," Gorbachev told the deputies of Reagan's response, not without bitterness and disappointment.[12]

The Soviet leader knew that his country had neither the resources nor the technology to match SDI, which was still in the design stage but, if realized, would mean another round of the arms race that the Soviet Union could not afford. He needed the money

and technical expertise that were being used by the designers of mis-
siles and nuclear arms to modernize the lagging Soviet economy.
The scientific establishment was on his side in principle. What its
members wanted was more money and continuing reliance on domestic know-how, even if what they had to offer was more expensive and inferior to the technologies and equipment available on
Western markets. The continuing Cold War, which led the West to
impose embargoes on the sale of advanced technologies to the Soviet Union, lent weight to their argument. The government-funded
military-industrial complex was eager to move into the economic
sphere while maintaining its monopoly on high-tech industries and
products. Many, including Gorbachev, saw this as the most effective
solution to the country's economic troubles.

THE DESIRES, fears, and aspirations of the Soviet military-industrial
complex and its scientific wing were articulated to the congress
by Anatolii Aleksandrov, the president of the Soviet Academy of
Sciences. The fact that he was the first representative of the Soviet intelligentsia to address the congress underscored the symbolic importance of his position in the party hierarchy as well as
the hopes that the new leaders were now placing in the scientific
establishment.[13]

A tall man with a long face, a big nose, and a shaved, egg-shaped
head, Aleksandrov had turned eighty-three earlier that month. He
was significantly older than most members of the Politburo, and
older than all three of the general secretaries who had passed away
in the previous three and a half years. But no one dared to suggest
that Aleksandrov was not fit to do his job or that "stagnation" had
become apparent either in the Institute of Nuclear Energy, which
he headed, or the Academy of Sciences, over which he presided. He
was fit, vigorous, and full of ideas. As one of the founders of the
Soviet nuclear program, he also enjoyed considerable respect in the
party, in industry, and in scientific institutions. When it came to
scientific and technological progress, Gorbachev's "wonder weapon"
for overcoming Soviet economic backwardness, everyone looked to
Aleksandrov and his scientists to show the way. They were the ones
who were expected to deliver the miracle.[14]

Aleksandrov began with a reference to Lenin and the attention he had allegedly paid to the development of the Soviet sciences. But his main historical emphasis was on the development of the Soviet nuclear program, led by Igor Kurchatov, the founder of the institute that Aleksandrov now headed. Under Kurchatov's leadership, said Aleksandrov, "the first atomic bombs were built, and then, earlier than in the USA, hydrogen bombs as well. The security of the Soviet Union was guaranteed." Aleksandrov neglected to mention the role played in the creation of the first Soviet nuclear bomb by atomic spies who had reported to Moscow on the progress of the American Manhattan Project. He particularly stressed the peaceful uses of nuclear energy: "In 1954, soon after the creation of atomic weapons, the first atomic energy station in the world was established in the USSR. I wish to applaud its creators." The audience responded with loud applause.

Aleksandrov recalled historical milestones not only to praise his predecessor and, indirectly, his own role in the development of the Soviet nuclear program, but also to remind his listeners of dangers emanating from the West. He argued against buying technologies and equipment abroad, reasoning that contracts could be canceled at any moment for political reasons. He wanted to invest in scientific development at home. The automation of a small production unit at his institute offered an apposite example. "We announced to the whole ministry: comrades, if you order parts, please order them from us," said Aleksandrov, eliciting a new wave of applause.[15]

Gorbachev, who had previously interrupted Aleksandrov's speech with supporting and reassuring remarks of his own, now fell silent. He did not ask what ministry Aleksandrov had in mind. He knew the answer himself. Aleksandrov was referring to a top-secret ministry with the awkward name of "Ministry of Medium Machine Building." Its minister, Yefim Slavsky, was sitting in the presidium behind Aleksandrov. A giant of a man, five years older than Aleksandrov and bigger and taller than he, Slavsky was one of the most powerful ministers the Soviet government had ever had. A pioneer of the Soviet nuclear program who had begun working on it with Kurchatov in the late 1940s, Slavsky was now in his twenty-eighth year of running the Ministry of Medium Machine Building, a

huge government enterprise responsible for the production of nuclear bombs, and, later, of nuclear power for peaceful purposes. Soviet leaders came and went, but Slavsky remained. When it came
to political power and resources, he virtually owned Aleksandrov's
Institute of Nuclear Energy and, through him, the Academy of Sciences. Aleksandrov's deputies were constantly knocking on Slavsky's
door, asking him to fund their projects. He assented if he was so
inclined.[16]

Slavsky and Aleksandrov were longtime allies. Both came from
Ukraine, where Aleksandrov, the son of a prominent judge in the
Kyiv region, had fought in the ranks of the White Army against
the Bolsheviks after the 1917 revolution, while Slavsky, the son of a
Cossack, had joined the Red Cavalry. The fact that they had fought
on opposite sides did not prevent them from forging a lasting alli-
ance. The story was told that in the early 1960s, Nikita Khrushchev
had summoned Slavsky and Aleksandrov to his office and, switching
to Ukrainian, demanded that they catch up with America in the
construction of nuclear power plants. Inspiration for how to produce a new nuclear reactor presumably came from the well-known
Soviet comedian Arkadii Raikin, who, in a standup act on television, joked that it was a shame to allow a ballerina to twirl without
producing any energy for the socialist economy, so a rotor should be
attached to her body. Slavsky and Aleksandrov, allegedly after seeing
the skit, decided to take a nuclear reactor designed for the production of weapons-grade plutonium and attach an enormous turbine
and rotor to it in order to use the reactor's excess heat to produce
electricity.[17]

Whatever the actual source of their inspiration, a new reactor,
called the RBMK (High Power Channel Reactor), was born out of
cooperation between Slavsky's ministry and Aleksandrov's institute.
Its chief designer was Nikolai Dollezhal, another native of Ukraine
who had made it big in the Soviet nuclear industry and served as
the director of the Research and Development Institute of Power
Engineering. He had designed the reactor that had produced plutonium for the first Soviet nuclear bomb and had then worked on
the reactors powering Soviet submarines. Aleksandrov, who also
worked on nuclear submarines, served as chief scientific consultant

for the RBMK design. The first RBMK units were tested and run by Slavsky's ministry. Aleksandrov kept telling everyone who would listen that his reactors were safe and sound. They were like samovars, he said, and could not possibly explode. Rumor had it that he went so far as to declare that his reactors were safe enough to be installed on Red Square.[18]

That never happened, but after the new reactor was tested, at a plant in Slavsky's ministry, it was deemed safe enough to be transferred to the Ministry of Energy and Electrification, which had no experience with nuclear energy. Few doubted the positive effect that the fusion of science and technology would have on the country as a result of the military-industrial complex's stewardship of the nuclear industry. Aleksandrov's RBMK reactors were placed all over the European part of the Soviet Union, producing much-needed clean energy for the country. With a capacity of 1,000 megawatts of electrical power per unit, they were more powerful than their Soviet competitors, the VVERs (or Water-Water Energy Reactors, meaning water-cooled and water-moderated), which had been produced starting in the early 1970s. By 1982, more than half the electrical power produced in Soviet nuclear plants came from Aleksandrov's reactors. Three of them were built at the nuclear power plant near Leningrad, two at the Kursk plant, one in Smolensk, and three in Chernobyl. The fourth RBMK unit was launched there by Briukhanov in 1983.[19]

Before coming to Moscow, Briukhanov had been under great pressure to help complete the construction of the fifth unit, which was 70 percent ready. The pressure increased in January 1986, when the local party committee officially reprimanded Briukhanov's deputy for failing to meet construction deadlines. The news got into the local media, and Briukhanov knew that if the situation did not improve, he would be next in line for a party reprimand. In his report to the congress, Soviet Prime Minister Nikolai Ryzhkov warned his underlings against any more delays in launching new reactor units, announcing, "Considering the strain on the country's fuel balance and the growing role of atomic energy, such disruptions are inadmissible in the future." The appetite for nuclear energy not only at the top but also at the bottom of the party pyramid was enormous.

Briukhanov could not help noticing that the regional leaders were eager to jump on the nuclear bandwagon, asking for investments of nuclear rubles in their regions. A party secretary from the Gorky (Nizhnii Novgorod) region on the Volga River argued in his address to the congress for the construction of a nuclear power plant in his district. A delegate from Siberia attacked Moscow officials for killing plans to build a new nuclear power station in his region. Everyone wanted to go nuclear.[20]

The man who served as gatekeeper to the Soviet nuclear paradise was Viktor Briukhanov's immediate superior, the fifty-six-year-old minister of energy and electrification, Anatolii Maiorets. New in office, he was eager to prove himself. Faced with the task of increasing the production of electricity at nuclear power stations by two and a half times within the next five years, he sought ways of fulfilling the task when the entire construction cycle of a nuclear power plant—from the start of architectural design to the completion of the reactor—took seven years. Maiorets told the congress that the cycle could be reduced to five years if the design and construction could take place concurrently. Briukhanov knew how difficult it was to deal with half-baked architectural designs not adapted to local conditions. Since few reactors were actually completed within the seven-year time frame, reducing it to five years seemed impossible. But if the party so ordered, and the state managers demanded it, the plant managers had no choice but to fall in line.

Maiorets finished his report on a high note: "Let me assure you that electrical power engineers and builders, inspired by the decisions of the 27th congress of the CPSU, will carry out the party's grand plans and make a worthy contribution to constructing the material basis of communism."[21] It seems to have escaped him that the party was no longer building communism, but then, as a new minister, he could be overenthusiastic.

The whole atmosphere at the congress was one of jubilation. Everyone wanted to think big and believe that anything was possible. Among the most optimistic delegates was Gorbachev himself. His report was received very well, his vision for the acceleration of economic development based on scientific and technological progress was endorsed by the congress, and he was now elected general secretary not only by the plenum of the party's Central Committee but by the congress as well. His standing had improved, and his mandate to carry on his acceleration policies was strengthened.

Moreover, Gorbachev was able to bring his own people into the Politburo. Among them was the energetic party boss of Moscow, Boris Yeltsin, who asked, rhetorically, "Why do we keep raising the same problems from one congress to the next? Why, even now, do demands for radical reform get bogged down in the inert stratum of opportunists with party cards?" His words sounded like a bombshell in an auditorium still full of Brezhnev's appointees. The word "perestroika," or restructuring, was mentioned in Gorbachev's report to the congress, but only once. The key word was still "uskorenie," acceleration, which had first been introduced in official party discourse immediately after Gorbachev's ascension to power in the spring of 1985. Most delegates believed that they were on the right track, the problem being the stagnation of the Brezhnev era. The solution lay in a return to Lenin's ideals of true communism.[22]

The congress ended on March 6. Viktor Briukhanov and his colleagues in the Ukrainian delegation packed their bags and headed home. The future looked bright, not only for the nuclear industry but also for the country as a whole. But there was one thing that bothered the director of the Chernobyl nuclear plant. In an evening telephone interview from his Moscow hotel, Briukhanov had shared his concern with the Kyiv reporter who had interviewed him a few weeks earlier on the occasion of his fiftieth birthday. As expected, he praised Gorbachev's report and embraced the new tasks assigned to the Soviet nuclear industry. But he also sounded a warning: "We must hope that this will also promote greater attention to the reliability and safety of atomic energy generation at our Chernobyl station in particular. That is most urgent for us." The interview with Briukhanov appeared in the newspaper without that warning.[23]

2

ROAD TO CHERNOBYL

O N THE evening of March 6, 1986, as the invigorated Mikhail Gorbachev hosted a reception in the Kremlin Palace for the foreign guests of the congress—most of them representatives of communist parties who had come to Moscow at Soviet expense—the Soviet delegates were leaving Moscow on planes, trains, and cars. Viktor Briukhanov and his fellow members of the Kyiv delegation boarded a night train to Ukraine's capital.

The next morning they were in Kyiv, where they were greeted by local party officials. There were hugs, handshakes, and flowers for the women in the delegation—the next day, Saturday, March 8, was International Women's Day, widely celebrated in the Soviet Union. A news photo taken at the main Kyiv railway station on the morning of March 7 featured Briukhanov in his fur hat and sheepskin coat, surrounded by fellow delegates, one of them a woman holding a bouquet of carnations. Briukhanov would have to get flowers for his wife, Valentina. But the drive home would take more than two hours—the distance between Kyiv and Prypiat was 150 kilometers.[1]

The driver of Briukhanov's company car came to meet him at the railway station and took Moscow Avenue to Highway P02. It led north of Kyiv along the Dnieper Reservoir, which had been created by the hydroelectric power station built there in the 1960s, then turned northeast toward the town of Ivankiv, passing through birch groves and, closer to Chernobyl, entering a zone of pine forests.

Briukhanov had first traveled the Kyiv–Prypiat highway by bus in the winter of 1970, when the city of Prypiat did not yet exist. He was young and full of enthusiasm. Becoming director of a nuclear power plant at such an age was quite an accomplishment, but for a while there had been no plant to talk about. Briukhanov had yet to build it—the plant, his own offices, along with a home for his family, which included his wife, Valentina; their nine-year-old daughter, Lilia; and one-year-old son, Oleg. He rented a room in a run-down hotel in the small town of Chornobyl—in Russian, Chernobyl—which would give its name to the power station Briukhanov yet had to build. He spread out the papers on his bed and began working through draft construction plans and contracts for the first temporary buildings on the spot chosen for the new plant and the new city. Construction would begin in a year.[2]

Meanwhile, Briukhanov's young family stayed in Sloviansk, then a city of 125,000 in the Donbas region of eastern Ukraine, where he had worked previously. The city would become known in 2014 as the place where the Russo-Ukrainian armed conflict began and the first people were killed. The fighting was so fierce there because Sloviansk was the local highway and railway hub as well as a major industrial center. That was why Briukhanov had ended up there in 1966—he had worked at a power plant that produced electrical energy by burning local coal.

Briukhanov's first power plant was in Angren, not too far from Tashkent, the capital of Uzbekistan, where he had been born on December 1, 1935. He was the oldest child in a large Russian working-class family that had come to the city from Saratov on the Volga. He did not remember much from the World War II years except that he was always hungry. At the age of twenty-four, he graduated from the local polytechnic institute and began his career at nearby Angren. It was there that he met Valentina, who also worked at the plant; in addition, she took evening classes at the local college. He was captivated by her eyes—as he remembered later, he felt that he could drown in them.

When Valentina first encountered Viktor's surname in a local magazine, he was already making headlines as a qualified and conscientious engineer. He had become head of his unit in just one year, and Valentina thought to herself, "God forbid having such a name"—"Briukhanov" was either derived from or was simply consonant with the Russian word for "belly." She would soon forget her worries after meeting the bearer of the name, the young, slim, and energetic Viktor. He won her heart, showing his love by inundating her with flowers. Cars coming from the nearby Qurama Mountains brought wild tulips, and Viktor would bring home enough to cover entire windowsills. They got married in a year and were very happy in Angren.

The Briukhanovs' tulip paradise came to an end early on the morning of April 26, 1966, twenty years to the day before the Chernobyl catastrophe. On that Tuesday, a powerful earthquake destroyed most of downtown Tashkent, Briukhanov's hometown, located 112 kilometers from Angren. More than 230 administrative buildings and over 700 shops and cafeterias were either completely destroyed or rendered unusable. Miraculously, only eight people died as a result of the earthquake, but many more were injured, and close to 300,000, almost one-third of the city's population, woke up without a roof above their heads. Among them were Briukhanov's parents, whose brick home was badly cracked and on the verge of collapsing. For Valentina Briukhanova, it was too much to take. What if another earthquake devastated Angren as it had Tashkent? What would happen to them and their young daughter? She wanted the family to move. Encouraged by his wife, Viktor began making inquiries at power plants in other parts of the Soviet Union. It turned out that they were looking for people just like him in Ukraine. The Briukhanovs packed their bags and went to Sloviansk, where Viktor soon rose through the ranks to become chief of the turbine division and then chief engineer of the plant.[3]

The Sloviansk plant was still growing when the Briukhanovs arrived in 1966. A new unit—the largest in the Soviet Union, as Viktor later recalled—was under construction. He embraced the challenge and soon showed himself a capable engineer and organizer once again. The start-up phase for new units was especially challenging, but Briukhanov managed the stress with aplomb, dealing simultaneously with construction crews that missed critically important deadlines and electricity production quotas. A hard-working, competent, and calm man of few words, Briukhanov seemed to have been born for such situations. That got him noticed in Kyiv, and in the spring of 1970 he was offered a job that would require the kind of qualities he had shown in Sloviansk, but on a much larger scale. The authorities wanted to put him in charge of the construction and then the operation of a new power plant to be built in the environs of Chernobyl, far away from the coal fields of either Uzbekistan or Ukraine. The new plant did not need coal. It would run on nuclear fuel instead.

It was a difficult decision for the young engineer. He asked Valentina for advice. She was afraid: it was a nuclear power plant, after all, and Viktor was a turbine specialist who knew nothing about reactors and nuclear power. But a power plant was a power plant, they told him in Kyiv. The superiors in Moscow agreed. There were few nuclear engineers available to build plants when the nuclear energy industry was just getting on its feet. Briukhanov accepted the challenge. But before becoming an expert in nuclear energy, he would have to become well versed in construction—a difficult task and initially a thankless job. He would regret his choice at first, but change his mind later. "I regret nothing," he would tell a correspondent on the eve of his fiftieth birthday in December 1985.[4]

THERE WERE even fewer reasons to regret anything in March 1986, when Briukhanov returned from the party congress in Moscow and got into his company car at the Kyiv railway station to drive back home to Prypiat. The Kyiv–Prypiat highway was a narrow two-lane road with a large amount of traffic. Both the nuclear power plant and the satellite city depended on it for many of their supplies.

Brikhanov's driver knew the Kyiv–Prypiat road almost by heart—his boss was constantly going back and forth between the two cities. Party bosses, ministers, and heads of departments were all in Kyiv, and the director had to travel there for numerous meetings. Then there were thousands, if not tens of thousands, of permits and other documents that needed signatures and stamps that could be obtained only in Kyiv. After a drive of almost two hours through forested countryside still covered with snow, Briukhanov's car finally approached Chernobyl. On the left was a concrete structure with the name of the city and a monument to Vladimir Lenin. Ahead was a central square that was rather large for a town that did not exceed 14,000 inhabitants.

Despite the construction of the nuclear power plant and the rapidly growing city of Prypiat a dozen kilometers to the north, Chernobyl had managed to stay almost the same as it had been ten, twenty, or even thirty years before it gave its name to the plant. While Prypiat had become a symbol of the industrialized socialist future, Chernobyl remained an embodiment of the rural presocialist past. The city on the Prypiat River and the port that had provided livelihoods for generations of its inhabitants featured numerous buildings dating to pre-Soviet times.

The settlement of Chernobyl was first mentioned in a Kyivan chronicle in 1193. It stood on the hunting grounds of the Kyivan princes, who ruled a vast medieval realm extending from the Carpathians in the west to the Volga towns in the east. The chronicle never explained the origins of the town's name, but scholars would eventually point to an abundance of common wormwood, or *Artemisia vulgaris*, a shrub recognizable by the black or dark red color of its branches. *Chornyi* is the Ukrainian word for "black." Thus the shrub gave its name to Chornobyl or Chernobyl, allowing future generations to associate the Chernobyl catastrophe with the biblical prophecy about a star called "Wormwood."

"The third angel sounded his trumpet," reads the Book of Revelation, "and a great star, blazing like a torch, fell from the sky on a third of the rivers and on the springs of water—the name of the star is Wormwood. A third of the waters turned bitter, and many people died from the waters that had become bitter." The common wormwood, after which the town of Chernobyl was named, is not exactly the same shrub as the wormwood (*Artemisia absinthium*) mentioned in the Bible, but it was close enough for many, including President Ronald Reagan, to conclude that the Chernobyl accident was prophesied in the Bible.[5]

The biblical prophecy aside, Chernobyl remained the capital of the north Ukrainian wilderness for most of its history. In early modern times, the rule of Kyivan princes over the region was replaced by that of Lithuanian grand dukes, and then of Polish kings. The

Cossacks claimed the territory in the mid-seventeenth century, but after a few years they had to cede it to the Poles. The town became the private property of local nobles and magnates. Mainstream history has forgotten most of Chernobyl's rulers and residents, with the exception of one young woman, Rozalia Lubomirska, a daughter of the town's owner, who had the misfortune of traveling to Paris at the time of the French Revolution. Maximilien de Robespierre put her on trial for close ties with the royals and alleged conspiracy against the revolution. Rozalia died on the guillotine in the French capital in June 1794. She was twenty-six years old. Her image survived on a wall tile in her old Chernobyl palace, which was later turned into a neurology ward of the local hospital.[6]

Whereas the French Revolution killed Chernobyl's internationally best-known citizen, the Bolshevik Revolution of 1917 did away with much of the town's rank and file. Some 60 percent of the town's population of 10,000 were Orthodox Jews, first invited there in the late seventeenth century by its Polish owners. Before the revolution, Chernobyl acquired a reputation as one of the centers of Hasidism in Ukraine. The spiritual leaders of the Chernobyl Jews were rabbis of the Hasidic dynasty established in the second half of the eighteenth century by Rabbi Menachem Nachum Tversky, a student of Baal Shem Tov—the founder of Hasidism—who was himself one of the pioneers of the movement. Rabbi Tversky's book *Me'or Einayim* (Light [of One's] Eyes) became a classic Hasidic text, and his sons and grandsons became rabbis in numerous towns of Ukraine.

The Chernobyl rabbis were famous for collecting money for charitable causes. In the early twentieth century there were a number of Jewish prayer houses, a school for Jewish girls, and an asylum in Chernobyl. The Jews of Chernobyl suffered disproportionately at the time of the revolution and the civil war that followed it, not only from passing military units but also from marauding gangs, the members of which were often recruited by local warlords from the area's Ukrainian and Belarusian villagers.[7]

Quite a few local Jewish youths had sided with the Bolsheviks—the political and military force that turned out to be the most friendly toward the poor Jewish masses and offered them the shortest path to emancipation. One of the leaders of the revolutionary transformation, Stalin's right-hand man Lazar Kaganovich, came from the Chernobyl region. In the mid-1920s, Kaganovich became the communist boss of Ukraine and presided over the policy of *korenizatsiia*, or indigenization, which put a temporary halt to the cultural Russification of the local population and promoted the development of Ukrainian and Jewish culture.

As Stalin's policies changed, however, so did the role that Kaganovich played in Ukraine. In the early 1930s, he became one of the main architects of the Holodomor, the great Ukrainian famine that took the lives of close to 4 million survivors of the revolution and the civil war and of the children who had been born to them in the years that followed. Some 1 million people died in the Kyiv region alone. In Kaganovich's native district of Khabne, the death toll was 168 individuals per 1,000—the area was more than decimated. Tens of thousands did not survive the famine and live to see the renaming of Khabne to Kaganovychi-1, and the village of Kabany, where the party boss was born in 1893, to Kaganovychi-2, in 1934, probably in recognition of the native son's loyalty not to his homeland but to his boss in the Kremlin.[8]

Then came the horrors of World War II. The Germans entered Chernobyl on August 25, 1941. Less than three months later, on November 19, the occupation authorities ordered the approximately 400 Jews still remaining in the town to gather near the synagogue. They were then marched to the premises of a Jewish collective farm called New World. There they were gunned down by automatic fire in the antitank ravine that some of them had helped dig as part of a futile effort by the Red Army commanders to stop the invaders. That was almost the end of the Jewish community of Chernobyl. By the time Briukhanov rented a hotel room there in the winter of 1970, there were only 150 Jewish families in the town where Jews had once been the majority. One of their synagogues had been turned into the headquarters of the local military commissariat.

The Chernobyl Jews who survived the Holocaust found refuge with partisan units in the surrounding forests. The communist-organized guerrilla groups, which enlisted local Ukrainian and Belarusian peasants, were active in the region from the fall of 1941. But the low-intensity war between the communist-backed partisans and the German-organized police (including local cadres) turned into a bloody vendetta. The executions of partisans, and, once the tide of the war turned, of policemen, took place in public, further brutalizing the local population. The settling of accounts among the relatives of those involved in the conflict would continue long after the end of the war.[9]

The Red Army recaptured Chernobyl and its environs from the Germans in the fall of 1943. It was a long and bloody battle. At stake was Chernobyl's port on the Prypiat River, the hub of the city's economic activity, as well as the bridges over the river and the nearby railway station. The Red Army suffered enormous casualties. It lost ten of its bravest—soldiers and officers decorated with the country's highest award, Hero of the Soviet Union. For the locals, the long-awaited liberation from the Nazis brought more death and hardship. With the Red Army in control of the region, the local male population was immediately called up and enlisted in the army. Many of those who survived the occupation were thrown into battle without arms, training, or even uniforms, dying on the outskirts of their villages and towns.

As Briukhanov's car passed the Chernobyl city limits, he recognized a familiar silhouette on the right side of the road. It was the statue of a Soviet soldier memorializing natives of the village of Kopachi who had died in the war and Red Army soldiers who had died in battle for the village in 1943. The first list was significantly longer than the second. In Chernobyl, the soldiers who had died in the six-week battle for the town were buried in a place that later became known as the Park of Glory. Its Alley of Heroes led up to an obelisk with an eternal flame at its base. "To the warrior liberators from the working masses of the Chernobyl district, May 1977," reads the inscription on one of the monuments in the park. Next to the inscriptions are plaques with names of Red Army generals and the units they led in the battle for the town.[10]

Over the years, Briukhanov had attended quite a few commemoration ceremonies at the Chernobyl Park of Glory on May 9, Soviet Victory Day. The cult of heroes of the Great Patriotic War, as the Soviet-German conflict of 1941–1945 became known in Soviet parlance, commemorated only those who had died wearing the Red Army uniform. The rest were largely forgotten. There would be no monuments to victims of the Holocaust or of the Holodomor. Both atrocities went unacknowledged.

A FEW minutes after the director's car passed the monument to the war heroes, Briukhanov could already see the huge white tube of the Chernobyl plant's cooling tower on the horizon. There, according to the official Soviet narrative, the shadows of the past were lifting: the miracle of technological progress was about to deliver a bright future. To the right of the canal that the car was following, there appeared the walls of Unit 5, which was still under construction and surrounded by tall, powerful cranes. Then the white walls of the operating units came into view: Units 3 and 4 were joined together in one huge building; Units 1 and 2 were separate.

The area near the village of Kopachi had been chosen as the construction site of the nuclear plant, in December 1966. The search for the right place had begun a year before that, with a memo from the deputy head of the Ukrainian government, Oleksandr Shcherban, to the Central Committee of the Communist Party of Ukraine (CPU). A former vice-president of the Ukrainian Academy of Sciences and an early enthusiast of nuclear energy, Shcherban had decried the lack of electrical-power-generating facilities in Ukraine, predicting a possible slowdown of the republic's economic development if new sources of energy were not found promptly.

Shcherban knew that two nuclear power stations had been launched in Russia in 1964 and advocated the construction of three such stations in Ukraine: one in the south, another in the west, and a third in the area around Kyiv. He was soon backed by his superior, the head of the Ukrainian government, Volodymyr Shcherbytsky, and Shcherbytsky's boss, the first secretary of the CPU, Petro Shelest, who was also a member of the all-Union Politburo. Shelest wrote to Moscow demanding that Shcherban's proposal be included in the Union's plans for the construction of new nuclear power stations. In response, the Union government approved the building of one nuclear power plant in Ukraine. Kyiv was not unduly disappointed by the reduction in the number of plants: the important thing was that the republic was getting on the nuclear bandwagon and acquiring what was then considered cutting-edge technology.[11]

In the fall of 1966, Volodymyr Shcherbytsky issued a decree ordering the start of the exploratory work for the construction of what was then known as the "Central Ukrainian Nuclear Power Plant." The commission formed in Kyiv in November of that year soon concluded that there was no better place to build the plant than the area near Kopachi. It was a fairly large settlement, with more than a thousand inhabitants, but surrounded by a sparsely populated area. The village was distant enough from big cities and towns, as well as from the resort areas—another site was turned down because it was too close to such an area. It was also close to the Prypiat River, which was essential for the functioning of a nuclear plant, but its location was not a marshland. No less important was the site's proximity to a railway station—the railway had been built during the first Soviet five-year plan, its construction begun under the Chernobyl area's native son, Lazar Kaganovich.

True, there were also some problems with the location. Underground waters were coming too close to the surface, and a great deal of soil would have to be brought in to ensure a solid foundation for the buildings. Kopachi was distant from sources of construction materials, including rocks and granite; only sand was available locally. But the surveyors considered those problems manageable. Because the area was not very agriculturally productive, converting it to an industrial site would have little impact on the rural economy. The future plant's cooling pond—the reservoir needed for the functioning of the plant—would take up the biggest chunk of the territory allocated for the construction of the plant and the satellite city—altogether more than 1,400 hectares, or 3,460 acres, of pastureland, 130 hectares (320 acres) of forest, 96 hectares (240 acres) of cultivated land, and 50 hectares (120 acres) of gardens planted by the local population.

Kopachi was ruled to be the best of the sixteen locations considered by the commission. The name of the facility was eventually changed from the Central Ukrainian Plant to the Chernobyl Plant. The "center" had moved north to the very border with Belarus. There is no indication that the Belarusians were ever consulted.[12]

The construction of the plant began on Briukhanov's watch. In the summer of 1970, he moved his headquarters from the rented hotel room to an office of less than six square meters in a mobile construction workers' barrack. From there he commanded his growing cohort of engineers, checked on the construction crews, and visited high offices in Kyiv and Moscow. Grigorii Medvedev, who came to the Chernobyl plant in the winter of 1971 to take the position of a deputy chief engineer, had idyllic memories of Chernobyl at the start of construction: "There was a sparse forest of young pines all around, with an intoxicating atmosphere unmatched anywhere else. There were sand hills covered with a low-growth forest and patches of pure yellow sand against a background of deep green moss. No snow to be seen. Green grass warmed by the sun dotted the area. Silence and a sense of primeval creation."[13]

The silence did not last very long. Excavators working around the clock soon removed some 700,000 cubic meters of soil for the foundations of the new unit. In August 1972, Petro Neporozhny, the all-Union minister of energy and electrification, visited the site himself to witness the first pouring of concrete. Speeches were delivered and pledges made, but it would take much longer than the authorities expected to complete construction and make the reactor operational. It was supposed to be up and running in 1975, but there were problems with the supply of reactor parts and related equipment. In April 1975, with the original deadline all but missed, Volodymyr Shcherbytsky, who had been promoted from head of government to party boss of Ukraine, appealed directly to Moscow. Things finally started moving, and the requisite equipment arrived. In August 1977, the first nuclear fuel was loaded into the core of the reactor. In September it began operation and was connected to the grid, and in December Briukhanov signed the documents declaring the reactor fully operational.[14]

It was only then that Briukhanov made the transition from director mainly responsible for the construction of the plant to director predominantly responsible for its operation. "The year 1977 will go down in the history of Soviet atomic energy as the year of the birth of an energy giant on the Prypiat," he wrote with satisfaction in one of Ukraine's leading newspapers at the end of the year. A new era had indeed begun. In December 1978, Unit 2 was connected to the grid. Three years later, in December 1981, Unit 3 produced its first electricity, and Unit 4 followed in December 1983.[15]

The December dates of completion of units or their connection to the grid were not coincidental. The pressure to launch reactors before the end of a given year was enormous—party leaders and ministry officials wanted to present achievements in their annual reports, while construction crews and operational personnel would not get their hefty bonuses if they did not complete planned tasks by year's end. "Interestingly enough, no one could speak aloud of the impossibility of going onstream in a given year until December 31," recalled Anatolii Diatlov, a nuclear engineer who first arrived in Prypiat in 1972.

None of the units were completed on time. Upon his arrival, Diatlov noticed a slogan above the entrance to the dining hall calling on construction workers and engineers to launch the first reactor in 1975. As 1975 passed without a working reactor, they changed "5" to "6" and then to "7." Every year representatives from the ministry would come and insist on a new unrealistic date that everyone knew could not be fulfilled. "And at first there would be jitters because of strict insistence on fulfillment of a timetable that was incapable of fulfillment from the moment of its creation," remembered Diatlov. "Tough production meetings and summons to work at night. Delays inevitably increased, vigilance fell off, and normal work began. Until the supervisor's next visit."[16]

Briukhanov vividly remembered the launch of every successive unit. He was often critical of the construction crews. At a meeting of the party committee for the city of Prypiat, he chastised them, saying, "Incompetence in production, that is, in the plant workshops, makes itself felt on construction sites in poor-quality parts and poor work. Take something as simple as calculating corners. Crooked openings for doors and windows, crookedly nailed finishing elements, wrong angles in the installation of plumbing." Briukhanov was in a difficult position. He was the one who had to sign papers attesting to the satisfactory completion of the work. The bosses, for their part, wanted to report the fulfillment of plans, and the construction workers wanted their bonuses, but Briukhanov was responsible for the operational soundness and safety of the units. The problem was that the government was at once contractor and client. The plant and the construction directorate both reported to the same bosses at party headquarters in Kyiv and Moscow. If Briukhanov complained too much about problems with the work done by the construction directorate, he could easily lose his job.[17]

FINALLY BRIUKHANOV'S car approached Prypiat, the city that had been built before his own eyes and with his active participation. Sometimes he thought he had had enough of the place. He would feel tired and want to try something else. They had asked him in Moscow whether he would be interested in going abroad, for example, to help build a nuclear power plant in Cuba, where Soviet architects and engineers had begun construction of the country's first nuclear reactor in 1983. But those were moments of weakness. Briukhanov stayed in Prypiat.

The city was only 3.5 kilometers north of the power plant. The road from the station turned into Prypiat's main street, the broad Lenin Boulevard, with trees and flowers planted in the median between the wide lanes leading to the town's main square. Located there were the administrative building housing party headquarters and the city hall, the palace of culture, called the "Energy Worker"; and the hotel, named Polissia, meaning "forested area," a reference to the huge ecological region covering most of northern Ukraine from the Dnieper in the east to the Polish border in the west. The main square was at the intersection of the city's two main boulevards, one named after Lenin and the other, perpendicular to it, after the founder of the Soviet nuclear program, Igor Kurchatov. On the corner of the two boulevards, facing the square, stood the town's White House, a nine-story apartment building that housed the city's elite.[18]

Briukhanov's apartment was on the fourth floor. It was his second home in Prypiat. The first was located farther down Lenin Boulevard in the first apartment building to be built in Prypiat, in 1971. Back then, he had needed permission from a party secretary in Kyiv to move in—so scarce was housing in the city during its first years of existence, and so careful was the party not to offend the working

class by creating the impression that it favored managers over workers when it came to goods and privileges. The new town was supposed to be an embodiment of socialism, with no private cottages
allowed. The planners projected that by 1975 it would house close
to 12,000 plant and construction workers. By 1980, as the third
and fourth reactors were to come onstream, the number of people
dwelling in the city was expected to grow to 18,000; the population
would then decline to slightly more than 17,000 and stay at about
that level for the next five years. In fact Prypiat grew much faster, attaining a population of almost 50,000 by 1986. Housing remained
a problem.[19]

The leaders of the Komsomol—the Young Communist League,
the youth branch of the Communist Party—had given the city of
Prypiat and the Chernobyl power plant the status of a Komsomol
construction site, recruiting young people all over the Soviet Union
to join its workforce. But most of those who came to work in Prypiat needed no special encouragement. Problems with housing were
common in Soviet cities, and in Prypiat housing was being built
faster and better than anywhere else. The city enjoyed special status
in the country and the republic with regard to supply of consumer
goods and agricultural products because the nuclear industry was
still very close to the military-industrial complex, with its special
privileges.

By the mid-1980s, it had become impossible to buy cheese or
sausage in most of the towns and big cities of the Soviet Union, but
those products were freely available in Prypiat. The signs in the local
supermarket did not lie in that regard. More difficult was the situation with fresh meat, which was often supplemented with lard and
bones, but then, at least there were villages around the town where
meat and milk could be bought. Life was relatively good in Prypiat, and many, especially from the neighboring villages, wanted to
move there. Once in the city, often as construction workers, they all
wanted to work at the nuclear plant, where bonuses could be earned
for fulfilling and overfulfilling production quotas.

Most people who came to the new city were young, and many
were unmarried. The average age of those living in Prypiat in 1986
was twenty-six. There were eighteen dormitories in the city for

singles, and most apartments were designed for young families. Not only were most inhabitants young, but their children, when they had them, were young as well. In the five local elementary schools, there were as many as fifteen parallel classes, each with no fewer than thirty students. Most rural schools, by contrast, had barely enough students for one class. In most urban schools, there were only three parallel classes running at most. And there was no sign of this trend slowing: more than a thousand newborns were welcomed to the city every year.[20]

There were two stadiums and two swimming pools, one of which qualified for international competitions. Briukhanov was proud of what he had helped to build in the city but also annoyed that he often had to divert funds and resources from the plant for construction projects in Prypiat. The city was run separately from the plant by the party and local authorities, but, as the city's coffers were often empty, while the power plant had an enormous budget, the local party authorities kept importuning Briukhanov to chip in and build new facilities for the city. Briukhanov could keep the local party bosses at bay, but he could not say no to the party officials on the regional and republican levels—they outranked him in the party-state hierarchy. Especially pushy was the first secretary of the Kyiv regional party committee, Hryhorii Revenko, who would become Gorbachev's chief of staff in 1991. In the mid-1980s, Revenko convinced Briukhanov to build the second swimming pool, which would meet international standards. Then he came up with plans for a skating rink. Briukhanov bristled at this: "There was nothing of that class in all of Ukraine, but I was supposed to build it in my little town?" he recalled later. Still, he went along with the request.[21]

Briukhanov realized that his own workers needed sports facilities and would benefit from them. The same was true of the shops, as the architects had planned for only one supermarket. The city needed more, and Briukhanov found money for them, occasionally misleading the banks—he would borrow funds for the plant and spend them on the city. "We get used to the abnormal and begin to accept it as something like the norm. That is what's so terrible!" he complained to a reporter a few months before the party congress in Moscow, referring to the need to take care of the city's problems when his time and energy were needed at the plant—to ensure its safety, among other things. "In those second-rate circumstances, the main thing is to ensure the reliability and safety of our work," said Briukhanov, continuing his litany of complaints. "Whatever you say, we are no ordinary enterprise. God forbid that we suffer any serious mishap—I'm afraid that not only Ukraine but the Union as a whole would not be able to deal with such a disaster."[22]

But for now, Briukhanov could put such disturbing thoughts aside. He was finally home. International Women's Day lay ahead, a chance to greet his wife and spend time with friends and colleagues. His daughter no longer lived with them: she and her husband were about to graduate from medical school in Kyiv. The Briukhanovs would have to greet her over the phone, but they soon hoped to host the young couple in Prypiat. They were expecting a child, and Bri-ukhanov would become a grandfather. He had worked hard for everything that he had achieved in life, but he had to admit that life was treating him pretty well. The year 1986 seemed to be off to a particularly good start. First, there was his participation in the party congress in Moscow. Then there were rumors that he was poised to be awarded the Star of Hero of Socialist Labor. Of course, it would mean fulfilling and overfulfilling his production quotas, but that was nothing new—he had done it before.[23]

On the morning of the next day, March 8, the main regional newspaper published a photo of a smiling Briukhanov surrounded by other delegates to the congress upon their return to Kyiv. He looked reserved, perhaps a little tired, but generally happy and satisfied—a man in control of his destiny and that of the people around him.[24]

POWER PLANT

INTERNATIONAL WOMEN'S Day notwithstanding, the single biggest event scheduled for Viktor Briukhanov and his colleagues and subordinates for the month of March 1986 was an all-Union conference of representatives of enterprises that supplied construction materials and hardware for the Chernobyl nuclear plant. The three-day conference, planned for the last week of the month, was intended mainly to coordinate the actions of the plant administrators, the construction directorate, and the vendors and solve problems that had emerged with the construction of the newest power unit of the plant, reactor No. 5. Construction quotas for 1985 had not been met, and the prospects for bringing the reactor onstream in 1986 were dubious, but the Moscow party congress had set a goal of more than doubling the construction of nuclear units in the course of the next five-year plan.[1]

In Prypiat, the person with the greatest stake in the success of the conference was Vasyl Kyzyma, the fifty-four-year-old head of the construction directorate that was charged with building the power plant. As far as power and prestige went in the city of Prypiat, Kyzyma outshone Briukhanov. Whereas the latter lived in an apartment building, though a prestigious one, Kyzyma and his family occupied one of four cottages built in a city that was supposed to have no cottages at all. When the Ukrainian party boss, Volodymyr Shcherbytsky, visited the site of the power plant in the mid-1970s, he was so impressed with the young director of construction that he ordered his aides to ensure Kyzyma's election to the Ukrainian Supreme Soviet. That would not only raise Kyzyma's salary but also boost his prestige and independence from local party and state officials. In 1984 Kyzyma was given the highest Soviet award—Hero of Socialist Labor. Briukhanov, despite being elected a delegate to the party congress in Moscow, was still waiting for that kind of recognition.[2]

It all made little sense if one looked at the results of the two managers' work, especially their fulfillment of production quotas. Kyzyma and his crews failed to build any unit of the nuclear power plant on time, while Briukhanov and his engineers invariably fulfilled or overfulfilled their quotas. The previous year, 1985, was no exception. That year, the four units of Briukhanov's power plant produced 29 billion kilowatts of electrical energy, as much as all the power stations of the then united Czechoslovakia, enough to power 30 million Soviet apartments, housing roughly half the Soviet population of 280 million, for the whole year. They had overfulfilled their quotas by more than 9 percent, reported the Prypiat newspaper *Tribuna énergetika* [Tribune of the Energy Worker] in January 1986. The same issue featured a brief note titled "Why Has the 1985 Plan for the Nuclear Power Plant Not Been Fulfilled?" It was not apparent from the title that the article concerned the failure to meet construction quotas.[3]

Still, Vasyl Kyzyma was not only spared public criticism but also showered with attention from his superiors. There was a simple reason for that: given the constant shortages of labor and materials, the higher-ups considered the supervision of construction a much more challenging task than that of running the plant itself. And Kyzyma had shown his ability to do the job better and faster than most of his rivals.

A native of the Kyiv region, Kyzyma was born in January 1932 into a peasant family in Tarashcha Raion, south of the city of Kyiv (a *raion* is a district similar to a county). He was only one year old when the Holodomor—the Great Ukrainian Famine—struck the region. He was lucky to survive: every third citizen of the raion died of hunger. In neighboring Volodarka Raion, the death toll was even higher—466 of every 1,000 residents did not live to see the year 1934. The mortality rate was especially high among children and the elderly. Kyzyma also survived the brutal Nazi occupation of Ukraine from 1941 to 1944. After graduating from the local school, he studied engineering in Kyiv and cut his teeth on the construction of coal power plants in western Ukraine. Kyzyma returned to his home region in 1971 to take charge of the flagging construction of the city of Prypiat and the nuclear plant nearby.[4]

From the very start, Kyzyma told his superiors that there could be only one boss on the construction site: emissaries from the Kyiv ministries could pack their bags and go home. They did so. But in the city itself, Kyzyma had to share power and prestige with Briukhanov and his people at the plant. While Briukhanov was regarded as competent, but reserved and not very sociable, Kyzyma combined the expertise and toughness of a construction foreman with the magnetic personality of a politician. Kyzyma's bosses praised his strategic thinking, but they were somewhat concerned about his canniness; construction workers considered him one of their own, and his subordinates emphasized his toughness.

Briukhanov was respected, but Kyzyma was either idolized or feared, depending on one's circumstances. Eventually the two men learned to work together and respect each other. Ultimately, they had a common cause, with the plant administrator being the customer and the construction directorate the vendor, each depending on the other in many ways.

Personal relations between the two masters of the city aside, there were also managerial, social, and cultural tensions between their teams. Most of Briukhanov's people were top-notch, highly educated engineers who came mostly from Russia, bringing to Prypiat not only their expertise in running the nuclear plant but also their big-city habits, culture, and, occasionally, arrogance. The locals who manned Kyzyma's construction crews were often former villagers with their own habits, culture, and prejudices. The newcomers spoke Russian, the locals largely Ukrainian, with *surzhyk*—a mixture of Russian and Ukrainian—emerging as the lingua franca of Prypiat's streets.

Managing tensions between the two groups and preventing occasional clashes among local youths and the young construction workers brought to the city by Kyzyma, known as "Rexes," were among the responsibilities and headaches of the Prypiat police. Most challenging were the two days when the workers received their pay, first an advance and then the rest of the salary. Celebrating both paydays by getting together and drinking was a long-established Soviet custom. The police took no chances and went after any young men who appeared on the streets in groups of more than five. It helped only to a degree. In 1985, officials had to deal with a youth riot that resulted in turning over cars and breaking windows.[5]

Nuclear power plant workers of all ages were often envied by those who worked in the construction directorate—their salaries were higher and working conditions better, and they moved from dorms into apartments faster than those who were actually building the apartments. The construction workers believed that the atmosphere at their building sites was better, and that their workmates were open and straightforward, whereas those at the plant were canny and calculating. The plant workers generally disagreed.[6]

Kyzyma ran his directorate like a huge peasant household. The heads of the directorate's units dreaded his invitations to the Romashka, one of the plant's dining halls, for a meal of Ukrainian borscht. At the head of the long table, in the spot reserved for the family patriarch, sat Kyzyma, with his "children" (subordinates) on both sides. "All sat silently spooning their borscht when suddenly the quiet was broken by [his] voice, . . . not angry, but decisive and insistent, saying that this group of deliverymen had not brought a crane to the building site at the appointed time, or that concrete factory had not completely filled the order of the concrete-pouring brigade," recalled a journalist who had been invited to one of Kyzyma's working lunches. Kyzyma had a superb memory and not only knew his numerous subordinates by first name and patronymic, but also remembered, in detail, all the particularities of their tasks and the resources at their disposal.[7]

Kyzyma was known to be a true patriot of the city, able to find money and resources to construct not only the plant but also the public buildings, the housing, and the infrastructure of Prypiat.

He also resisted pressure to build quickly, preferring to build well. When party leaders reprimanded him for slow work on the Prypiat palace of culture, he referred to the St. Volodymyr Cathedral in Kyiv as an example. The cathedral had taken decades to build but was still standing, he told a surprised party official, who could hardly imagine a worse ideological sin than that of comparing a socialist palace of culture to a church. Kyzyma did not mind. He not only took his time, but insisted that the palace he built would be the best in Ukraine. He obtained scarce marble for its walls, aluminum parts (almost impossible to acquire in the USSR) for its infrastructure, and rare varieties of wood for its floors. The crews that installed the wooden floors in the Prypiat palace of culture were the same ones that had restored the imperial-era tsarist palace in Kyiv.[8]

THE CONSTRUCTION of Unit 5 at the Chernobyl power plant presented Kyzyma with a new set of problems. He had built reactors before and figured out how to do so with minimal delays. But now, with the party authorities in Moscow rushing to launch as many reactors as possible, the time frame for the construction of the new unit had been cut from three years to two. Construction had begun in 1985 and was supposed to be finished by the end of 1986. The new all-Union minister of energy and electrification, Anatolii Maiorets, who had suggested, at the party congress in February 1986, that the time line for the construction of multiunit nuclear power plants could and should be cut from seven years to five, was eager to make his mark on the industry.

In December 1985, Maiorets had come to Prypiat to see for himself what was going on at the construction site; he wanted to speed up the work, which was already falling behind. In his presence, the Kyiv regional party committee issued a resolution calling for an uptick in the pace. The party used the main tool at its disposal—a barrage of reprimands to the officials deemed responsible for the delays. Some were oral, others written, and still others recorded in the personal party files of the victims. If those did not work, the official would be fired. If the situation improved, the reprimands would be withdrawn, the manager rewarded, and the game would start again. In January 1986 Prypiat managers became the targets of party reprimands. Among the victims was R. L. Soloviev, Briukhanov's deputy responsible for construction projects—that is, for working with Kyzyma and his people. At a meeting of the Prypiat city party committee, Soloviev was issued a reprimand of the severest kind—it was recorded in his party file. Kyzyma and his deputies were spared, for the time being.

Kyzyma was a rarity among Soviet officials in unhesitatingly opposing party officials who wanted him to divert scarce resources from the plant he was building and the city he adored to anything other than the purposes for which they were intended. Ever since the forcible collectivization of farms in the Stalin era, agriculture, as a sector of the Soviet economy, had been chronically lagging. The party, in order to increase production, had resorted to a system of "partnerships" between large urban industrial and scientific enterprises and collective farms; in practice, that meant sending urbanites to the villages to help with farm work. Party officials in Prypiat demanded that nuclear plant personnel and construction workers do their part, and the local newspaper would regularly report on the results. Briukhanov complained, but he sent his people to the neighboring collective farms. Kyzyma, however, simply refused.

On one occasion when a party boss demanded agricultural labor from the nuclear plant personnel, Kyzyma, red in the face, asked his secretary to connect him with party headquarters in Kyiv. I have your man here in my office, Kyzyma told the official at the other end of the line. He is telling me what to do, but my plate is full. Please figure out for yourselves who is boss here, him or me. He hung up the receiver, and that was the end of the matter. Given the importance of nuclear energy, party officials had little choice but to back down. But they were furious: as they saw it, Kyzyma was getting away with murder.[9]

One of Kyzyma's tools in dealing with the growing party pressure was his use of the media to make sure party officials helped him by insisting that suppliers deliver products to the construction site in a timely fashion. Kyzyma had learned media tactics long before the era of glasnost, or openness, declared by Mikhail Gorbachev and his reformers. In the spring of 1980, Kyzyma and Briukhanov were summoned to Moscow to report to the deputy prime minister of the all-Union government on the construction of the third unit of the power plant. Representatives of other plants attended as well, but none of them were doing worse than Chernobyl. If average fulfillment of construction quotas was approaching 90 percent, Kyzyma and Briukhanov reported the fulfillment of only 68 percent of their quota. The problem was lack of qualified personnel, equipment, and supplies. They were ordered to catch up within a month—an impossible task, since they were at least two or three months behind the original schedule.[10]

With the revised deadline in jeopardy and new reprimands on their way from the very top of the Soviet party and state pyramid, Kyzyma invited Oleksandr Boliasny, a reporter for the regional communist newspaper, *Kyïvs'ka pravda* [Kyiv Truth], to his office for an interview. The editor in chief expected an upbeat report and a pledge that the quotas would be fulfilled in the shortest possible amount of time. Instead, Kyzyma kept Boliasny in his reception room for hours while he himself took pen and paper and personally wrote responses to the interview questions. Kyzyma explained that the unit would not be finished on time because the requisite supplies simply were not there. Miracles did not happen in the socialist economy. He wanted party officials to put pressure on the vendors and make them fulfill their obligations. The terrified newspaper editor called the party officials, who were not happy but eventually gave their go-ahead for publication of the interview. After all, it shielded them as well against unjust criticism from Moscow.[11]

In early 1986, the problem with the construction of Unit 5 was similar: the construction crews were there, but the supplies were not. With the reduction of the construction cycle, decreed long before similar resolutions were passed by the party congress in Moscow, Kyzyma and his people did not get their design documentation until July 1985. That delayed their orders for construction materials and equipment until the fall, and the vendors began to deliver construction blocks and hardware only at the end of that year. Some arrived on time, but others did not, halting the installation process.

When Liubov Kovalevskaia, a reporter for the Prypiat newspaper *Tribuna ėnergetika*, approached Kyzyma with the idea of publishing a special issue discussing the problems with vendors, he readily agreed. More important, in a violation of security regulations, he gave Kovalevskaia permission to consult data collected by the nuclear plant's computer center on the quantity and quality of hardware supplied by the vendors. That was a major coup for her, but others were less obliging than Kyzyma. Mindful of the watchful eye of the KGB, the computer center's managers allowed Kovalevskaia to work with printouts of the spreadsheets for only fifteen minutes, hoping that Kovalevskaia, a teacher of the Russian language and literature by training, would not be able to understand much of what they contained. She managed to work on the spreadsheets for half an hour, though, taking a lot of notes.

The figures that Kovalevskaia copied added up to a terrifying picture. As much as 70 percent of the hardware supplied by one of the vendors had serious defects. Of the metal structures for the reservoir that would store used nuclear fuel, 356 metric tonnes (392 US tons) had major defects as well. Concrete panels supplied by another vendor were of the wrong size and had to be adjusted on the construction site. But the main problem was that even if some parts were up to specifications, others had not yet arrived—altogether, 2,435 metric tonnes (2,684 US tons) of metal structures were still missing. Kyzyma gave his go-ahead for publication of the article, which appeared under Kovalevskaia's pen name, L. Stanislavskaia, in *Tribuna energetika* on Friday, March 21, 1986. The all-Union meeting of representatives of the suppliers was scheduled to start on the following Monday.[12]

ANATOLII DIATLOV, the deputy chief engineer of the Chernobyl nuclear plant, believed that Kyzyma's workers were doing a good job of building the plant. They were also very successfully withstanding pressure from party and industry officials, who were making unrealistic demands and setting impossible deadlines. According to Diatlov, Kyzyma and his subordinates "did not take all those assaults seriously, although that attitude was not on display. . . . Otherwise, it would be impossible to survive for long under those working conditions."[13]

As far as Diatlov was concerned, the main problem was not with Kyzyma but with the lack of a proper manufacturing base for the construction of nuclear plants, and hence the problems with vendors. Nuclear power stations like the one in Chernobyl were no longer the responsibility of Yefim Slavsky, the all-powerful head of the Ministry of Medium Machine Building, which was the hub of the Soviet nuclear program and its military-industrial complex. Slavsky's ministry was a virtual empire, a state within a state with its own manufacturing plants capable of producing most of the equipment needed for the nuclear industry. Those plants had been used to build the nuclear power station at the Sosnovyi Bor settlement near Leningrad, where the first reactor began producing electrical energy in December 1973. But soon after that the construction of nuclear power stations had become the responsibility of the Ministry of Energy and Electrification, which was not part of the military-industrial complex; it had a poor manufacturing base of its own and none of the political clout that came with Slavsky's power and prestige.

As Diatlov later recalled, "the government resolution indicated that nonstandard equipment for the four blocks of the first of those stations [Nos. 1 and 2] would be produced by the same factories that produced them for the Leningrad station. But the Ministry of Medium Machine Building did not take the government resolution as an order." No prime minister could control Slavsky, whose main responsibility was the Soviet nuclear weapons program. Nuclear power plants like the one in Chernobyl were on their own. "They said you have your own factories, so go ahead and make the equipment; we'll provide the plans," continued Diatlov. "I had been to several factories making auxiliary equipment for the Ministry of Energy—the machine tools were on the level of poor workshops. Commissioning them to produce equipment for a reactor was like making a carpenter do the work of a joiner. So there was constant difficulty with production for every block."[14]

Complaints about difficulties with the construction of the new units were all but ignored at the top. After all, the Chernobyl power plant was assigned reactors of the type that in theory could be built by unspecialized machine-building factories almost anywhere by almost anyone at minimum cost. The Chernobyl plant originally was supposed to use Water-Water Energy Reactors (VVERs), the Soviet equivalent of the Pressurized Water Reactors (PWRs) in the United

States. Like its American counterpart, the Soviet VVER originated in the 1950s as a byproduct of building reactors for nuclear submarines. In those reactors, energy was produced by placing fuel rods, which generate heat through the fission of uranium atoms, into pressurized water. Water is also used as a coolant, to prevent the whole system from overheating. The design was extremely safe. In the unlikely event of the failure of coolant circulation, the increased heat would effectively shut down the reaction (the less water there was in the core of the reactor, the smaller the neutron moderation effect slowing the fast energy neutrons produced by water on the radioactive fuel in the rods, which could not continue the reaction without water). The VVER reactors tested quite well at a number of Soviet nuclear power plants, which was why they were initially chosen for the Chernobyl plant.

In the corridors of power, however, the VVER reactors lost in competition with the RBMK, or High Power Channel Reactor, which used graphite to moderate the reaction and water as a coolant. The RBMK reactors had an output of 1,000 megawatts of electrical energy, twice that of the VVERs. And they were not only more powerful but also cheaper to build and operate. Whereas VVER reactors required enriched uranium, RBMK reactors were designed to run on almost natural uranium-238, with an enrichment level of a mere 2 to 3 percent of uranium-235. Last but not least, the RBMK reactors could be constructed on the spot from prefabricated components produced by machine-building plants that did not specialize in the production of high-precision equipment for the nuclear industry. As far as the party leadership in Moscow was concerned, it was a win-win situation. While the rest of the world chose VVER reactors, the USSR went mostly, but not exclusively, with the RBMK type. The Chernobyl power station was caught up in the new Soviet trend.

By the time the decision was made to switch the Chernobyl reactors from VVERs to RBMKs, the latter had not yet been fully tested. But behind it stood the powerful figure of Yefim Slavsky. The director of the Igor Kurchatov Institute of Atomic Energy, Anatolii Aleksandrov, had been the scientific director for both types of reactors, and he knew their strong and weak points. But, while making the claim, mentioned earlier, that the RBMK was as safe as a samovar, Aleksandrov went along with the general preference for a cheaper and more powerful reactor over a safer one. The designers argued that the RBMKs were so safe that costs could be reduced even more by building them without the concrete structure that could contain radiation in the event of a reactor failure. Thus Chernobyl got the reactors but not the containment.

There were some dissident voices, but they were silenced or ignored. The most powerful of these voices belonged to Nikolai Dollezhal, the chief designer of the RBMK. Although Dollezhal did not disown his brainchild, he considered it a mistake to build nuclear power stations in the European part of the USSR. The whole nuclear industry, in his opinion, was not safe enough for that. Dollezhal lobbied his colleagues and petitioned the government, but to no avail. He then decided to publish an article spelling out his concerns in an academic journal. The authorities offered him publication in the party's main ideological journal, *Kommunist*, instead. Publication there would require him to pull his punches, but the article would reach a broader audience, with the potential of inspiring a public discussion. Dollezhal took the deal.[15]

The article, titled "Nuclear Energy: Achievements and Problems," which Dollezhal coauthored with a fellow scientist, Iu. I. Koriakin, appeared in *Kommunist* in the summer of 1979, a few months after the Three Mile Island nuclear accident in the United States. In March of that year, a malfunction of the cooling system led to the partial meltdown of one of the reactors and the release of radioactive gas, which led to the voluntary evacuation of close to 140,000 people from the area. Dollezhal and his coauthor wrote that in the United States, the cost of building nuclear power plants had increased seven to eight times over because of safety concerns—a trend not followed by the Soviet nuclear industry. They were especially concerned about the quality of nuclear power equipment and safe transport of nuclear fuel and waste in the Soviet Union. With the growth in the number of nuclear energy plants, the probability of accidents would also increase, they argued. They were worried, too, about climate change as a consequence of the construction of nuclear power stations—in the Soviet Union, each plant emitted huge quantities of heat into the atmosphere—two or three units of energy for every unit turned into electric power. Dollezhal proposed that instead of building power stations in the European part of the country, as current policy dictated, nuclear stations should be located in the sparsely populated northern reaches of the USSR, close to uranium deposits.[16]

Dollezhal's article presented a huge challenge to the existing Soviet nuclear energy industry. Stations were already under construction, with billions of rubles and the reputations of leaders in the field invested in ongoing projects in the European part of the USSR, and research on long-distance transmission of electricity (which would be required if plants were built far away from populated areas) was not of high priority. According to conventional wisdom, a nuclear power plant should be a maximum of 500 to 600 kilometers from consumers. That was the logic used in choosing the site of the Chernobyl station. The Soviet academic establishment fought back. Anatolii Aleksandrov published an article criticizing Dollezhal's ideas in the journal *Problems of Peace and Socialism*, intended for foreign, largely Eastern European audiences: since the Soviets were building reactors abroad, he said, safety was a paramount concern, just as it was for any other Soviet technological or industrial export. Curiously enough, the Soviet-constructed plants in Eastern Europe were of the VVER type and used water, not graphite, as a moderator of neutrons.[17]

IN UKRAINE, Dollezhal's concerns about building nuclear reactors in general and RBMK reactors in particular in the European part of the Soviet Union attracted the attention of the republic's minister of energy, Aleksei Makukhin. His ministry, together with the all-Union Ministry of Energy and Electrification, oversaw the performance of the Chernobyl nuclear plant. Makukhin, like most newcomers to the field of nuclear energy, was concerned about the safety of the newly constructed reactors, but he had no independent opinion on the matter. Long before the publication of Dollezhal's article, but already aware of Dollezhal's reservations, Makukhin had asked Grigorii Medvedev, then a deputy chief engineer of the Chernobyl plant, for his opinion. Medvedev responded that Dollezhal was right. The RBMK reactors were indeed "dirty." "What are the projected emissions of the Chernobyl reactor?" asked the now concerned minister, according to Medvedev's memoir. "Up to 4,000 curies per 24 hours," came the answer. "And of the Novovoronezh reactor?" asked the minister, referring to a nuclear plant near Novovoronezh in Central Russia employing VVER reactors. "Up to a hundred curies," responded Medvedev. "A substantial difference." "But the academics, after all . . . ," responded Makukhin. "The Council of Ministers has approved the use of [the RBMK] reactor. Anatolii Petrovich Aleksandrov praises that reactor as the safest and most economical one." He then added: "You're laying it on thick. No problem, we'll take care of it."[18]

Without the expertise and manufacturing base of Slavsky's Ministry of Medium Machine Building, the construction of the Chernobyl nuclear plant turned out to be a major challenge. The construction and plant managers, along with party officials, tended to report achievements, whereas the KGB, charged with maintaining the security of what was still considered to be secret technology and the safety of its operation, was busy pointing out design and construction flaws. Occasionally the local KGB directorate that oversaw the safety and security of the plant served as Vasyl Kyzyma's secret weapon—another channel through which he and his staff could alert party and state officials to problems with suppliers and lobby for their own interests. For instance, in August 1976, with the deadline for the launch of Unit 1 nearing, the Kyiv regional KGB office reported to republican headquarters on the failure of suppliers to deliver parts and hardware, as well as on their delivery of substandard or damaged items. KGB headquarters then reported the findings of their subordinates to the party's Central Committee in Kyiv.[19]

But the KGB was not only after the suppliers who failed to meet deadlines or comply with quality standards. It also targeted the shabby work done by Kyzyma's construction crews and Briukhanov's willingness to approve construction that was not up to industry standards. In February 1979, KGB reports about problems with the construction of Unit 2 reached KGB headquarters in Moscow, forcing none other than Yurii Andropov, the future general secretary of the party and then head of the KGB, to report to the Central Committee about the poor quality of the work. The report held one of Kyzyma's deputies responsible for building the foundations of the unit without proper hydro isolation, installing pillars that were up to 10 centimeters (almost 4 inches) off the mark, and raising walls that were 15 centimeters (almost 6 inches) off their intended location.[20]

Surprisingly, there were no accidents at the plant that could be directly attributed to construction problems. The most serious accident during this phase took place on September 9, 1982. On that day, planned repairs to reactor No. 1 were completed, and the operators began bringing it back to full capacity. Everything went without a hitch until the reactor reached the power level of approximately 700 megawatts of thermal energy (MWt), more than two-thirds of the projected capacity and a dangerously unstable condition in all RBMK reactors. One of the fuel channels burst, releasing enriched uranium into the core of the reactor, and it took operators close to thirty minutes to figure out that something was wrong and shut down the reactor. According to KGB reports, the release caused the beta radiation (high-energy electrons) in contaminated areas to rise to ten times normal levels.[21]

The commission investigating the accident concluded that the repairmen were at fault—one of them had allegedly shut the valve of the water coolant channel going to that part of the reactor, causing the fuel channel to burst. Briukhanov's second-in-command, the chief engineer of the plant, lost his job, but Briukhanov survived. After all, by industry standards, the Chernobyl plant was doing rather well, with a lower accident level than other plants of the same type. In the following year, Briukhanov was awarded the prestigious Order of the October Revolution.[22]

But problems persisted. In February 1986, as Briukhanov attended the party congress, and party leaders called for doubling the construction of nuclear units in the next five-year period, the KGB reported on continuing difficulties at the plant, pointing to a breach of technical specifications in the construction of Unit 5. Because the vendors did not deliver the right kind of crushed stone for making concrete, the managers used the stone they had, with particles twice the size of the type that should have been used. The concrete that they poured into the forms therefore failed to get into tight spots, leaving holes in the structure. "The estimated coverage area unfit for use is 300 meters," wrote the KGB officer. "As Unit 5 comes into use, the technical deficiencies discovered in the production of concrete may lead to accident-threatening situations, including the possibility of human losses." There was no response to this warning.[23]

IN MARCH 1986, the all-Union conference of vendors—suppliers of construction materials and specialized hardware—went on for three days and was declared a success in Prypiat. Vasyl Kyzyma, the head of Prypiat's construction directorate, was among the key speakers. Representatives of twenty-eight firms that provided supplies for the construction of Unit 5 were in attendance, and only two enterprises failed to attend.[24]

Viktor Briukhanov was not at the meeting, but the nuclear plant was represented by his chief engineer, Nikolai Fomin. Cooperation with the construction directorate and the launching of new reactors were among his responsibilities. Fomin had become an international celebrity of sorts the previous month, when an article about the Chernobyl power station in the February issue of the English-language magazine *Soviet Life* quoted him as saying that the station's cooling pond was being used for breeding fish—an indication of the station's complete safety. Even if an accident happened, argued Fomin, the automatic security system would immediately shut down the reactor.[25]

Fomin never mentioned the 1982 accident at the power plant. It would have been illegal for him to do so. In the summer of 1985, censorship of information about nuclear accidents had been reinforced by an instruction from the new energy minister of the USSR, Anatolii Maiorets. "Reports on adverse consequences of ecological effects on service personnel and the population, as well as of energy sources (the effect of electromagnetic fields, irradiation, or pollution of the atmosphere, bodies of water, or soil) on the environment, are not subject to open publication in the press or in radio and television broadcasts," read the minister's instructions to the industry's personnel.[26]

At the conference, the usually active and energetic Fomin was largely silent, not because of the ministry's restrictions on speech but because of injuries he had sustained in a car accident in late

1985. He had just returned from medical leave, and his very appearance at the conference was something of a miracle: those in attendance could see how difficult it was for him to participate in discussions—indeed, how hard it was for him to speak at all. Grigorii Medvedev, who was no longer deputy chief engineer of the plant, but had moved to Moscow to serve in the all-Union Ministry of Energy and Electrification, attended the conference to represent the ministry. He found Fomin a shadow of his former self. Medvedev almost did not recognize his colleague, whom he had known as a healthy man with a charming smile and a pleasant voice, someone always "coiled like a spring and ready to leap." "There was some kind of restraint in his whole appearance, the mark of the suffering he had borne," Medvedev later recalled. He said to Fomin, "Maybe you have to take a few more months and get better. Those were pretty serious injuries." But Fomin would not listen: "Got to get on with the job," he said.

Viktor Briukhanov gave Fomin all the support he could offer. "I do not think it's particularly serious," he said of Fomin's physical condition. "He's made good recovery," he assured Medvedev. "He'll get back to normal faster by working." The party secretary of the power plant had convinced Fomin to come to work earlier than scheduled because Briukhanov was attending the party congress in Moscow, and it was up to Fomin, as second-in-command, to take charge of the plant. To Medvedev, Briukhanov himself looked overworked and tired. The director was particularly concerned about units of the power plant leaking radioactivity. Altogether the leaks amounted to about 50 cubic meters (1,765 cubic feet) of radioactive water per hour coming from the drainage channels and air vents. The steam extraction units were barely able to deactivate the radioactive water. They were holding so far, but had reached the limit of their capacity. The only effective way to deal with the problem was to stop the reactors and carry out the repairs, but that would jeopardize the fulfillment of the annual plan of electricity production, and Briukhanov was not ready to face the rage of the party officials, whose main concerns were deadlines and quotas. Briukhanov told Medvedev that he was considering a move somewhere else—thoughts of a possible job abroad that he had previously dismissed were now probably testing his resolve to stay in Prypiat.[27]

Alarm bells were going off not only in Prypiat but also in Kyiv. On the day after the all-Union conference of vendors, Liubov Kovalevskaia, the *Tribuna énergetika* reporter, managed to publish a version of her earlier article about problems with the construction of Unit 5 in the Kyiv newspaper *Literaturna Ukraïna* (Literary Ukraine), the mouthpiece of the Ukrainian Union of Writers. Much of it was taken verbatim from her Russian-language article in *Tribuna énergetika* and translated into Ukrainian.

But the article in *Literaturna Ukraïna* was addressed to a wider audience and also made points of more general significance. While observing the obligatory idealization of Soviet socialism and praising the party for its accomplishments and concern about the Soviet people, Kovalevskaia listed arresting examples of the problems faced by construction crews at the Chernobyl power station. According to her, out of 45,500 cubic meters of prefabricated concrete that the construction directorate had ordered in 1985, 3,200 cubic meters of concrete never arrived, and 6,000 cubic meters of concrete turned out to be defective. Also defective were 326 tonnes of sealant for the nuclear waste depository and 220 tonnes of columns for the turbine hall of the reactor under construction. Kovalevskaia allowed herself to criticize not only the vendors who had failed to deliver parts and hardware on time, but also her own construction directorate—the owner of the newspaper she worked for.

"Disorganization weakened not only discipline but also the responsibility of each and all for the results of joint efforts," wrote Kovalevskaia:

> The impossibility or even incapacity of engineering and technical personnel to organize the work of brigades weakened standards. 'Exhaustion,' the deterioration of equipment, machines, and mechanisms, the lack of mechanical implements and tools, and the like all made themselves apparent. In a word, all the shortcomings—typical, alas—of the construction mechanism manifested themselves in aggravated form. And the time

coincided with the start of economic restructuring, which, as is well known, requires first and foremost the restructuring of human consciousness.[28]

Kovalevskaia waited for a reaction to her article, but there was none, either from Prypiat or from Kyiv. As far as the Prypiat party officials were concerned, Kovalevskaia's style of investigative journalism and record of uncovering abuses by senior officials had long marked her as a troublemaker. Rumor had it that the bosses were preparing to expel her from the party, which would make it all but impossible for her to continue her journalistic career. For the time being, however, no one seemed to care. The world media would not discover the article and its author until a month later, after April 26, 1986.[29]

II

INFERNO

FRIDAY NIGHT

O N FRIDAY, April 25, Prypiat residents were looking forward to the weekend. The previous one had been anything but restful, as the party authorities had turned Saturday, April 19, into an unpaid workday—a yearly ritual to mark the birthday of Vladimir Lenin. Had the father of the Soviet state possessed the life everlasting that the media claimed for his ideas, he would have turned 116 that April 22—not an anniversary of any significance, but that did not matter.

Ostensibly, work on the weekend closest to Lenin's birthday was a volunteer activity, but in reality, party officials demanded it. Mikhail Gorbachev skipped the occasion, as he was on an official visit to the German Democratic Republic, pushing his "acceleration" ideas there, but his compatriots did not let him down. The Politburo in Moscow reported that 159 million people all over the Soviet Union took part in the event—more than half the recorded population.[1]

According to the Prypiat newspaper, the ubiquitous *Tribuna ėnergetika*, citizens were eager to participate in "Red Saturday," as the authorities called it. A city official reported, in an article titled "The Holiday of Labor," that more than 22,000 of his fellow citizens had gone to work that Saturday for free. Workers at the nuclear plant and other city enterprises had produced goods and services valued at more than 100,000 rubles, while the contribution of those

involved in construction exceeded 220,000 rubles. The main action was of course on block No. 5, where, according to internal
party reports, delays were increasing. The newspaper claimed that
the construction workers were doubling their efforts—one crew had
managed to pour 30 cubic meters of concrete, which apparently was
an achievement significant enough to be praised; it received its due
in an article titled "Return in Full."[2]

The next weekend was supposed to be free of such duties.
Quite a few young couples in Prypiat were getting ready to hold
their weddings on April 26. Sundays were normally reserved for the
registration of newborns. Komsomol (Young Communist League)
organizers were happy to reserve premises and lend ideological approval to the wedding ritual, formerly part of church tradition, but
now not only secularized, but heavily ideologized—newlyweds were
expected to lay flowers on monuments to Lenin and memorials to
the heroes of the Great Patriotic War. Since Gorbachev had initiated an anti-alcohol campaign, party and Komsomol officials were
pushing for alcohol-free weddings. There were few takers in Prypiat.
The authorities would count it a success if weddings did not end in
drunken brawls between plant workers and "Rexes"—that is, the
construction workers from the villages.

Wedding or no wedding, everyone was glad that the weather
had finally improved. It was unseasonably warm, with temperatures
in the 70s (Fahrenheit). For many, that meant two or even three days
of picnics, hiking, and fishing in the neighboring forests along the
Prypiat River and its tributaries. Not for nothing had Chernobyl entered the annals of history as a princely hunting ground. The hunting season was still distant, but fishing had already begun. On that
Friday *Tribuna ènergetika* ran a photo on its back page featuring a
young Prypiat resident with a huge catfish in his arms. The caption
read, "Such a catch deserves to be in the 'red corner.'" The reference
was to makeshift propaganda displays organized by party officials
at workplaces. Judging by the image, the catfish weighed at least 40
pounds, and the caption suggested that such a prize catch deserved
public recognition.

The newspaper carried all-important information about the
fishing season. While fishing was generally prohibited until early

June—the end of the spawning season—the authorities allowed limited fishing in areas where spawning was not taking place. The article accompanying the photo identified spots on the Dnieper, Prypiat, and Uzh Rivers where onshore fishing was legal. Members of fishing and hunting associations could catch up to 3 kilograms of "valuable" fish daily, including carp and bream, and up to 10 kilograms of "less valuable" or common fish, such as crucian carp or bleak. Those not belonging to associations were allowed only 2 kilograms of valuable fish and 5 kilograms of common fish. The newspaper said nothing about fishing in the Chernobyl power plant's cooling pond, which the chief engineer, Nikolai Fomin, had recently advertised as a perfect place for breeding fish. Fishing there was strictly prohibited but still widely popular. Quite a few plant workers fished from the pond at night as the fishing inspectors slept—not from the shore but from boats.[3]

While the Prypiat *Tribuna énergetika* catered to the tastes of urbanites getting ready for a pleasant weekend outdoors, the Chernobyl newspaper, *Prapor peremohy* (Flag of Victory), with a largely rural readership—the town of Chernobyl, located about 16 kilometers southeast of Prypiat, continued to serve as the administrative center of the traditionally rural district around it—was preparing peasants for a busy weekend of collecting birch sap and planting potatoes, one of the region's main agricultural exports and the peasants' most important staple. The Saturday issue had articles of interest on both topics. It turned out that in the birch forests near Dytiatky—a village south of Chernobyl that would later give its name to the main entrance point into the Chernobyl Exclusion Zone—a local forestry brigade had managed to collect 90 tonnes of birch sap and ship it to customers. Local collective farms were competing with one another in planting potatoes. Very appropriately, given its name, the Peremoha (Victory) farm was leading in the competition. Its headquarters were located in the village of Stechanka, which turned out to be in its last days—its inhabitants would soon be resettled.[4]

But by far the most important information provided in the Chernobyl newspapers concerned the different types of potatoes available for planting. Two experts, one from a Kyiv research institute and the other from the Chernobyl center for potato seeding, explained the pros and cons of each. This subject was also important to many Prypiat residents, some of whom had *dachas*, or small summer houses, in the neighborhood, and planted potatoes and other vegetables there. But a much larger group, mostly young men and women from Vasyl Kyzyma's construction crews, were also getting ready to go back to their villages to help their parents plant potatoes. Good information on the best types to plant was the best gift they could offer their parents and themselves—during the long winter months they would rely on products sent to their city apartments from their parents' plots.[5]

Best of all, the coming weekend was the beginning of a long stretch of holidays. May 1, International Workers' Day, was a statutory holiday in the Soviet Union. Orthodox Easter, officially ignored by the authorities but widely celebrated, was coming up on Sunday, May 4. The week after that would also be a short one—May 9, marking victory in World War II, was a statutory holiday widely celebrated by the authorities and average people alike. Friday, April 25, was the last opportunity to take care of unfinished business before the holiday mood took over the entire city, making it difficult, if not impossible, to get anything done before mid-May.

LIKE EVERYONE else in Prypiat, Viktor Briukhanov was looking forward to a restful weekend. He badly needed one. Ever since his return from Moscow, Briukhanov had worked nonstop, coming home only to sleep. Now he would have an opportunity to enjoy the warm weather and the outdoors. Once, when he and his wife, Valentina, were bathing in the river in early April, they had encountered two elks in the water—an unforgettable experience. Maybe they would see the elks again. But the most important thing was that their daughter, Lilia, and her husband were coming from Kyiv for a visit.[6]

As always, Friday was a hectic day for Briukhanov, but there was no indication of trouble that might spoil the weekend. Radioactive leaks were still occurring, but they managed to keep them under control. At the moment, there was no need to shut down any of the reactors in order to deal with the leaks, as had happened a few weeks earlier at the Zaporizhia nuclear power plant in southern Ukraine, which had two operational units and one under construction. There, on April 7, controllers had detected radioactivity in cooling water coming from the reactor that exceeded the norm fourteen times over. To fix the problem, they had to shut down the reactor for two weeks, stop work on the unfinished unit, and stop generating power entirely, as their third unit was under repair. That meant no electricity, no bonuses, and a lot of questions from party and government officials.[7]

The Chernobyl plant was working to schedule. It was considered one of the best in the industry, with an average of five technological accidents and equipment malfunctions per year. They were about to shut down reactor No. 4, but that was for regular system checks and repairs according to industry regulations. Depending on the condition of the reactor, such repairs could take several months. The frequency of shutdowns depended on the ministry. The new all-Union minister of energy, Maiorets, was determined to make history and get noticed by his bosses in the government by lengthening the intervals between shutdowns and decreasing the amount of time spent on repairs. Increasing the flow of energy would please the higher-ups. In 1985, the Chernobyl power plant had overfulfilled electricity production quotas by almost 10 percent, partly by cutting time allocated for repairs. With much-needed shutdowns coming in 1986, the plant was scheduled to produce less energy than the previous year, and the local party authorities were not happy about it.[8]

Nevertheless, there were industry standards that neither party officials nor the minister could ignore. Unit 4's turn to stop for repairs was coming up in late April, and Briukhanov's engineers were ready to do the job. As usual in such cases, the shutdown of the reactor would be used to test its numerous systems at a low level of radioactivity. One of the tests to be conducted on the unit before shutting it down concerned the steam turbine and was designed to find a way to make the reactor more secure during SCRAM—the Safety Control Rods Activation Mechanism, which automatically inserted all the control rods into the active zone of the reactor and halted the reaction in case of emergency. The idea behind the test was fairly simple. In case of an emergency leading to the shutdown of the reactor, it was expected that the electricity would go off when the unit still needed it to pump coolant into the overheated reactor and prevent its meltdown. Emergency diesel generators were supposed to take care of that problem and provide badly needed electricity to continue pumping the water, but at present they would kick in only forty-five seconds after the shutdown, creating a supply gap and thus a potential safety problem. It had to be fixed.

Engineers from a research institute in the eastern Ukrainian city of Donetsk had come up with a solution to the problem. They pointed out that just as the reactor did not cool down immediately after the shutdown, so the turbine driven by residual steam pressure would also keep rotating for some time. Energy produced by that continuing rotation could be used to produce enough electricity to cover the forty-five-second gap. How long the rotation produced by the momentum of the turbine generator would continue and how much energy it would produce was the question that the Donetsk engineers wanted to answer with the help of their Chernobyl partners during the shutdown of reactor No. 4. That was the essence of the test.

The trick was that in order to conduct the test, which, ultimately, would improve the automatic shutdown mechanisms, those mechanisms would have to be disabled to simulate a power failure and blackout of the plant. Consequently, there was a risk that the reactor might go out of control during the test itself. But no one believed it was a huge risk. The plant managers were interested in conducting the test because it would allow them to activate one more emergency safety system envisioned by the builders of the reactor. Moreover, ministry instructions required them to do so. They had attempted to run the same kind of test earlier, but it had failed because the steam turbine generator had malfunctioned. That problem had been fixed, and everything now seemed good to go.[9]

Preparations for the test had begun in March and went into full gear in mid-April. The steam turbine test was potentially the most complex part of the system checks to be completed at reactor No. 4, but there were several others that had to be conducted. The task of preparing a timetable to coordinate all the tests went to one of the most experienced engineers around, Vitalii Borets. Then in his late forties, Borets had worked at power stations all over the Soviet Union. He had first come to Prypiat in March 1974. By that time he had close to twelve years of experience in the nuclear industry, mostly in the closed city of Tomsk-7, a nuclear facility and town near the Siberian city of Tomsk. Unlike Tomsk itself, Tomsk-7 was not marked on any Soviet map. It was home to the first Soviet industrial nuclear power plant, launched in 1958, and its main product was weapons-grade plutonium, not electricity. In December 1963, Borets belonged to the crew that launched the station's fourth reactor, ADE-4. Like the Chernobyl reactor, the one at Tomsk-7 used a graphite moderator to slow down the neutrons that bombarded the nuclei of enriched uranium. Borets worked at the Chernobyl plant for more than ten years before taking a job with a contractor responsible for launching reactors. Reactor shutdowns were also within his area of expertise.[10]

When asked to prepare a timetable for the tests, Borets readily agreed. He knew the plant exceptionally well. Unit 4 was the newest and, as many in Prypiat believed, the safest unit in the entire plant. It was part of the second stage of construction: it was housed not in a separate structure, like Units 1 and 2, but in a "duplex" that it shared with Unit 3. Both units had a production capacity of up to 1,000 megawatts of electrical energy (MWe). In order to reach that capacity, they had to generate at least 3,000 megawatts of thermal energy (MWt). Their generating capacity was rated as 3,200 MWt. Unit 3 had become operational in December 1981, and Unit 4 in December 1983. The protocol of the commission that inspected and launched Unit 4 had been signed by Nikolai Fomin on December 18, 1983.

The protocol described the reactor's main characteristics. Its vessel was a steel cylinder over 10 meters in diameter and 7 meters high made of high-purity graphite placed in a concrete pit almost 22 meters long, 22 meters wide, and 26 meters high. It was filled with graphite blocks to slow down the fast-moving neutrons and help them sustain a nuclear chain reaction by splitting the atoms of uranium into two smaller atoms and releasing kinetic energy. At the top and bottom of the cylinder were two massive metal plates that served as biological shields. The top one, called "System E" and dubbed "Elena" by the operators, was penetrated by numerous standpipes used for two kinds of channel assemblies—control and fuel rods. The reactor had 1,661 fuel rods or pressure channels, each approximately 3.5 meters long, filled with pellets of 2 to 3 percent enriched uranium-235 and natural uranium-238. There were 211 movable control rods made of boron carbide that absorbed neutrons, and which could therefore slow down the fission reaction if they were inserted into the active zone (or core) of the reactor, or increase the speed of the reaction if they were removed from the active zone. Two coolant loops circulated water overheated by the energy produced by the chain-reaction fuel rods, conducting it to steam drums that separated the steam from the water and fed it into the turbine to produce electrical power.

As with all Soviet reactors, there was no secure containment of the Unit 4 reactor except its concrete pit, but the commission found the functioning of all systems of the reactor satisfactory. As always, there were caveats. The commission noted problems with the reactor that would have to be fixed in the future. They recommended, among other things, that the construction of the control rods be modified. It turned out that they could produce an effect of positive reactivity, or a spike in the fission reaction and increase of the reactor's power, when inserted into the graphite core at a depth of less than 2 meters. (The overall length of the rods was 6 meters.) The adjustments that the commission recommended had already been made to Unit 3. These improvements were supposed to take care of problems such as those experienced in 1975 at the Leningrad power plant, where a drastic rise of radioactivity and destabilization of the reactor had resulted from a positive void effect—an increase of radioactivity within the reactor caused by the loss of coolant. The details of that accident, which almost destroyed the reactor, had been withheld from the personnel of other nuclear plants, but Borets knew firsthand how dangerous the situation had become at the Leningrad location.[11]

On November 30, 1975, Borets, who had been sent to the Leningrad plant from Chernobyl to undergo training on an RBMK reactor, inadvertently became a witness of the worst accident in the reactor's history. That day he decided to stay at the plant after his shift to see how the reactor would behave during the process of "shifting gears"—stopping the reactor and moving from one mode of operation to another. It soon became clear to Borets that there was something wrong with the reactor. Running at a low power level, it began to increase the radiation rate, even though the operator tried to slow down the process by inserting additional control rods into the active zone of the reactor. Normally, the nuclear radiation rate decreased as operators added control rods to the active zone of the reactor, where the nuclear reaction took place and nuclear energy was released from the fuel channels. But the RBMK reactor at the Leningrad power station behaved differently. Even when the experienced operator, whose skills impressed Borets, inserted additional rods manually, it did not slow down the rapidly increasing rate of radiation. The reactor was not behaving as expected.

As far as Borets was concerned, the reactor was out of control. Having operated reactors previously, he knew that if the rapidly rising level of radiation was not checked, it could cause an explosion. "Imagine yourself behind the steering wheel of a car," said Borets the following day to a security official who had little understanding of nuclear physics. "You start the motor. You start moving. You accelerate smoothly. You shift gears. Your speed is 60 kmph. You take your foot off the gas pedal. And suddenly the car independently begins to increase speed to 80, 100, 130, 150 kmph. You brake to no effect; the speed keeps increasing. How would you feel?"[12]

The ungovernable reactor was stopped twice by means of SCRAM, the emergency system, which halted the reaction. An explosion was averted, but because of spikes in the strength of the reaction, one of the fuel channels inside the reactor core melted, releasing uranium into the core. The reactor was shut down. It was "cleaned" the following day with nitrogen, and the waste, amounting to 1.5 million curies of radionuclides, was released into the environment through the exhaust pipe. One curie is equivalent to the amount of radiation released by the fission of 37 billion atoms. It can contaminate 10 billion quarts of milk, making them unfit for human consumption. According to the International Atomic Energy Agency, a safe level of nuclear contamination of a territory is 5 curies per square kilometer. It is anyone's guess what impact 1.5 million curies had on the people and the territory around the station, including the city of Leningrad, less than 50 kilometers away.[13]

Vitalii Borets never received any explanation of what had gone wrong with the reactor, and he knew nothing about this major technical weakness in the reactor design. The information was kept under wraps. The designers of the reactor made no major changes in the RBMK model; instead, they issued instructions on how the control rods should be improved, failing to explain why that had to be done. Those instructions eventually made their way into the recommendations of the commission that inspected reactor No. 4 at the Chernobyl nuclear plant. But everyone there believed that the problem with the rods was a minor one. The lessons of the Leningrad reactor's failure were not learned. There were many improvements that could be made to a reactor, but the main task of the plant's operators was to produce energy, not to design new reactors or improve existing ones. The repairs could wait.

The KGB, constantly keeping a watchful eye on the plant, was generally satisfied with the condition of the two newest units. Its agents keeping track of the progress at Chernobyl believed that Units 3 and 4 were generally safer than Units 1 and 2. By 1984, they had concluded that despite serious safety issues at the plant, the general situation was improving: in 1982, there had been three accidents and sixteen breakdowns of equipment at the three working units, but in the first nine months of 1984, there had been no accidents and only ten breakdowns.[14]

There seemed to be no reason for Vitalii Borets or anyone else to recall the accident at the Leningrad station when they were getting ready for the shutdown of reactor No. 4 for testing. Borets did as he was asked. He collected information on all the proposed tests from the various units and consultants and submitted the timetables he prepared to the expert group responsible for planning the shutdown. Borets suggested that the shutdown begin on Thursday, April 24, at 10:00 p.m. The whole operation, including the shutdown test, was to end by 1:00 p.m. on Friday afternoon, April 25.

The expert group replied that the shutdown should end by 10:00 a.m.; otherwise, the radioactivity level in the reactor would fall below admissible levels. Everyone agreed. The chief engineer, Nikolai Fomin, signed off on the program. He later recalled that originally the shutdown had been planned for April 23, but then they decided to do it over the weekend. They never submitted the program for approval by representatives of the Ministry of Energy and Electrification in Moscow or the manufacturer of the reactors, as was prescribed by instructions but rarely done in practice. The Chernobyl nuclear plant was supposed to enter the last weekend of April with Unit 4 safely shut down.[15]

PREPARATIONS FOR the shutdown of reactor No. 4 began not on the evening of April 24, as suggested by Borets, however, but with the start of the new shift of operators in the early morning hours of April 25. By 4:48 a.m. the power level of the reactor had been reduced by half and secured at 1,600 MWt.[16]

The shutdown procedures were continued by Igor Kazachkov, the head of the morning shift in Unit 4 that took over from the night shift at 8:00 a.m. on April 25. In his mid-thirties, Kazachkov was one of the most experienced shift leaders at the plant. He had come to Prypiat in 1974, immediately after graduating from the Odesa Polytechnic Institute, and risen through the ranks to his current position. In December 1985, the regional newspaper even ran a profile of him with a photo—in his white cap and gown, sporting glasses and a goatee, he was standing by the screen of one of the plant's computers. He appeared to be a thoughtful young man, and the brief article praised him for his "precise organization and work discipline."[17]

Kazachkov had inherited from the night shift a reactor that had almost all its control rods inserted into the core in order to reduce the intensity of the nuclear reaction. Fewer than fifteen rods were left uninserted and were still at his disposal to regulate the reactor's behavior. The manufacturer's instructions suggested the need for a shutdown when this point was reached, but neither the personnel on the night shift nor Kazachkov began the shutdown procedure. That would have been a violation of the program they had been given by their superiors. The power generator test, along with the numerous other scheduled tests and measurements, had not yet been carried out. But even more importantly, short of an extreme emergency, the reactor could be shut down and disconnected from the power grid only on orders of the plant superiors, who needed the approval of the power-grid inspector in Kyiv.

Kazachkov later said, "Why did neither I nor my colleagues shut down the reactor when the number of protective rods [those not yet inserted into the active zone] was reduced? Because none of us imagined that that could bring on a nuclear accident. We knew that it was forbidden [by the manufacturer] to do so [delay the shutdown], but did not think [about it]. But if I [had] shut down the reactor, I would [have gotten] a severe tongue-lashing. After all, we were racing to fulfill the plan." Asked what the consequences of his shutting down the reactor might have been, Kazachkov responded: "I think they would have fired me. They would certainly have fired me. Not for that, of course. But they would have latched onto something. That particular parameter—the number of rods—was not something we considered serious."[18]

So, not overly concerned about the inadequate number of rods at his disposal, Kazachkov followed the prearranged program and prepared for the test by disabling the system of emergency water supply to the reactor. The process for stopping the reactor was supposed to last only a few hours, so Kazachkov considered the odds of the regular water supply system failing about as likely as that of a plane falling from the sky and hitting someone on the head. Shutting down the emergency water system was a lengthy and laborious process in which the operators had to close the valves of huge pipes by hand. That took up to forty-five minutes, with teams of two or three men working on each valve. The system was shut down by around 2:00 p.m. They had another fifteen or twenty minutes before the start of the actual shutdown of the reactor. But then came a call from the plant's administration, which wanted the reactor to maintain its current production of 1,600 MWt. The shutdown and the test, which had to be carried out at a power level of 700 MWt, had to be postponed.

The plant's administrators changed their plans because of a call they had received from the headquarters of the Kyiv area power grid—the office charged with managing the distribution and consumption of electrical power produced by the plant. It was, in fact, the Chernobyl plant's only customer. Unless there was an emergency at the plant, the power grid operator had to be reckoned with. It turned out that at the southern Ukrainian nuclear station in the Mykolaiv region, one of the units had unexpectedly gone offline, and the operator wanted the Chernobyl unit to maintain its current power level until evening, when the demand for electricity would decrease and the plans for the shutdown of the reactor could finally proceed. No one at Unit 4 was happy with the request, especially as it came a mere fifteen minutes before the start of the shutdown, with the emergency water supply system already shut off. But there was little they could do other than follow instructions from the authorities at the Kyiv power grid. The electricity produced by just one unit of the Chernobyl power station was enough to keep the entire city of Kyiv going, so they could not go on or off line just as they pleased. They had complained in the past about similar conflicts, but to no avail.[19]

In February 1986, the dissatisfaction of the Chernobyl power plant personnel with the grid operators had been noted in KGB reports that found their way to Moscow. In 1985 alone, there were twenty-six cases in which the Chernobyl operators had had to change the power output of the units in response to orders from the grid dispatchers. In the first three weeks of 1986, there had been nine such cases, leading to an overall decrease of power output. The reactor operators complained that RBMK-1000 reactors were designed to work at a constant level of power output, and a decrease could lead to malfunctions. Moreover, changes of power output released radioactive particles into the atmosphere. Moscow's response to the reports indicated that even the KGB was powerless to do anything about the existing practice—the ministries in Moscow responsible for nuclear energy had simply taken the problem under advisement.[20]

By 4:00 p.m., the evening shift had taken over the control room of the unit. Its leader, Yurii Trehub, was not familiar with the testing program: the shutdown of the reactor was supposed to be over by the time he started his shift, and he was clearly upset by the dispatcher's demand, which he considered unreasonable. "I'm surprised that there could possibly be such a turn of events—a dispatcher taking command of an atomic power station," said Trehub a few months later, venting his frustration. "After all, even if we had an accident or a power interruption, the dispatcher might not give permission for a shutdown. But we're not talking about a thermal power station, a mere boiler exploding on the premises. . . . It's always very difficult dealing with dispatchers . . . there are plenty of arguments."[21]

Trehub and his group of engineers tried to come to terms with the situation. By the time he took over the shift, the reactor's power level had been reduced by half, from its full power output of 3,200 MWt to 1,600 MWt. Trehub was surprised that the safety system had been shut down. "What do you mean, they turned it off?" he asked Kazachkov. "On the basis of the test program, although I objected," was the response. Kazachkov added that the power grid dispatcher was expected to give his approval for the shutdown of the reactor around 6:00 p.m. Accordingly, the steam turbine experiment that required the safety water supply to be turned off was postponed, not canceled altogether. Given the difficult and laborious process of enabling and then shutting off the safety system again, Trehub decided to leave it as it was and wait for the grid dispatcher's approval, then begin the shutdown of the reactor. He discussed the situation with his immediate boss, the leader of the plant's evening shift, and they agreed that they had no choice but to stick to the approved program, which Trehub studied carefully. Not everything was clear to him, but he had no one at hand to consult with, and he was busy running the other tests on Borets's program. Most of them did not require a complete shutdown of the reactor. In front of him was a display with close to 4,000 indicators of the reactor's activity to be monitored and controlled.

Six o'clock passed with no news from the dispatcher. Around 8:00 p.m., the worried Trehub called the plant's shift leader—still nothing. The leader advised Trehub not to proceed with the shutdown of the reactor until a deputy chief engineer, Anatolii Diatlov, arrived. Diatlov was in charge of operating the reactors, and when it came to a shutdown, he was the ultimate authority for the personnel. Trehub called Diatlov. It turned out that he had left his office around 4:00 p.m. to get some rest. Trehub eventually found Diatlov at home and told him, in distress, "I have questions, many questions." "This is not a phone conversation—don't start without me," came the brisk answer. Then, out of the blue, Trehub received a call from Diatlov's boss, Nikolai Fomin himself. He also told Trehub not to start without Diatlov. But Diatlov would not come to the plant until they received approval from the dispatcher of the power grid to shut down the reactor. Finally, shortly after 9:00 p.m., word came from the dispatcher that the shutdown could start at 10:00 p.m. Trehub immediately called Diatlov, whose wife, Isabella, answered that he was already on his way.[22]

They were finally ready to start shutting the reactor down. With the test scheduled to take less than two hours, Trehub expected that it would be conducted on his shift, before midnight on April 25. They had to hurry. But where was Diatlov?

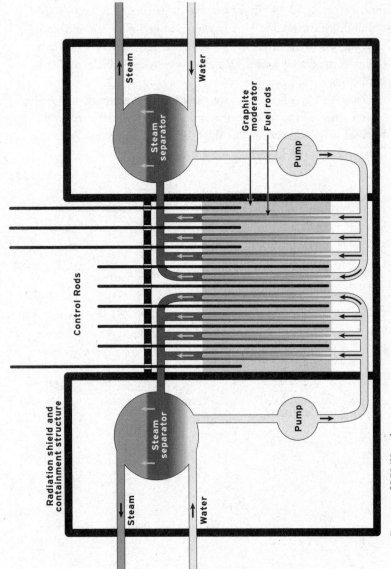

Diagram of RBMK nuclear reactor

EXPLOSION

A s usual, Anatolii Diatlov, a fifty-five-year-old deputy chief engineer of the Chernobyl power plant, reached his workplace on foot. A tall, well-built man with a broad, open face and silver-gray hair and moustache, he cared about keeping fit, and the walk to the station from his apartment building on Lenin Boulevard and back was part of his daily routine. He calculated that by walking 4 kilometers each way, he covered close to 200 kilometers per month. Along with all the fast walking he did at the plant from one reactor to another, and then within the units themselves, that added up to 300 kilometers per month—sufficient exercise, in his opinion, to keep him in good physical form. He found his daily walks good for his mental balance. "Walking, you clear your mind of all unpleasant thoughts. And if something does find its way into your head, speed up," he wrote later.[1]

On the night of April 25, his walk was like any other. No untoward thoughts entered his head—at least, he did not remember any later. Everything seemed normal and under control. The shutdown schedule had been slightly adjusted, but that was nothing to worry about—it had happened before. Like everyone else in Prypiat, he was looking forward to the weekend—relaxing with his family, which did not see much of him during the week, and spending some time with his granddaughter. A lover of classic Russian poetry who could recite entire poems by Aleksandr Blok and Sergei Yesenin by heart, he planned to spend at least part of the weekend with a book in his hands. But first he had a task to complete. When it came to the plant's senior management, the shutdown of Unit 4, planned for that day, was mainly his responsibility.

Diatlov was one of the top nuclear experts at the plant and an early arrival in Prypiat. He had come to the city in September 1973 at the age of forty-two. Born in Siberia, he had studied in Norilsk in the far north and at the Moscow Engineering Physics Institute (MIFI), the Soviet Union's leading school for nuclear physicists, engineers, and practitioners in other technological fields. For more than a decade, Diatlov had worked at the shipbuilding yard in Komsomolsk-on-Amur, a far eastern industrial city of more than 200,000, where he headed the group that tested nuclear reactors designed for Soviet submarines. He decided to change jobs after he and his family got tired of his regular tours of duty on nuclear submarines, where he installed and tested nuclear engines. While Diatlov had no previous experience of working on powerful reactors like the ones designed for the Chernobyl power station, he had no problem training himself on the job. He would later bring quite a few of his colleagues from Komsomolsk-on-Amur to work at the Chernobyl station.

In Prypiat, Diatlov rose through the ranks from deputy chief of the reactor group to deputy chief engineer. He received two high state awards for good work. With Nikolai Fomin injured in a car crash, some believed that Diatlov had a good chance of replacing him in the chief engineer's office. Many thought that would be a good thing: both the director of the plant, Viktor Briukhanov, and its chief engineer, Fomin, had come to Prypiat from conventional thermal plants that ran on coal, and neither was an expert on nuclear power. Diatlov was. In fact, he was the highest-ranking nuclear power expert at the plant, and he was responsible for the operation of the reactors, including launches and shutdowns.

With reactors, as with airplanes, "takeoff" and "landing" are the most challenging moments. Diatlov had to be there for the shutdown of Unit 4 to see that everything went according to the program. There would also be a turbine test, which had been approved the previous day by Diatlov's boss, Nikolai Fomin. Preparations had begun in March, but it was not until mid-April that Fomin called a meeting of engineers from the plant and representatives of research units and consulting firms to discuss the procedures and establish a combined program for the tests. It was Diatlov who gave Vitalii Borets the task of coordinating the schedules. He was also the first to approve the proposed test programs and timetable.[2]

At the plant, Diatlov was known for being difficult, occasionally even rough. "Diatlov was a complicated man to deal with, direct, with his own point of view that he never changed according to a superior's wish. He argued and did not agree; ultimately he obeyed but stood by his opinion," recalled one of his acquaintances. "In just the same way, he did not reckon very much with the views of subordinates. You understand that not everyone likes such a person." A more favorable assessment of Diatlov's management style came from another colleague: "Diatlov would instantly chew out anyone who tried to deceive, to weasel out of fulfilling a task, to hide behind far-fetched explanations, or, even worse, to conceal a violation of instructions. And then the punishment would fit the crime. Many reacted with anger and indignation, while understanding the justness of the assessment in their hearts."[3]

If Diatlov was a strict disciplinarian, he was also good at getting things done—the quality most valued by his bosses. His argumentative nature and his subordinates' complaints about his occasional rudeness could be brushed aside. Those who knew Diatlov appreciated his sense of humor. He had a wonderful memory and could memorize whole pages not only of poetry but also of technical instructions. In his line of work, that was a valuable asset.

ON THE night of April 25, Diatlov clearly took his time. With the power plant only a few kilometers away from his apartment building, and Diatlov himself in good physical shape, Yurii Trehub estimated that it shouldn't have taken him longer than forty or fifty minutes to reach Unit 4. Trehub had called Diatlov's apartment around 9:00 p.m. and learned that he had already left. But now it was well past 10:00 p.m., when the operator of the Kyiv power grid had given permission to shut down the reactor, yet Diatlov was nowhere in sight. It was only around 11:00 p.m. that colleagues called Trehub from the neighboring Unit 3 to inform him that Diatlov was there. "He stopped off at Unit 3 on his way and evidently found something wrong with regard to discipline. He dressed them down," remembered Trehub. That was typical of Diatlov. "He punished personnel severely for mistakes and disobedience, bellowing at them and fraying nerves," recalled another colleague.[4]

But Diatlov was in no hurry—he had delayed his arrival for one simple reason: when the request from the power grid operator earlier in the day had held up preparations for the reactor shutdown, he had asked Hennadii Metlenko, the head of the group from Donetsk conducting the turbine test, to check with the unit shift at 10:30 p. m. to see whether the test could proceed. It would take the consultants at least half an hour to reach the unit, so there was no point in rushing to get there before 11:00 p.m. When Diatlov showed up in the control room of Unit 4, it was already after that, and the consultants, led by Metlenko, came in a few minutes later. They could finally start. Trehub had a number of questions about the shutdown procedure, but Diatlov refused to discuss them. Trehub understood that Diatlov had already decided to conduct the turbine test on the next shift, which made good sense. At 11:10 p.m., Trehub began to reduce the power level of the reactor, then at 1,600 MWt. By midnight, the official end of his shift, the level was down to 760 MWt, as prescribed by the program for the turbine test.[5]

The new shift that took over from Yurii Trehub was headed by Aleksandr Akimov, a thirty-three-year-old engineer with ten years of experience at the Chernobyl plant. Bespectacled and sporting a fashionable moustache, Akimov was regarded as competent, friendly, and susceptible to pressure from the higher-ups. A member of Prypiat's Communist Party committee, he was clearly on his way up— he had been appointed head of shift only four months earlier. Also new on the job was the twenty-five-year-old Leonid Toptunov. Like Akimov, he was nearsighted, wore glasses, and had a moustache. It was Toptunov's third month as senior engineer in charge of the reactor's operational regimen, a demanding job of operating dozens of switches and tumblers. An extra worker was added to do that job during the first days after the engineers in charge of reactors returned from their vacations. Otherwise, quipped Trehub, who had performed that function earlier, it would be like expecting a pianist to perform without rehearsals. Other members of the shift, engineers responsible for the operation of the unit and the turbines, had more experience in their respective positions.[6]

Akimov and his crew had worked the previous night shift, and it was on their watch that the slowing down of the reactor had begun in the early hours of April 25. By the time they arrived for the new shift, the process of shutting down the reactor was supposed to be complete, and they expected a relatively peaceful period of watching over a reactor that had been stopped earlier in the day by more experienced crews. But now the task of shutting down the reactor was being passed back to them. Akimov, who had arrived half an hour before the start of his shift, was trying to figure out what he had to do. Shutting down a reactor was a difficult task, and the handover of responsibility was rushed. The control room was full of people—old shift, new shift, managers and engineers from other divisions who were involved in the turbine test or who simply wanted to see how their equipment behaved during the shutdown—altogether, up to twenty people.

Trehub had had some time to figure out how the shutdown and test program should work, but Akimov had none. Akimov sat next to Trehub, who explained to him what he had figured out about the test program—he had been studying it for most of his shift. But some questions remained unanswered, such as what to do with the power produced by the reactor once the supply of boiling water and steam was cut off from the turbine. Diatlov had not been interested in discussing that matter with Trehub, so Trehub simply offered Akimov what he believed to be the best solution. Trehub stayed on for the next shift, but only as an observer, since he was interested in seeing how the test would go. Akimov was now formally in charge of the control room, while informal authority was held by Diatlov, the senior person in the room—who immediately reminded everyone of that.[7]

Razim Davletbaev, the deputy chief of the plant's turbine unit, was in the control room that night, too. He recalled:

> Immediately upon the start of the shift, Diatlov began demanding that the program continue to be carried out. When Akimov

sat down to study the program, Diatlov began reproaching him for working too slowly and failing to pay attention to the complexity of the situation that had arisen in the unit. Diatlov shouted at Akimov to get up and started insisting that he hurry up. Akimov, holding a sheaf of papers in his hands (evidently the program), began going around to the control room operators and establishing whether the equipment was in appropriate condition for the program that was being carried out.[8]

They started the shutdown. By that time, the reactor's power level had been reduced from 1,600 MWt to 520 MWt. Following the program, Leonid Toptunov continued gradually reducing power output by inserting control rods into the core of the reactor. The deeper the rods penetrated, the more neutrons they neutralized, and the slower the reaction would become. Everything went as planned until an emergency signal indicated that the supply of water to the reactor had dropped to an unacceptably low level. Yurii Trehub, who saw the emergency light go on, rushed to help Toptunov. Toptunov was confused and did not call technicians to check whether the signal was correct and whether he should open the valve to increase the supply of water. Trehub did that instead. He began switching tumblers to check the water level when suddenly he heard Akimov's voice: "Keep up the power!" He looked at the indicators on the control panel and realized that the power level of the reactor was rapidly falling.

It turned out that when Toptunov switched from one regulator of the control rods to another, which turned out to be out of order, he caused a sudden reduction of the power level, bringing the reactor near to shutdown long before the test had been completed. At twenty-eight minutes past midnight on April 26, the reactor's computer recorded the level falling to 30 MWt—a huge drop, given that it had begun descending at a level close to 520 MWt. According to one eyewitness, the power level dropped to almost zero. With Akimov's assistance, Toptunov switched off the automatic system controlling the rods and began withdrawing them manually to breathe life back into the dying reactor. Trehub was also there to help. "Why are you withdrawing [the control rods] inconsistently? This is where you should be withdrawing," he told Toptunov, who was initially taking them out of some sectors of the reactor but not from others. Trehub eventually replaced Toptunov at his work station and brought the power level up to keep the reactor going. In four minutes they raised the level from 30 MWt to 160 MWt. Everyone breathed a sigh of relief. "That moment of keeping up the power level was somewhat nerve-wracking," remembered Trehub. "As soon as we got up to a level of 200 MWt and switched to automatic, everything settled down."[9]

The question was what to do next: shut down the reactor in orderly fashion or raise the power to a level at which the test could continue. If the reactor was shut down, then the turbine test, so long in preparation, would have to be abandoned. No one wanted that. Diatlov, who had left the control room when the loss of power took place, was now back, and, as he later recalled, he gave his permission to raise the power level. Those in the control room saw him wiping sweat off his brow. Trehub remembered him in discussion with Akimov, who, with papers in hand, apparently was trying to convince Diatlov of something. Trehub overheard Diatlov demanding that the power level be kept at 200 MWt. That was much lower than the 760 MWt prescribed for conducting the test, and deviation from the specified test parameter could spell disaster. But at 200 MWt they thought they could stabilize the reactor and start the test. Whether Diatlov suggested that level first and then ordered Akimov and Toptunov to attain it or simply accepted their proposal remains a mystery. But he never denied that he approved conducting the test at that power level. Since he was the senior official in the control room, everyone followed his orders.[10]

Diatlov was determined to go ahead with the test. Those present in the control room remember him urging people to hurry up. At forty-three minutes after midnight, he ordered the operators to block the emergency signal from the two steam turbines that would be involved in the test. Twenty minutes later, at 1:03 a.m., in order to increase the flow of water through the reactor, they activated one of the two reserve pumps, then another one four minutes later. It was all part of the test program, but, given the reactor's low power level, the addition of two pumps to the six already in operation further

destabilized the reactor. Additional pumps increased the flow of water and decreased the amount of steam in the steam/water separator
drums. The power level took another hit: water, unlike steam, absorbs neutrons and slows down the rate of nuclear reactions. At 1:19
a.m., the low steam pressure alarm sounded. The operators switched
off the alarm and shut down the reserve pumps.

Keeping the reactor going, at the low power level of 200 MWt,
turned out to be a problem that became ever more difficult to solve
as they moved closer to the moment when they could begin the test.
The power level kept falling. The rapid decline of the power level
earlier that morning, and the reactor's continuing operation at a level
of 200 MWt or less, caused a rapid accumulation in the fuel rods
of xenon-135, a byproduct of nuclear fission that slowed down, or
"poisoned," the reaction by absorbing neutrons. To keep the power
levels from falling further, Toptunov had to keep removing control
rods from the active zone of the reactor. Numerous emergency sig-
nals indicating the unstable state of the reactor were ignored. Soon,
only 9 rods out of close to 167 available remained in the core of the
reactor—all the rest had been withdrawn, making the reaction difficult to control and the reactor highly unstable.

Then, at 1:22 a.m., the computer system indicated that the reaction rate had begun to pick up—yet another uncontrolled factor.
The water in the cooling system, with only four pumps operating,
reached boiling point and turned into steam, drastically reducing
the ability of the water, acting as a coolant, to absorb neutrons. Less
water and more steam meant more unabsorbed neutrons and a spike
in the intensity of the reaction. Toptunov read the computer data
and reported it to Akimov. The power level was rising at an alarming speed. But Akimov was wholly focused on the logistics of the
turbine test, which was supposed to start in a few seconds.[11]

Razim Davletbaev, the deputy head of the turbine division, who
was in the control room, recalled that "the head of the shift in Unit
4, Aleksandr Fedorovich Akimov, . . . went up to every operator,
among them the senior engineer in charge of turbines, Igor Kirshenbaum, whom he briefly instructed that after the command was
given to start the test, he should shut off the steam on Turbine 8.
Then Akimov asked the operators to get ready, after which the test

representative of the Dontekhenergo enterprise, Metlenko, gave the command 'Attention, oscillograph, start.'" It was 1:23:04 a.m. "After that command," continued Davletbaev,

> Kirshenbaum closed the stop-valves of the turbine, while I stood next to him and watched the tachometer monitoring the speed of Turbine 8. As was to be expected, its speed fell sharply as a result of the electrodynamic braking of the generator. . . . When the speed of the turbogenerator had been reduced to the level specified in the test program, the generator started up again, that is, the "coasting down" segment of the test program worked properly, and we heard the command of the shift leader, Akimov, to shut down the reactor, which was done by the operator of the unit's control panel.

The time was 1:23:40 a.m.[12]

The test had lasted thirty-six seconds. Those seconds turned out to be crucial for the fate of the reactor and the people in the control room of Unit 4. The power level of the reactor was out of control. The situation was worsened by the growing number of steam voids in the water coolant, which was now failing to absorb neutrons, and the lack of control rods in the active zone of the reactor, which had been removed in an attempt to revive the reaction. The automatic control system was trying to reduce the speed of the reaction with its twelve rods, but the rest of the rods had been withdrawn manually and remained out of the active zone.

Toptunov, who had access to the computer data, shouted that the power level was rapidly rising. Diatlov later remembered that at the end of the test he heard the voices of Akimov and Toptunov. "I was about 10 meters away from them and did not hear what Toptunov said," wrote a deputy chief engineer in his memoirs. "Sasha Akimov ordered a shutdown of the reactor and pointed with his finger— press the button." The button that Akimov ordered Toptunov to press was AZ-5, the one used for SCRAM, or emergency shutdown of the reactor. Toptunov removed the paper cover from the button and pressed it. Diatlov and the rest of the crew in the control room could finally breathe a sigh of relief. The difficult test was over. The red AZ-5 button was supposed to do its job and shut down the reactor. It was an unusual measure, but this was an emergency.[13]

Once the button was pressed, 178 control rods began to move in the active zone of the reactor. They were 7 meters long, moved at a speed of 40 centimeters per second, and were made of boron, which absorbed neutrons and reduced the rate of the reaction. The tips of the rods, however, were made of graphite, and the graphite tips appear to have tipped the already highly unstable reactor toward catastrophe. As the rods began to descend into the core of the reactor, the tips replaced neutron-absorbing water in the top part of the active zone, thus not decreasing, but further increasing, the rate of the reaction. This was the positive void effect—the deadly design problem of RBMK reactors that had almost destroyed one of them at the Leningrad power station in 1975. Now the positive void effect was once again at work.

The introduction of the control rods with their graphite tips caused a spike in the level of the reaction and a dramatic rise of the core's temperature. The rise in temperature, in turn, caused the cladding of the fuel rods to fracture. These tubes, less than 14 millimeters, or approximately half an inch, in diameter, have zircaloy walls less than 1 millimeter, or 0.04 inches, thick, making them thinner than a strand of hair. The fractured fuel rods jammed the control rods, which by that time had been inserted to only one-third of their length. The core and the bottom of the reactor's active zone remained out of reach of the rods, and the reaction there spun completely out of control. The power output of the reactor, which had hovered around 200 MWt, jumped to more than 500 MWt within a few seconds and then shot up to more than 30,000 MWt—ten times the norm. The rapidly increasing number of unabsorbed neutrons had burned away the xenon-135 that had been preventing the reactor from picking up speed a few minutes earlier. Now there was nothing to slow down the nuclear reaction. The fuel rods disintegrated, and the uranium fuel tablets in the zircaloy tubes of the fuel rods were released into the water of the cooling system, causing an enormous spike in the production of steam, which had nowhere to go.[14]

Those in the control room heard a sudden roar. "That roar was of a completely unfamiliar kind, very low in tone, like a human moan," remembered Razim Davletbaev. For Yurii Trehub, it was at first "as if a Volga [automobile] proceeding at full speed had started to brake and began to skid. It sounded like du-du-du-du." Trehub then heard a roar like the one described by Davletbaev. After that came the shocks. "But not as in an earthquake," remembered Trehub. "If you count[ed] ten seconds, there came a roar, and the frequency of the shocks diminished. But their strength increased. Then came the sound of the blast." Those were the effects of the steam explosion that destroyed the casing of the reactor, throwing the concrete plate—the upper biological shield of the reactor that the operators called "Elena," which weighed 200 tonnes—through the roof of Unit 4 and into the air. The plate, to which the entire infrastructure of the reactor was fastened, landed back on top of the reactor but did not cover it completely, leaving an opening through which it could freely "spit" radiation into the atmosphere. The time was 1:23:44.

Two seconds later, the operators heard another, much more powerful blast. "The floor and walls shook violently, dust and bits of debris fell from the ceiling, the luminescent lighting went off, semidarkness descended, and only emergency lighting was on," recalled Davletbaev. Those inside the control room heard and felt the explosions but did not know what had happened. An exploding reactor was the last thing on their minds. It was a difficult shift, and numerous alarms were going off, but such things had happened before. If something was going wrong, it could only be the cooling system or the steam turbine, not the reactor. As far as they were concerned, the reactor and its panoply of safety systems were idiot-proof. No textbook they had ever read suggested that reactors could explode. "Everyone was in shock," recalled Trehub, describing the scene after electricity was restored in the control room. "Everyone stood around with long faces. I was very frightened. Complete shock."[15]

They thought there had been an earthquake. It took them a while to realize that it was a man-made earthquake—one that they themselves had produced. The first blast was caused by the steam explosion as the excess steam produced by the breaking of the fuel channels escaped into the external cooling system, detonating it. That explosion blew the Elena biological shield into the air, further damaging the fuel channels and tearing off the coolant lines attached to the shield. Without water to cool the active zone, the thermal power of the damaged reactor increased even more, causing a second, even more powerful explosion.

The second blast destroyed a good part of the containment building and threw graphite blocks into the air—the moderator core of the reactor, along with part of its fuel. The highly radioactive pieces of graphite landed on the roof of the neighboring Unit 3, and were scattered all over the land on which the power plant was built. The graphite also caught fire inside the damaged reactor, sending radioactive particles high into the sky.

The first to see what had happened were the scores of men who were fishing in the cooling pond of the power station on that warm April night—the pond that was used to breed fish and served as proof of the safety of the power plant. Two of the fishermen were very close to Unit 4, a mere 260 meters from the turbine hall. Suddenly they heard the dull sounds of the explosions, one and then the other. The ground shook under their feet, and the flames following the explosion suddenly illuminated the area, betraying their location. But no one came out to investigate. As fire rose higher and higher from the ruins of the reactor, they kept on fishing. They were hardly in a position to appreciate the significance of what they had witnessed: a nuclear star had fallen onto the earth, poisoning the land and water nearby, their catch, and the two fishermen themselves. They saw everything but realized nothing. They were the first but not the last to fail to grasp the reality.[16]

FIRE

For the firefighters at the Specialized Military Fire Department No. 2 on April 25, it was a day like any other, except that it was Friday. Everyone was discussing plans for the coming weekend. Many were getting ready to visit their families in neighboring towns and villages, where they would help with sowing potatoes. Those living in Chernobyl had private plots beside their cottages. Potatoes were on everyone's mind. Most of the engineers and technical personnel working at the power plant were newcomers from Russia and other parts of Ukraine, but the firefighters, like the construction workers, were locals—Ukrainians and Belarusians from across the border. These village boys had taken jobs and salaries that they could only have dreamed of in their rural birthplaces.

The shift lasted twenty-four hours, from 8:00 a.m. one morning to 8:00 a.m. the next. Including the extra half-hour for their shift change with the next group of firefighters, they were supposed to be free to go home no later than 8:30 a.m. on April 26. They would have forty-eight hours off, with another shift starting on Monday, April 28. They had spent the day studying, exercising, and training around Unit 5, which was still under construction, but in the evening they had taken it easy. Some of the men had taken naps, while others had watched television. The evening news came on at 9:00 p.m.

The main news item was the recent Politburo meeting, at which Gorbachev and the party leadership had discussed ways of increasing production of consumer goods—an all-important issue in a society that suffered shortages of basic consumer items. Cars, an especially coveted item, could be bought only after spending years on the waiting list. Some of the older firefighters were driving cars, but the younger ones had motorcycles. In the international news, the hot topic was the conflict between the United States and Libya. President Ronald Reagan had ordered air strikes on Libya on April 15 in response to the bombing ten days earlier of the *La Belle* discotheque in West Berlin, which was frequented by American military personnel. The move was highly controversial not only in the Soviet Union but also in the West, and Soviet television had a field day covering protests against the American action throughout the world.[1]

On Friday night, people waited impatiently for the end of the news and the start of entertainment shows. At 11:10 p.m., one of the two all-Union television programs broadcast a popular contest, "Song 86." The second all-Union program was a report on a gymnastics competition, another favorite of Soviet viewers and a source of pride throughout the country. That year, the star of the competition was Elena Shushunova. Two years later she would win gold at the Summer Olympics in Seoul—the first Olympics since 1976 in which Soviet and American athletes took part together. The Americans had boycotted the Moscow Olympics in 1980 after the Soviet invasion of Afghanistan, and the Soviets retaliated by refusing to attend the 1984 Olympics in Los Angeles.[2]

As others watched television, chatted, or napped, the officer in charge of the firefighters, a twenty-three-year-old lieutenant named Volodymyr Pravyk, was sitting in his office writing. Some thought he was working on his study notes, as he was planning to continue his education and enter a firefighters' school for senior officers. But Pravyk was also in the habit of writing letters to his wife, Nadiika, because they had lived apart for a year: Volodymyr was in Chernobyl, while Nadiika was finishing her studies in Cherkasy, a city on the Dnieper 320 kilometers south of Prypiat. They had met there when Volodymyr was studying at the firefighters' academy and she was a student in the local music school. They fell in love instantly, but Nadiika was only seventeen—too young to marry. Volodymyr returned to his hometown of Chernobyl, while she remained in Cherkasy. They married in 1984. In the following year, Nadiika moved to Prypiat, but Volodymyr was still more comfortable expressing his deepest feelings for her in writing. He used his long shifts away from his family to write her letters.

Two weeks earlier, Nadiika had given birth to their daughter, Natalka. Lieutenant Pravyk asked his superiors to move him to a different position without night shifts so he could spend more time with his family. They promised to arrange the move, but there was no one to replace him yet, so he had to stay for the time being. He loved his job and his crew, and he was always working on projects to improve things in the fire department. With the help of one of the firefighters, he designed and installed remote-control doors in the department's garage, a rare feature at the time. That day he had brought his tape recorder to work with the idea of preparing a musical greeting to his crew for the coming holidays. At 2:00 a.m. he was supposed to be replaced on duty and take a nap before the end of the shift at 8:00 a.m. Later that morning, he, Nadiika, and their daughter were planning to visit his parents in Chernobyl. Like everyone else in his unit, he was preparing to help his parents with the gardening.[3]

Lieutenant Pravyk and his crew were by no means the darlings of the Fire Department command. If anything, they were a high-maintenance unit that demanded considerable attention from the department's commander, the thirty-five-year-old Major Leonid Teliatnikov. "It was a highly distinctive unit. You could say that it was a unit of individuals," Teliatnikov recalled. "Because everyone was on his own. There were a great many veterans there, a great many mavericks." Pravyk was the youngest of them: every subordinate of his was older than he was. The firefighters were paid quite well, and it was not easy to get a job in the department—many had gotten theirs through family connections. Fathers, sons, and brothers formed networks that were hard to penetrate even for a seasoned commander. Pravyk believed in leading by example. Teliatnikov wanted him to be tougher with his men, who took advantage of him and occasionally let him down.

Pravyk brought their requests for better living conditions, days off, and so on directly to Teliatnikov. On one occasion he publicly opposed his commander on the issue of punishing a subordinate who had missed his shift—the firefighter had confused the dates when he was supposed to be on duty. Pravyk believed that he should not be punished too harshly. Teliatnikov disagreed and had numerous discussions with Pravyk about it. He even spoke with his wife, who he believed had influence over her husband, but to no avail—Pravyk continued to put his people first. It cost him delayed vacations and deferred promotion to the next rank. But the rough veterans loved their young commander. "Pravyk was a very good guy," said one of the men, the thirty-five-year-old Leonid Shavrei, in describing his commanding officer. "Brainy and competent. He was very knowledgeable about radio engineering, which he loved very much. He was something of a master with light shows or repairing receivers or tape recorders. And he got along well with the men. A fine commander. He could settle any question; if you approached him, he would see to it promptly."[4]

Shavrei was the oldest of three brothers, all of them firefighters from a Belarusian village on the border with Ukraine. The city of Prypiat was only 17 kilometers away from the village, while the administrative center of their county in Belarus was 50 kilometers away. Leonid and Ivan Shavrei were both members of Pravyk's crew, and both were on duty that night; their younger brother, Petr, a lieutenant, was at home. After watching television, Leonid went to take a nap—he was supposed to replace Pravyk, who was still on watch at 2:00 a.m. Ivan was in front of the fire station, chatting with other firefighters, when they heard a sound. Ivan recognized it immediately as a release of steam from the power station. That had happened before, so he did not pay much attention to it.[5]

Ivan had just walked into the fire department building when he heard an explosion, then another one. What had happened? He rushed to the window and saw a fireball rising above Unit 4. The siren sounded, waking up Leonid. "Look, it's burning," came the voices of his fellow firefighters, pointing to the power station. Above the station, usually illuminated at night by numerous spotlights, a mushroom cloud of smoke was rising from the top of the reactor building—a red column turning blue as it ascended and exploded into a black cloud.[6]

BEFORE THEY knew it, they were already in the fire truck, Leonid Shavrei sitting next to Lieutenant Pravyk, Ivan Shavrei following in the next truck. Three trucks were speeding toward the power plant. As they approached the gates, they could see nothing of the fire and smoke they had just observed. They reached the administration building, located at some distance between Units 1 and 2. Behind the units and connecting them was a turbine hall—a long structure approximately 32 meters in height that connected all four reactors of the power plant. Units 3 and 4 were next to each other and shared a tall exhaust pipe that would end up in almost all the photos of the Chernobyl plant. The height of the units, and the level at which the pipe began, was 72 meters—the height of a seventeen-story building, comparable to a medium-size American skyscraper of the 1920s and 1930s. As they looked in the direction of the pipe, they were stunned by the scope of the disaster—the roof of Unit 4 was gone, as was a good part of the wall. Tongues of flame were licking the remaining walls.

The shocked Pravyk radioed alarm signal no. 3, the highest level of alert, which meant that all firefighting units in the entire Kyiv region were immediately called into action. Pravyk was acting as he always did, taking responsibility. "Well, Mikhailovich, it's going to be hot for us," said Pravyk, addressing Leonid Shavrei respectfully by his patronymic. "We'll really have to work here." Shavrei understood immediately that the situation was grim: "My hair stood on end," he remembered later. The time was 1:28 a.m., five minutes after the explosion.

Pravyk and his crew left the truck and ran along the transport corridor through Unit 3, trying to figure out what had happened. They found a telephone in the corridor, but no one responded to their call. Finally they noticed two distressed technicians running toward them from Unit 4. "What happened? Where's the fire?" asked the firemen. The technicians were not sure but suggested that the roof of the turbine hall might be on fire. That was bad news, Pravyk realized. There were many combustible items—including extremely valuable equipment—in the turbine hall that connected all four units of the power plant, and the fire could spread to all of them.[7]

Pravyk had to act, and act fast. He ordered Leonid Shavrei to get back to his fire truck and drive it up to the wall of the turbine hall. Pravyk himself would stay inside the building, try to find out more about the accident, and devise a plan for further action. Shavrei did as he was told. Together with another firefighter, Volodymyr Pryshchepa, he then climbed onto the roof of the turbine hall—no easy task, as they were wearing full gear and had to climb the ladder, which was 12 meters long and moving as they ascended it. What they saw at the top was much more than a fire. "When I made my way onto the roof, I saw that parts of the ceiling were damaged, and some had fallen," recalled Pryshchepa a few days later. "Closer to the edge of the roof on Unit 4, I saw the spot where the ceiling had started to burn," he continued. "I wanted to approach it to put out the fire, but the ceiling was shaking. I returned and walked along the wall, following the pipe that supplied water for fighting fires, approached the heart of the fire and covered it with sand [stored on the roof], as it was impossible to attach a hose."

Leonid Shavrei also remembered that they extinguished the fire without using water: "We tried to beat out the flames with canvas fire hoses. There were water pipes for fighting fires on the roof, and hoses in containers, so we used those hoses to fight the fire. . . . There were holes in the roof, and once we started pouring water, there might be short circuits. . . . We beat the flames with the hoses and stamped on them with our feet." Contrary to all safety regulations, the roof on which the firefighters were working was covered with bitumen, a highly inflammable petroleum product. "It was hard to walk, as the bitumen on the roof had melted," recalled Shavrei. "Such heat. . . . With the slightest increase of temperature, the bitumen immediately caught fire. . . . If you stepped on it, you couldn't put one foot in front of the other; it tore off your boots. . . . And the whole roof was littered with luminous, silvery pieces of debris of some kind. We kicked them aside. One moment they just seemed to lie there, the next moment they would catch fire."[8]

What Shavrei and Pryshchepa were kicking aside were pieces of graphite and radioactive fuel. These radioactive materials were irradiating everything around them, first and foremost the crew members, who had no instruments to measure the radiation or proper gear to protect themselves against it. They were trained to extinguish everyday fires and enter rooms and buildings full of smoke; no one had ever explained to them how to deal with radiation, even though their fire station was near the power plant. They had little, if any, understanding of how the fire they were fighting was different from regular fires, and that it might be radioactive. With the temperature rising, Shavrei and Pryshchepa removed parts of the regular gear they were wearing. "The temperature was high, making breathing difficult; we unbuttoned our gear, took off our helmets, and laid them down," recalled Shavrei. Unbeknownst to them, they were being applauded by observers on the ground. The fishermen on the nearby pond, who had witnessed the scene, were impressed. "He's taken his helmet off!" exclaimed one of them. "Fantastic! He is a real hero!"[9]

IN PRYPIAT, Liudmyla Ihnatenko, the wife of Vasyl Ihnatenko, a crew member of Fire Department No. 6, which was responsible for the fire safety of the city, woke up when she heard a noise under her windows. Like some other young firefighters and their families, Vasyl and Liudmyla lived in an apartment above the department's garage. Vasyl was on duty that night, and she looked out the window, searching for him. He was there, getting into his truck. "Close the window and go back to sleep," Vasyl called out to his wife. "There is a fire at the reactor. I'll be back soon." Liudmyla looked toward the power plant and saw flames above Unit 4. "Everything was radiant," she remembered. "The whole sky. A tall flame. And smoke."

The trucks pulled out of the garage and sped off toward the power plant. In charge was Vasyl's commander, a twenty-three-year-old lieutenant named Viktor Kibenok. Trained at the Cherkasy firefighters' academy—he had graduated in 1984, a year after Pravyk—Kibenok was in charge of an exemplary unit. All his subordinates were more or less of his age, and he was friends with many of them, including Vasyl Ihnatenko. Ihnatenko, a prize-winning athlete who had won the title of "Master of Sports of the Soviet Union," was two years older than Kibenok. Their families spent holidays together, and their wives were friends. The two men worked on the same shift and were now on their way to the nuclear plant.[10]

Lieutenant Kibenok reached the plant at 1:45 a.m., seven minutes after the arrival of Pravyk's unit. Since Pravyk and his people were dealing with the fire on the roof of the turbine hall, the roof of the reactor hall of Unit 3 became the new priority. It had caught fire from the explosion on the nearby Unit 4. Building the two units next to each other had saved money—they shared an exhaust pipe and other facilities—but now their proximity had turned dangerous. Kibenok, Ihnatenko, and most of their fellow crew members connected fire hoses to hydrants and to pipes on the walls of the reactor building; they then climbed the ladders attached to the building up to the roof. It was a dangerous task, as they were dressed in full gear, and the height of the unit was close to 72 meters. From the roof of Unit 3 they could see the terrifying sight of the exploded reactor hall and the fire emerging from below. They located the hoses stored on the roof, connected them to the wall pipes, and began to extinguish the fire.

Lieutenant Pravyk soon joined Kibenok and his crew on the roof of the reactor hall. His men, including Leonid Shavrei, were keeping the roof of the turbine hall under control. Pravyk could now help Kibenok. Those remaining on the ground saw Pravyk climbing the fire ladder up to the machine hall, and from there to the top of the reactor hall of Unit 3. The location around the base of the exhaust pipe became their battleground. Some of the firefighters, including Ihnatenko, were using hoses; others were busy kicking graphite blocks off the roof. They had no idea that the blocks came from the reactor and were killing everything around them by irradiation. Their primary concern was to contain new fires that were being ignited by extremely hot pieces of graphite on the roof.[11]

Major Leonid Teliatnikov, the head of Fire Department No. 2 and Pravyk's commanding officer, was fast asleep when the telephone rang in his Prypiat apartment. The officer on duty told him that there was a fire at the nuclear plant. Teliatnikov was on vacation, but that did not matter now: he had to join his men at the plant. With all the fire trucks gone, he called the local police and asked them to send him a car. They obliged, and he reached the plant at 1:45 a.m., around the same time as Kibenok's unit. He saw the ruined Unit 4 and the fire spreading on the roof of Unit 3—the tongues of flame there were up to 2 meters high. What else was on fire? He spent his first minutes at the plant as Pravyk had done, running around and assessing the situation.[12]

Near the walls of the turbine hall he ran into Leonid Shavrei from Pravyk's crew, who had climbed down from the roof to get a fire hose. "Leonid Petrovich, the power cables are broken here. They could have killed me," Shavrei told his boss, referring to the destroyed power lines and electric wires hanging from what remained of the walls of Unit 4, from which Teliatnikov was coming. "Well, they didn't kill you, so you're alive," responded Teliatnikov, who asked Shavrei where Pravyk was. Teliatnikov was concerned about the young lieutenant. But Shavrei's reply was encouraging. "When the accident happened, regardless of any friction in the crew, regardless of anything, the whole crew followed Pravyk without hesitation," recalled Teliatnivok. "And no one flinched."[13]

Teliatnikov finally saw Pravyk when the lieutenant and the group of firefighters from Kibenok's crew climbed down from the roof of the reactor hall of Unit 3. Pravyk reported that the fire on the roof had been largely extinguished. But he was clearly in a bad way. "There were seven men with him, all in bad shape and feeling sick," recalled Teliatnikov. He spotted an ambulance nearby and ordered Pravyk and his men into it. It was 2:25 a.m. They had spent less than half an hour on the roof but were now feeling very sick. Everyone now realized that it was not just the effect of the fire. As Pravyk got into the ambulance, he asked those around him to tell his wife, Nadiika, to close the windows in their apartment. The ambulance sped off to the Prypiat hospital. Ivan Shavrei, who was then on the roof of the turbine hall, saw the firefighters from Kibenok's crew starting to climb down as well. They all felt sick. Vasyl Ihnatenko, who was lying on the edge of the roof, was in particularly bad shape. Kibenok, his shirt torn, was moving slowly, holding onto the wall. They were helped down to the ground, and the ambulance took them to the hospital.[14]

The only doctor on duty in Prypiat that night was twenty-eight-year-old Valentyn Belokon, who had been responding to emergency calls: sick children, patients with chronic diseases needing help, and a man who had jumped out of a window after drinking too much.

Then came the call from the nuclear station. Belokon arrived at the power plant with painkillers, expecting to treat burns, but there were none. He ran into Kibenok before the latter joined his men on the roof of Unit 3. "Is there anyone with burns?" asked Belokon. Kibenok responded: "No, there isn't. But the situation is not quite clear. Something is making my boys feel a bit sick." Belokon recalled that the lieutenant was "somewhat agitated, worked up."[15]

After Belokon had spoken with Kibenok, firefighters brought him a young workman, approximately eighteen years old, who had been working at Unit 3 and had walked into Unit 4 from there. Now he complained of an excruciating headache and nausea. Belokon asked what he had been eating and drinking. It was a Friday night, and his first suspicion was alcohol poisoning. The boy told him that he was sober. Meanwhile, his condition worsened. His speech was slurred and he grew pale, repeatedly saying, "Horrors! Horror!" Still, there was no smell of alcohol coming from his mouth. Belokon administered two relaxants—relanium, known in the West as diazepam, and Aminazin, whose generic name is promenazin.[16]

More people were soon brought to Belokon—all operators from the plant, and all complaining of headaches, dry throat, and nausea. Belokon applied the same treatment and sent them to the hospital in an ambulance. Then came the firefighters, who were in even worse condition. It was only when Belokon sent the firefighters to the hospital that he realized what was happening. He called his superiors and asked for potassium iodide, a drug that protected the thyroid from radiation. They were skeptical at first, until they saw the first patients sent by Belokon, and then they sent the drug. There was no longer any doubt that they were dealing with radioactive poisoning. But no one knew how high the level of radioactivity was.

Belokon started to recall what he had learned about radiation in medical school—not much. But even the operators at the power plant were oblivious to the danger. The people being brought to Belokon were vomiting, but they refused to admit what was going on, attributing their condition to shock. Ashamed to be seen vomiting, they went outside the building when they felt sick so that other workers would not see them.[17]

MAJOR TELIATNIKOV was running out of time. He sent reinforcements to the roof of reactor No. 3. By 3:30 a.m., he, too, was experiencing the symptoms he had seen in the firefighters he had sent to the hospital only an hour earlier: nausea and retching. It was his turn to be taken to the hospital.[18]

By now, the Shavrei brothers Leonid and Ivan, who were both at the power plant fighting the fire, had been joined by their younger brother, Petr. Leonid was on the turbine hall, and Ivan had been sent to the top of reactor No. 3 with those who were replacing the Pravyk and Kibenok crews. Petr had arrived with another off-duty officer to help deal with the fire. He was not even wearing his gear when, approaching the turbine hall, he heard the voice of his brother Leonid, who was shouting, "Give me hoses; there are no hoses!" The hoses he was using had been burned by the melting bitumen. "I immediately took off my shoes and put on kersey boots, throwing my cap into the car," recalled Petr. "I put two hoses under my arms and climbed the ladder to the top. And that was all the protective gear I had—just my boots! Protection didn't matter—minutes counted to stop the flames from spreading."

The new hoses were finally on the roof of the turbine hall, but there was no water. With the electricity gone, the plumbing system that was supposed to supply water to the pumps was dead. Petr Shavrei made a decision on the spot—to use water from the cooling pond nearby. It was easier said than done. Around him was a scene from the Apocalypse—the concrete blocks, window glass, graphite, and fragments of fuel rods, all thrown up by the blast, created a hellish obstacle course for the trucks that had to get to the pond. "I ran in front of the truck—there was no lighting; everything was covered with debris," remembered Petr. "I dodged like a rabbit, with the truck following me. And still the wheels were punctured. I took metal rods out of the wheels with my hands and kicked them out with my feet. Then the skin peeled off my hands—the metal rods were radioactive." Eventually they managed to reach the pond. Water was delivered to the roof of the turbine hall, and the fire was again brought under control.

It was only closer to 7:00 a.m., when the fire was finally extinguished, that the Shavrei brothers were allowed to leave their positions around the damaged reactor. Ivan, who had been fighting the fire on the top of reactor No. 3, was taken away by an ambulance. He had a sweet taste in his mouth and could barely stand up. Leonid climbed down from the roof of the turbine hall on his own, but he was vomiting. Petr, who had arrived last, already did not feel well: "I was retching and felt terribly weak. My legs wouldn't respond, as if they were made of cotton," he remembered. What he wanted most was a drink. He reached for a water hose and drank from it, experiencing immediate relief. "What are you doing, it's dirty!" said a fellow crew member, referring to the water brought from the cooling pond. Petr responded that the water looked clean. "But the water was radioactive—I knew that, but it seemed to me that if I didn't get a couple of swallows, I would fall and wouldn't be able to get up," Petr recalled. He would pay dearly for those two sips of water from the Chernobyl pond—his digestive tract was severely damaged.[19]

Another young firefighter officer, Petro Khmel, had arrived at the plant at the same time as Petr. Like him, Khmel had relatives already fighting the fire. His father, Hryhorii, a truck driver for the Chernobyl city fire department, had been among the first to reach the plant after the no. 3 alarm radioed by Pravyk had sounded. Petro Khmel was an officer of Fire Department No. 2 and was supposed to replace Pravyk and his crew on duty at 8:00 a.m. The first thing he learned on arrival was that Pravyk had already been taken to the hospital. "I went up to the roof of the machine hall to reconnoiter: the roof was completely destroyed, and there were flames—true, not high ones. . . . They gave me a fire hose. . . . After a while, I was the only one left there. I asked on the portable radio what to do. They answered, 'Wait for your replacement.'" He stayed. He was not sure how long he remained there—in his haste to reach the plant, he had left his watch at home—but he would later learn the terrible truth that every second counted.[20]

While Petro Khmel was on the roof, his father, Hryhorii, spent most of the night near the walls of the turbine hall. He saw Pravyk climb the fire ladder, learning subsequently that not only Pravyk but also Teliatnikov had been taken to the hospital, and began to worry about his son. He had no doubt that Petro would be called as well.

Around 7:00 a.m., when Hryhorii and his fellow firefighters were ordered to leave their positions and potassium iodide was administered to them, he began asking people whether they had seen Petro. The answer was no. Then someone said: "Petro Khmel was taken there as a substitute." Hryhorii's heart sank. "There" meant the damaged reactor. "I thought it was all over, finished," he recalled later.

Hryhorii was told to surrender all his clothing and take a shower. Only after that did he see his son. "I went out onto the street, looked around—it was light, and everything was visible—and saw my Petro coming in uniform, with a coat on, a fire belt, a cap, and leather boots." "Are you here, Father?" Petro asked his dad before being taken away for decontamination. Hryhorii must have felt like Nikolai Gogol's Taras Bulba at the execution of his son Ostap, who shouted into the crowd, "Father, where are you? Do you hear me?" before he was put to death. Hryhorii refused to leave the premises and waited until his son had taken a shower. Petro was obviously sick. As he recalled later, "I started to feel bad in the shower. I came out; my father was waiting for me. 'How do you feel, sonny?' Hearing almost nothing by then, I heard only 'Hold on.'"[21]

In the dorm room above the garage of Fire Department No. 6 in Prypiat, Liudmyla Ihnatenko could not sleep after her husband, Vasyl, left for the plant before 2:00 a.m. She was pregnant with their first child, whose birth was expected soon, but it was more than that. She felt that something was wrong. The fire truck was not coming back. "Four o'clock. Five. Six," she recalled. "At six we were supposed to go to his parents' home to plant potatoes." Not until seven did Liudmyla learn that Vasyl was not coming back. He was in the hospital. She rushed there but could not get through— the police had cordoned it off, allowing only ambulances to go back and forth at great speed. They ordered people to stay away from the ambulances.

Liudmyla was desperate. She found a doctor whom she knew. "I grabbed her white coat when she came out of the ambulance. 'Get me inside!' 'I can't. He's bad. They all are.'" Liudmyla begged, and the doctor finally agreed. "He was all swollen and puffed up," recalled Liudmyla. She could hardly see Vasyl's eyes. Liudmyla asked her husband what she could do to help him. "Get out of here! Go!

You have our child!" he said weakly. "Go! Leave! Save the baby!"
Liudmyla remembered the doctors telling Vasyl and others that they
had gas poisoning, but he probably knew that it was radiation. That
realization came too late to Vasyl and his fellow firefighters, but
once they understood what was happening, they wanted their families out. They had extinguished the fire, but they could do nothing
about the radiation. It was out of control, and only starting its destruction of their bodies and the environment.[22]

Volodymyr Pravyk's parents waited in vain for hours for their
son to come and help them with the gardening. When they finally
learned that he was in the Prypiat hospital, they rushed to see him.
Volodymyr told them through the window to get on their motorcycle, grab his wife, Nadiika, and daughter, Natalka, wrap Natalka in
as many blankets as possible, and send both immediately to Nadiika's parents in central Ukraine, away from Prypiat and Chernobyl.
They did as he said. Before leaving their apartment, Nadiika left a
letter for Volodymyr on the table, telling him where she and Natalka
were. Their romance had been conducted largely by letter—this
would be the only one to remain unanswered.[23]

DENIAL

VIKTOR BRIUKHANOV was asleep in his elite apartment building on the corner of Lenin and Kurchatov Avenues in Prypiat when the telephone rang around 2:00 a.m. on April 26. "Viktor Petrovich, something has happened at the station. Do you know what?" came the distressed voice of the chief of the plant's chemical unit, who had decided to call the director at home in violation of every written and unwritten rule of conduct. Briukhanov knew nothing.

The chief chemist told him that there had been an explosion, but he knew little more. After hanging up the receiver, Briukhanov tried to call the power plant, but there was no answer. He dressed and rushed downstairs. On the street he boarded the company bus to the plant. What could have happened? All he could surmise was that something had happened to one of the steam pipelines. That was the kind of accident everyone knew could happen, though they hoped it would not. As the bus entered the grounds of the plant, he realized it could not just be the pipeline—the top of Unit 4 was gone. His heart sank. "This is my prison," he said to himself.[1]

Briukhanov realized immediately that life as he knew it—a successful career, participation in the party congress, government awards—was over. He would have to bear responsibility for the disaster, whether he was guilty or not. In the 1930s, thousands of managers had been accused of sabotage—or worse, of being foreign spies—and sentenced to long terms in the Gulag or even shot for accidents at their plants. Although such accusations were no longer in vogue, in the Soviet system a director was invariably blamed for a disaster that took place on his watch. Normally a man of few words, Briukhanov would become even less talkative. Those who saw him that night had no doubt that he was not only under stress but also depressed by what had happened: his face was rigid, his reactions slow, his whole appearance that of a man who had lost his way.[2]

The situation was obviously bad, but Briukhanov wanted to know how bad and what had caused it. From the director's office in the administration building, he dialed the number of the chief of the night shift, Boris Rogozhkin. But Rogozhkin's telephone was silent. Briukhanov then ordered all his top managers to be summoned to the plant. On Rogozhkin's orders, the telephone operator was already calling the managers one by one. Briukhanov asked why the automatic alarm system that used prerecorded tape to make telephone calls to the senior management was not activated. He was told that the operator was not sure how serious the accident was and which tape to use. Briukhanov ordered that the "comprehensive accident" tape be used, indicating the highest level of alert. It meant that the accident could have an impact beyond the premises of the plant. Although Briukhanov still did not know what had happened, what he had seen was alarming enough.

With no one available to advise him about what had actually happened at the plant, the distressed Briukhanov decided to investigate for himself and ran in the direction of Unit 4. As he approached it, he stumbled over pieces of graphite on the ground but dismissed the possibility that they might have come from within the reactor. He clipped one of them with his shoe and kept going. The building next to the reactor hall that housed the emergency cooling system was in ruins, destroyed by the explosion. That was bad enough. He did not want to see the rest. Briukhanov retraced his steps and returned to his office.

The first managers summoned in the middle of the night by the telephone operator finally started to arrive. Briukhanov ordered that the entrance to the underground bunker—a nuclear shelter designed as a command post in case of a nuclear war or emergency—be opened. He told the managers to find out what was going on in their respective units and report to him. Everyone hit the phones. The director took the most thankless task of all upon himself—to call the ministry officials and party bosses in Moscow and Kyiv to inform them of the accident. He told them what he had seen: the explosion had destroyed a good part of Unit 4, and he was studying the situation. They responded that they were sending their people to the plant and ordered him to figure out what had happened as soon as possible.[3]

Briukhanov finally got Rogozhkin on the phone. The chief of the night shift had just returned to his office after visiting the turbine hall. He had also been to the control room of Unit 4, where he had discussed the situation with the chief of the unit's shift, Aleksandr Akimov, as well as the operator of the reactor, Leonid Toptunov, and the man in charge of the turbine test, deputy chief engineer Anatolii Diatlov. They were shocked, depressed, and unable to understand what had happened. "Boria," said Diatlov to Rogozhkin, addressing him by the diminutive of his first name, Boris, "we pressed the AZ-5 button [emergency shutdown], and 12 to 15 seconds later the unit exploded." Rogozhkin had seen people burned by steam and helped evacuate one of them, Volodymyr Shashenok. They were looking for missing people—a few were still unaccounted for. That was all Rogozhkin could tell Briukhanov. He offered to put him in touch with Diatlov, but Briukhanov responded that he would call Diatlov himself.[4]

LIKE EVERYONE else, Anatolii Diatlov was at a loss to understand what had happened. When the emergency electrical generators started up and the lights came on again in the control room after the second blast, his first guess was that the emergency protection and control system tank, which contained about 110,000 liters of hot water and steam and was located at Level +71 (71 meters, or 233 feet, above the ground), directly over the control room, had exploded. If that was indeed the case, the control room would be flooded with hot water in no time. He ordered everyone to relocate to the emergency control room, but the operators were too distressed and preoccupied with trying to understand their indicators—which were flashing all over the control panel, with needles swinging erratically on instrument scales—to pay any attention to the order. Since no water was seeping through the ceiling, Diatlov did not insist. But if the emergency water tank had not exploded, what was going on?

Diatlov rushed to the control panel and began to examine the scales and indicators. By now they were either dead or putting out information that was hard to comprehend. The indicators showing water flow to the active zone of the reactor were on zero. That was terrible news. Diatlov believed that the reaction had already been shut down. But he also knew that with the reactor still overheated, the fuel rods, deprived of coolant, would disintegrate in no time. In a loud voice he gave the order: "Cool the reactor at emergency speed." The emergency water pumps that had been shut off in preparation for the test now had to be restarted, and the laborious task of opening the gate valves had to be done in record time. Diatlov was afraid that he did not have enough time to cool the rods, but he saw no other way of saving the reactor. He ordered Akimov to get in touch with the electricians as soon as possible and restart the pumps.

Then, as if the water situation were not bad enough, Diatlov found that something had gone terribly wrong with the control rods. The indicators showed that the rods, which were supposed to stop the reaction, had stopped after descending only one-third of their way into the active zone of the reactor. The fission reaction was still going on. And as far as Diatlov could tell, there was no water there to cool the fuel rods. Akimov cut off the power supply to the servo-drives that moved the jammed rods in the hope that they would continue the descent on their own. They did not. Diatlov ordered Viktor Proskuriakov and Aleksandr Kudriavtsev, the two interns who had been present in the control room during the test, to run to the reactor hall and try to use mechanical gears to insert the control rods by hand more deeply into reactor's active zone. They left before he realized the absurdity of his command—if the rods had not gone down with the electricity cut off from the servo-drives, then they could not be pushed down by hand either. He ran into the corridor to stop the interns, but they were gone. The corridor was full of smoke and dust. Diatlov returned to the control room and ordered the ventilation to be turned on.[5]

Diatlov's next destination was the turbine hall, located on the other side of the control room. The hall was on fire. That terrifying news had been brought to the control room by one of the mechanics soon after Diatlov ordered the reactor to be cooled at emergency speed. The first to hasten there was Razim Davletbaev, the deputy head of the turbine unit. What he saw was more than shocking. The roof over one of the turbines was breached and caved in. "Some of the girders were hanging down," remembered Davletbaev. "I saw one of them fall onto the low-pressure cylinder of Turbo Generator 7. From somewhere above I could hear the sound of escaping steam, although neither steam nor smoke nor fire was visible in the broken spaces of the roof; what could be seen instead were the stars shining brightly in the night sky."[6]

Diatlov was horrified by what he saw: "A picture worthy of the pen of the great Dante!" he recalled later. "Streams of hot water were bursting in every direction from the damaged pipes and falling on the electrical equipment. There was steam everywhere. And the crackling sounds of short circuits in the electrical system resounded as sharply as gunshots." Near Turbine 7, Diatlov saw turbine engineers with fire extinguishers and water hoses trying to douse burning machine oil. The oil feed pump had been broken by falling equipment, and 200 tonnes of machine oil were making their way onto the floor, threatening to turn the turbine hall into a burning hell that could consume not only Unit 4 but the whole plant, as all the units shared the same turbine hall. Davletbaev and his men began redirecting machine oil from the feed pipes and turbines into the underground reservoir. They also started to remove hydrogen from the turbines located next to the damaged turbine generator to avoid a new explosion.[7]

Outside the control room, Diatlov ran into the first victims of the explosion, engineers burned by steam from the damaged pipes. He told them to go to the medical station. After he returned to the control room, Volodymyr Shashenok, an engineer, was brought in from the startup unit. During the test he had been monitoring the indicators on Level +24 of the unit when the explosion burst the hot water and steam pipes, causing severe burns to his entire body. "Volodia sat in a chair just weakly moving his eyes, neither groaning nor crying out. Evidently the pain had exceeded any conceivable level and rendered him unconscious," recalled Diatlov. Rogozhkin, who visited the control room, helped to take Shashenok to the medical station on a stretcher. Several engineers who had been in the unit at the time of the explosion could not be located.

Meanwhile, Akimov was struggling to supply water to what he believed was a rapidly overheating reactor. The internal telephone system was dead, as cables had been severed by the explosion, but, miraculously, the city telephones were working. Akimov called the electricians, asking them to restore power to the pumps so that he could cool the reactor. They promised to do what they could. Akimov asked Yurii Trehub, the chief of the evening crew, who was still around, to go and try to open the gate valves of the cooling system manually. Trehub and another operator from his shift, Sergei Gazin, ran up the stairs to Level +27. As they arrived, they noticed that they had difficulty swallowing because their throats and tongues were swollen. But that seemed to be the least of their problems. When Trehub opened the door to the hall where the mechanical controls for the valves were located, he was hit with a blast of hot steam. He had to give up any hope of getting inside—it was impossible to breathe there. Trehub and Gazin had to return to the control room, their task unfulfilled. Not knowing that the reactor had already exploded, Akimov was terrified of what could happen to it because of overheating.[8]

More bad news came when Proskuriakov and Kudriavtsev, the two interns sent by Diatlov to try to push the control rods into the reactor by hand, returned to the control room. They had tried taking the elevator to reach the reactor hall, located on Level +36, 36 meters above the ground, but it was jammed, and they had to use the stairs. It was a difficult ascent through stairwells full of debris; it was wet and dangerously hot as well, with water and steam bursting out of the damaged pipes. Eventually they had to turn back, as access to the reactor hall was blocked by collapsed walls and parts of the concrete ceiling. They did not reach their objective, but had gotten close enough to the damaged reactor for that trip to cost them their lives. They told Diatlov and Akimov that their mission had failed. Those in the control room concluded that the reaction was still going on, with no water to cool the reactor. They tried not to think about the possible consequences.

Proskuriakov told Trehub that he thought the active zone of the reactor had melted down. What he had in mind was the melting of the core of the reactor, caused by the overheating and the possible escape of uranium fuel from containment. Trehub agreed—he had seen a glow when he stepped out of the unit a few minutes earlier and assumed that once the overheated reactor had melted down, the abrupt rise in temperature had caused "Elena," the 200-tonne concrete plate covering the reactor, to heat up and illuminate the surroundings. He could not imagine that the plate had actually been blown into the air by the first explosion. "Why is nothing being done?" asked the distressed Proskuriakov. Trehub went to see Diatlov and shared his concerns. "Let's go," responded Diatlov. They went outside. Trehub remembered telling his boss: "This is Hiroshima!" Diatlov remained silent at first but then told Trehub: "I've never dreamed anything like this even in a nightmare." That night, Diatlov walked twice around the ruins, first around 1:40 a.m. and then soon after 2:00 a.m. He still could not comprehend what had happened. His theory about an explosion in the emergency water tank had by now been discarded for another: for some reason, the water pumps had burst in the active zone of the reactor, causing it to explode and raise Elena into the air. The explosion allowed steam and radioactivity to be released before the concrete plate fell back into place and sealed off the reactor.[9]

Diatlov returned to the control room and began thinking about how to prevent the disaster from spreading beyond the confines of Unit 4. He told Akimov to switch off electrical cables wherever possible to avoid short circuits and the spread of the fire to Unit 3. Davletbaev and his men were releasing machine oil from the pipes and the turbine generator, and Diatlov told Major Teliatnikov, who showed up in the control room, that with the fire on the roof of the turbine hall extinguished, the priority was the roof of Unit 3. He walked to the control room of Unit 3 and ordered the shutdown of its reactor while that was still possible.

What else could be done? Diatlov realized that there was no need for all those people to be in and around the control room. He knew that the radiation levels had risen and were high. Those in charge had tried to measure the levels immediately after the blast but did not get very far. The dosimeter they had measured radiation in microroentgens per second. The name "roentgen"—a unit of measurement of exposure to gamma rays and X-rays—comes from the discoverer of X-rays, Wilhelm Roentgen. Roentgens, milliroentgens (0.001 of a roentgen), and microroentgens (0.001 of a milliroentgen) did not measure all ionizing radiation, which also included alpha and beta particles as well as neutrons, but they gave a good idea of overall radiation levels. The maximum value of the dosimeter available to Diatlov and his crew was 1,000 microroentgens per second, or 3.6 roentgens per hour. The dosimeter showed that in one part of the control room the radiation level had reached 800 microroentgens per second, while in another part the reading was off the scale. They assumed that the radiation level might be around 5 roentgens per hour. The emergency maximum for operators was 25 roentgens. Diatlov and others thought they were good for a few hours—it was an emergency, after all.

But if the situation in the control room was tolerable, radiation levels outside must have been extreme. People around Diatlov who had ventured outside the control room for a longer period were already feeling dizzy; their skin was getting darker and their headaches were worsening—all signs of radiation sickness. It was best to get them away from the source of radioactivity as soon as possible. Diatlov ordered everyone who was not a member of the shift crew to leave. Such people included the interns Kudriavtsev and Proskuriakov. He also saw no reason for Leonid Toptunov and Igor Kirshenbaum, the two operators from Akimov's shift, to stay, so he ordered them to go to Unit 3. To prevent radiation from penetrating that unit, he ordered the ventilation system in Unit 4 to be shut down and ventilation in Unit 3 to be switched into high gear. It seemed like a good idea at the time, but in fact Diatlov's order increased the intake of radioactive air from outside, and the outside air was far more contaminated than the air inside Unit 4.

Those crew members missing immediately after the explosion had by now been located, with the exception of Valerii Khodemchuk, the circulating pump operator, who was on shift that night.

The part of the engine room where he had been working, at Level +10, had collapsed. Diatlov and two other engineers decided to try searching once again. They reached the entrance to the engine room but could not move much farther: the concrete ceiling had collapsed, and the door to the operator's office had been smashed by a falling crane; water was pouring in from broken pipes on one of the floors above. Valerii Perevozchenko, the chief of the turbine shift, who accompanied Diatlov, climbed up to the office door but could not open it. He shouted through the door, but received no response. Soaked by the falling water, he had to turn back. The radioactive shower would cost him his life.

It was then that Diatlov began to feel exhausted. He was nauseous and could barely stand on his feet. Doctors would later estimate his dose of ionizing radiation at thirteen times above the emergency norm. At his level of biological damage, most people did not survive for more than sixty days. Diatlov had given up fighting for the reactor. As he subsequently wrote in his memoirs, he knew the reactor was dead, but he lacked the strength to say so aloud. He assumed that there was no need for words: as an experienced engineer, Akimov surely understood the consequences of interrupting the flow of water to the reactor. In fact, Akimov refused to admit that the reactor was damaged and kept pumping water. Some witnesses claim to have heard Diatlov giving instructions to do that very thing. Diatlov was probably of two minds himself, as he understood that there was no point in continuing to pump water but had nothing better to suggest. At least supplying water meant keeping some kind of hope alive. Around the time he decided he had done everything he could, Diatlov received a call from Viktor Briukhanov, who wanted him in the underground bunker. Around 4:00 a.m., Diatlov finally left Unit 4. Akimov remained behind: his shift was not over yet.[10]

"WHAT HAPPENED?" was the first question Briukhanov asked his deputy chief engineer when Diatlov arrived. Diatlov spread his hands as an indication that he did not know. He said to the head of the plant's Communist Party committee, Sergei Parashin, who was sitting next to the director: "I don't understand it at all." The power level had been rising rapidly before the explosion, he said, and something had happened to make the control rods stick in the intermediate position. Diatlov gave the director the automatically recorded data on the equipment in Unit 4. Briukhanov took the records but did not seem interested in continuing the conversation. Diatlov was clearly in bad shape, pale and nauseous, almost at the point of vomiting. After talking to Briukhanov and Parashin, he rushed out of the bunker, where an ambulance picked him up.[11]

Meanwhile, everyone in the bunker was working toward one goal—to supply water to reactor No. 4. Diatlov knew that it had exploded, but did not say that to his bosses. "There were about thirty or forty people in the bunker. There was constant noise as everyone spoke with his unit by telephone," recalled Parashin, a bespectacled thirty-nine-year-old nuclear engineer turned party organizer. With Briukhanov busy responding to endless calls from Moscow and Kyiv, Parashin took upon himself the role of crisis manager, talking to those coming into the bunker, suggesting solutions, and giving recommendations to the director, which Briukhanov was generally happy to accept.

A few hours before Diatlov reached the bunker, Briukhanov, still in the dark about what had happened at Unit 4, had sent another deputy chief engineer, Anatolii Sitnikov, and the chief of Unit 1, Vladimir Chugunov, to reconnoiter. They had previously worked at Unit 4 and knew it better than anyone else. "They have to be sent: no one can understand the situation and help Diatlov better than they can," suggested Parashin. Briukhanov agreed. The duo's main tasks were to assess the situation and make sure the reactor's emergency cooling system was on. They examined the half-destroyed unit: the explosion had blown away a good part of it, and radiation levels were high, but they did not know how high—dosimeters with 1,000 microroentgens per second scales were of no use. Sitnikov could see graphite blocks and fragments of red-hot fuel rods lying around. He climbed up to the roof of the reactor hall and looked into the exploded reactor. It was a glance that would cost him his health and his life. "I think it's been destroyed," Sitnikov told one of his fellow engineers a few hours later. "It's blazing away. Hard to believe, but it's true."[12]

Still, Sitnikov and Chugunov kept doing what they were told. They tried to open the gate valves on the water feed lines to supply water to the reactor: Sitnikov knew that it had been destroyed but refused to admit it. The task was impossible. The gate valves were closed too tightly: the two managers, exhausted and already dizzy from radioactive poisoning, needed help. Chugunov went back to get assistance. When he returned to Unit 4 around 7:00 a.m. with three young engineers, he found Sitnikov, his head down on a desk, suffering from nausea. Next to him were Aleksandr Akimov and Leonid Toptunov, the chief of the night shift and his reactor operator, who ignored Diatlov's order to leave the unit. They were also in bad shape. Arkadii Uskov, one of the engineers who had come to help with the valves, remembered Toptunov as "confused and dispirited. He stood silently." Uskov and another engineer managed to open one of the valves and could hear water rushing through the pipes. Akimov, Toptunov, and one of the new arrivals were supposed to work on another valve. When the Uskov team completed their task and went to check on Akimov and his crew, they found both him and Toptunov vomiting.

Viktor Smagin, another new arrival to Unit 4, bumped into Akimov and Toptunov as they were returning to the control room. He remembered them "looking extremely depressed, with swollen dark brown faces and hands. . . . Their lips and tongues were so swollen that they could hardly talk." The two had clear signs of radioactive poisoning, but that was not what bothered them most. Smagin found them confused, bewildered, and full of guilt. "I'll never understand it," said Akimov to Smagin, another shift leader, who was scheduled to replace him that morning in the control room of Unit 4. "We did everything right. Why did it happen?" He then continued, addressing Smagin by his diminutive: "It really hurts, Vitia. We really screwed up."[13]

When Uskov and his crew finished opening the second valve and returned to the control room, Akimov and Toptunov immediately went to the restroom, where they continued to vomit. "How are you?" Uskov asked Toptunov when he returned to the control room. "Fine, I'm already feeling better," responded Toptunov. "I can work some more." One of the newcomers helped Akimov and

Toptunov walk to Unit 3. Akimov had a copy of the turbine test program in his hands. He must have felt that it was his only hope of proving that what had happened at Unit 4 that night was not his fault. Akimov told one of his friends that he had just followed the program and asked him to give the document to his wife and no one else. He no longer seemed to trust anyone. Now that his shift was finally over, an ambulance took him to the Prypiat hospital. He would die on May 11, the same day as Lieutenant Pravyk. Toptunov would pass away three days later.[14]

Smagin, Uskov, and their crew stayed in Unit 4 helping to open new valves—water was pouring into the reactor zone, but no one knew exactly where. The water became radioactive and flooded the underground compartments of the reactor, which were crammed with switching gears and cables. When the engineers working the next shift looked out the window, they could not believe their eyes: the reactor hall of Unit 4 was in ruins, and among the debris were pieces of the square block of graphite from the active zone of the reactor. "What we saw was so terrifying that we were afraid to say it aloud," recalled Uskov. When they told the shift's deputy chief engineer, Mikhail Liutov, about the graphite upon his arrival, he dismissed their fears. Liutov's full title was deputy chief engineer for scientific affairs, and he declared that if anyone could tell him the temperature of the graphite inside the reactor, he would describe the state of the reactor precisely. They responded that the graphite was not inside but outside the reactor. Liutov did not believe them. "Although he was my supervisor, I began to scream at him, asking what he thought it could be if not graphite," remembered Smagin. Liutov finally had to agree that it was graphite.[15]

It was no longer possible to deny that graphite was out there. But where had it come from? They checked the graphite that was to be installed in the fifth reactor, still under construction. The piles of that graphite were intact. The only other possible source was the reactor of Unit 4. That would also explain the exceptionally high levels of radiation at the plant. But few dared to challenge the belief that the reactor could not possibly have exploded. "The mind refuses to believe that the worst that could happen has happened," wrote Uskov in his diary for that day. Like Smagin, he was upset by

Liutov's refusal to admit the obvious. "The stress was too great, and our belief that the reactor could not explode was also too great," said Sergei Parashin, summing up the situation a few months later. "Mass blindness. Many see what has happened but do not believe it."[16]

As the sun rose above the ruined Unit 4, the bunker shelter at the Chernobyl power plant began to fill with high officials who had arrived from Kyiv. The most senior of them was Volodymyr Malomuzh, the second secretary of Kyiv's regional party committee, who had headed the Kyiv delegation to the party congress in Moscow a few months earlier. Briukhanov had called him earlier to inform about the accident; at that time, his information had been very general. Now Parashin approached Briukhanov with one of the department heads of the Kyiv party committee, suggesting that Malomuzh would need an official statement on the causes of the accident and the situation at the plant.

Briukhanov delegated the task to Parashin. When the document, which was put together with the help of quite a few officials, was finally ready, Briukhanov signed it along with the head of the plant's dosimetry unit, Vladimir Korobeinikov. The memo provided general information about the accident, mentioning the collapse of the roof of reactor No. 4. It also gave the radiation levels: 1,000 microroentgens per second at the plant and 2–4 microroentgens in the city. The figures reflected more wishful thinking than reality, especially when it came to the plant. Briukhanov knew that the actual radiation levels were higher, but, with the scales of the available dosimeters limited to 1,000 microroentgens per second, he decided to list that figure. Korobeinikov, who had taken measurements in Prypiat, was at his side. His measurements indicated that the radionuclides, or atoms with unstable nuclei that were emitting ionizing radiation, displacing electrons in human DNA and disrupting its functioning, had very short half-life periods and were about to dissipate without causing any harm. It turned out that the radioactive cloud caused by the explosion had "missed" the city for the time being.[17]

With Korobeinikov insisting that radiation levels were relatively low, Briukhanov felt that he could ignore the alarming data coming from the plant itself. Serafim Vorobev, the head of the plant's civil defense department, first reported the radiation data a few minutes after following Briukhanov's order to open the underground bunker that then served as the emergency headquarters for Briukhanov and his team of managers. Shortly after 2:00 a.m., Vorobev had switched on his dosimeter, the only one at the plant that had a scale of 200 roentgens, and found that the radiation level in the bunker was 30 milliroentgens per hour, 600 times above the norm. The radiation could only have come from outside, so Briukhanov ordered Vorobev to switch on the ventilation filters. Vorobev did so, went outside, and switched on his dosimeter once again. The radiation level was five times the one he had registered in the bunker. He went around the plant with his dosimeter on. Near Unit 4, the dosimeter went off the scale, indicating that the radiation level exceeded 200 roentgens per hour.[18]

When Vorobev returned to the bunker and reported his measurements to Briukhanov, the director did not want to listen. "Go away," he told the only man around with the right dosimeter, pushing him aside with his hand. Vorobev then turned to Parashin. But the party chief was of little help. Like Briukhanov, he was not prepared psychologically to accept any more bad news. "Why didn't I believe him?" asked Parashin in retrospect, trying to explain his reaction to what he had heard. "Vorobev is highly emotional by nature, and when he said that, he was frightening to look at. . . . And I didn't believe him. I told him: 'Go and convince the director.'"

Vorobev went back to Briukhanov. Following the instructions in his manuals, he demanded that Briukhanov declare a state of radioactive emergency. "The requirements of the standard documents in such cases were as follows," Vorobev later said. "If a radiation level of 0.05 milliroentgens per hour was exceeded, the civilians had to be informed and given directions on actions to take under those circumstances." Furthermore, he added, "If the level exceeded 200 milliroentgens, the siren was to be sounded, indicating 'radiation danger.'" He reminded Briukhanov of the protocol, but Briukhanov would not listen. He told Vorobev to inform civil defense headquarters in Kyiv but not to tell anyone else about his findings.[19]

The radiation levels that Vorobev detected in and around the bunker were from ionizing radiation. Ionizing radiation is caused by high-energy, high-speed atoms and subatomic particles that are moving quickly enough to free electrons from atoms—or ionize them. Electromagnetic waves can be strong enough to have the same capacity. Gamma rays and X-rays belong to the latter category, while alpha and beta particles (along with neutrons) belong to the former. The ionizing radiation at Chernobyl was activated by the radioactive fission products blasted into the atmosphere by the explosion of the reactor. They included isotopes of iodine and cesium, including iodine-131 and cesium-137, as well as gases such as xenon-133. The radiation levels were extremely high after the blast, and Vorobev's dosimeter went off the scale at 200 roentgens per hour for good reason. It was later estimated that the debris around the damaged reactor emanated radiation at a level of 10,000 roentgens per hour. That radiation was killing human cells or making them malfunction. The more cells are destroyed, the lower the chance of survival. Five hundred roentgens sustained over a period of five hours meant certain death.[20]

Even as Briukhanov dismissed Vorobev, he was by no means certain that his civil defense chief was making things up. When Parashin asked Briukhanov about the situation after his conversation with Vorobev, Briukhanov responded with one word: "Bad." His main concern was about the city of Prypiat rather than the plant. He had first raised the possibility of evacuating civilians sometime in the morning, in conversations with local party and city officials. The head of the city council had scolded him: "Why are you in a panic? A commission will come; they'll come from the oblast, and they'll decide." Late in the morning, Briukhanov raised the same question with Volodymyr Malomuzh upon his arrival. The answer was the same: "Don't panic!" As far as Briukhanov was concerned, he was in no position to argue with such a high-ranking party official. Besides, the statement he had signed earlier that morning conformed to the general tendency to do nothing.

When Briukhanov was invited to a meeting of the Prypiat party committee, which was convened by Malomuzh at about 11:00 a.m., he kept silent. "It was mainly Malomuzh who spoke at the meeting," recalled Briukhanov. Malomuzh took the line that radiation levels were too low to justify any drastic measures. Later the city and party officials would blame Briukhanov for not providing them with accurate data. But he remembered: "The directive remained the same: 'Don't panic! The government commission will be here soon. It will investigate, and then measures will be taken!'" Soviet managers and bureaucrats tended to do what years of party rule had trained them to do—avoid responsibility. Everyone was afraid of being accused of spreading panic, and everyone was glad to defer to the higher authorities to make a decision. They were company men, the "company" being the Soviet system.[21]

ON APRIL 26, starting early in the morning and over the course of the entire day, 132 people—firefighters, operators, and engineers—were admitted to the Prypiat hospital with signs of acute radioactive poisoning. As fire trucks sped toward the damaged plant and ambulances brought people to the hospital, the KGB cut intercity telephone lines to prevent information about the accident from spreading beyond the city of Prypiat. The engineers and workers on duty at the plant on the night of April 26 went home in the morning with strict orders not to say a word about what had happened. With smoke over the destroyed central hall of reactor No. 4 clearly visible from balconies of Prypiat apartment buildings, and police patrols controlling access to the plant, the explosion became an open secret to residents of Prypiat. But what exactly was going on? Few thought it amounted to much. That day there were seven weddings in the city, with people making merry in the shadow of the smoldering Unit 4.[22]

G. N. Petrov, a manager at one of the Prypiat firms that installed equipment at the nuclear plant, woke up around 10:00 a.m. It had been a rough night for him. Driving back to Prypiat at about 2:30 a.m., he had seen Unit 4 on fire. He drove there, stopped the car about 100 yards from the unit, and spent a minute or so observing the damage and the activity of firefighters on the roof. He then felt a sense of alarm and left for home in a panic. There, a neighbor whose husband had been at Unit 4 confirmed what Petrov already knew: there had been an accident. She mentioned radiation and suggested a cure—a bottle of vodka. They drank vodka while making jokes about the whole thing. Now that Petrov was awake, things seemed as normal as could be.[23]

"I went onto the balcony for a smoke," he recalled. "There were lots of children in the street. There were kids playing in the sand, building houses, making mud pies. The older ones were racing about on their bikes. Young mothers were out pushing their baby carriages. Everything looked normal." Petrov's next-door neighbor had decided to take it easy that Saturday morning and lay sun-tanning on the roof of the apartment building. "At one point he came down for a drink and said how easy it was to get a tan that day; he had never seen anything like it. He said his skin gave off a smell of burning right away," remembered Petrov. "And he was tremendously jolly, as if he had just been boozing." He invited Petrov to join him on the roof, saying, "Who needs to go to the beach?" In the evening an ambulance picked up Petrov's neighbor—he was vomiting—but Petrov made no connection between his neighbor's sudden illness and the accident at the plant. "Everything about that day was normal," he recalled.[24]

Liubov Kovalevskaia, the author of the article about problems with the construction of Unit 5 of the Chernobyl plant, which had been published a month earlier, but largely ignored, woke up at 11:00 a.m. She had spent most of the night finishing a poem titled "Paganini." Now she was ready to go to a meeting of the writers' club she had helped organize in the city. The club was named after Prometheus, the Greek hero who had stolen fire from Mount Olympus and delivered it to mankind. The name seemed perfectly appropriate for the nuclear city of Prypiat, or at least it felt that way at the time. On her way to the meeting, Kovalevskaia noticed something unusual: "I looked and saw a policeman here, a policeman there—I had never seen so many police officers in the city." Becoming alarmed, she returned home and told her mother to keep her daughter and niece inside when they came back from school. Her mother asked what was going on. "I don't know anything; I just feel it," she replied.[25]

Her premonition was accurate. The background radiation level of 4 microroentgens recorded by Korobeinikov in Prypiat that morning was a thousand times higher than the natural background. By 2:00 p.m. it had increased ten times, to 40 microroentgens per second, and in the evening it rose to 320 microroentgens per second

(80,000 times the natural background). Later that day, experts from Moscow who came to Prypiat to assess the situation measured the biological damage done by the radiation exposure in the units known as rem (1 rem, or roentgen equivalent man, is equal to 0.88 of a roentgen, or 0.01 of a sievert, the unit more commonly used today). They estimated that the radioactive assault on the thyroids of children within 3 kilometers of the reactor amounted to 1,000 rem, and 100 rem in the city of Prypiat. The same experts pointed out that the emergency norm was 30 rem. Children playing on the streets of Prypiat were being exposed to at least three times the dose of radiation that was considered unsafe but acceptable for workers in nuclear plants under emergency circumstances.[26]

Staff at the Prypiat hospital, which was equipped to deal with almost anything but radioactive poisoning, were busy making room for new patients, who kept arriving with symptoms of acute radiation sickness. They were cleaning the floors in an attempt to keep the radiation levels down in the hospital rooms. Viktor Smagin, the power plant engineer who had helped Aleksandr Akimov and Leonid Toptunov open the gate valves, and had joined them in the hospital by midday, heard a dosimetrist in the corridor demanding that the cleaning ladies do a more thorough job. But there was little they could do: the radiation was coming from the patients themselves. "The truth is, nothing like this has ever happened anywhere in the world. We were the first after Hiroshima and Nagasaki, though that was nothing to be proud of," reflected Smagin. Those patients who could still walk gathered in the smoking-room of the ward. Among them were Diatlov and Akimov. Everyone wanted to know what had caused the explosion, but neither of them had an explanation.[27]

III

ATOP THE VOLCANO

HIGH COMMISSION

THE CALL came around 5:00 a.m. on April 26, awakening the most powerful man in the land, the general secretary of the Communist Party of the Soviet Union, Mikhail Gorbachev. The message was that there had been an explosion and fire at the Chernobyl nuclear power plant, but the reactor was intact.

Gorbachev's first reaction was to ask how the explosion could have happened. "After all, the scientists had always assured us, the country's leaders, that the reactor was absolutely safe," he recalled. "Academician Aleksandrov said, for example, that an RBMK reactor could be installed even on Red Square, since it was no more dangerous than a samovar." Apart from being puzzled and probably annoyed by being awakened at such an early hour, Gorbachev did not show much concern at the time. "In the first hours and even the first day after the accident there was no understanding that the reactor had exploded and that there had been a huge nuclear emission into the atmosphere," remembered Gorbachev later. There was no need to wake up other members of the Soviet leadership or interrupt the weekend and call an emergency session of the Politburo. Gorbachev approved the creation of a state commission to look into the causes of the explosion and deal with its consequences. That was standard procedure in the case of high-profile accidents, and the initial alert concerning Chernobyl noted the possibility of four different types of emergency: nuclear, radioactive, fire, and explosion.[1]

Nikolai Ryzhkov, the chairman of the Soviet cabinet, had been on the phone at least since 2:40 a.m., when he had called the premier of the Ukrainian government, Oleksandr Liashko, to inform him about the accident. Ryzhkov did not know much at the time and did not learn the details until later in the morning. He recalled later that he had not been particularly worried. The son of a miner from eastern Ukraine, Ryzhkov had spent his entire career in managerial positions in the Soviet machine-building industry, which was prone to accidents involving malfunctioning equipment or glitches in the technology. "Well, what could have happened there?" he asked himself. "Any station—atomic, steam, coal, or gas—must have a turbine, a wheel with all kinds of blades," said Ryzhkov, recalling his thoughts of that morning. "We had cases, for instance, of a wheel breaking. There was a defect in it somewhere. Sometimes the roof was penetrated: part of a turbine would be hurled right onto the roof." They learned how to deal with such accidents: "We replaced the equipment and carried on." It was a big country; accidents happened all the time, and this seemed to be just one more mishap.[2]

Ryzhkov began to realize that the accident was much worse than it had at first seemed when he arrived at his office and spoke with Anatolii Maiorets, the minister of energy, who oversaw the Chernobyl plant. It appeared that the damage was not limited to the turbine generator poking a hole in the roof. At a minimum, there was a lot of machinery to be replaced before things could return to normal. Maiorets, a new minister who had been in office less than a year, was reluctant to deliver bad news to his boss. At the Twenty-seventh Party Congress he had promised to double the number of reactors to be constructed in the next five years and offered an innovative way of achieving that goal—cutting the time allocated for the design of nuclear plants. By the time he had been in office for a few months, he had met several deadlines for the production of electrical energy and the stabilization of the Soviet electrical grid, which had previously experienced sudden surges and drops in the supply of electricity. And now this.

News of the accident caught Maiorets by surprise. He was at work by 5:00 a.m. trying to get a handle on developments in faraway Prypiat. Thank goodness, the radiation level seemed to be within the norm and the number of casualties low. But the accident would delay the fulfillment of the year's production plans and compromise the stability of the grid. They had to put the plant back into operation as soon as possible; otherwise the plan figures and Maiorets's career prospects would be in trouble. The report prepared by Maiorets's ministry for the Central Committee of the Communist Party reflected that generally upbeat mood.

The report was signed by Maiorets's first deputy, Aleksei Makukhin, the former energy minister of Ukraine, on whose watch the Chernobyl power plant had been built. It was based on reports submitted by Viktor Briukhanov, who had informed party officials that the explosion that occurred at 1:21 a.m. had destroyed the roof and part of the walls of Unit 4. It had caused a fire that was extinguished by 3:30 a.m. Twenty-five firemen and nine engineers had been hospitalized, but the medical officials saw no reason for any special measures to protect workers at the plant, or for evacuating the population of Prypiat. The operators were cooling down the reactor, and the ministry was examining the causes of the accident and its consequences. Everything seemed to be under control.[3]

Protocol called for the creation of a commission to look into the causes of the accident. Someone in the Central Committee or the Council of Ministers decided to appoint as high profile a commission as possible. The man chosen to chair it was Maiorets's boss, Boris Shcherbina, the sixty-six-year-old deputy chairman of the Council of Ministers in charge of the energy sector. He looked, spoke, and acted tough. The commission included Maiorets himself and a number of his ministry's officials and scientists, as well as experts from other ministries and the Academy of Sciences. The first members of the commission to be appointed, middle-ranking experts from various ministries, began to leave Moscow for Prypiat around 9:00 a.m. They were flown to Kyiv and taken from there either by another plane or by car to Prypiat and the nuclear plant. More planes would arrive that day, bringing scores of officials and experts from Moscow to the nuclear plant.[4]

MINISTER MAIORETS left on a plane that took off from a Moscow airport around 4:00 p.m. on April 26. He did not expect to spend more than two days in Prypiat. It was Saturday, and he hoped that by early the following week everything would fall into place. The Central Committee's top nuclear industry expert, Vladimir Marin, seemed to be upbeat as well. He was encouraged by the good news that the radiation levels allegedly had not risen. "Amazingly, there is no contamination," he told a member of the traveling party. "It's a really terrific reactor." Gennadii Shasharin, Maiorets's deputy in charge of nuclear plants, who had been recalled from his vacation in the Crimea, was making plans to create groups of experts who would assess the damage and advise on repairing it as soon as possible. Construction managers were thinking about how to fix the roof of the unit.[5]

The first signs that things might be worse than they appeared came when the delegation landed in Kyiv. The Ukrainian energy minister informed Maiorets that radiation levels at the plant were above the norm. From Kyiv they flew to Prypiat, where, to their surprise, they were not greeted by the plant's managers—neither Briukhanov nor his second-in-command, chief engineer Fomin, was anywhere in sight. But the always energetic and decisive Vasyl Kyzyma, the director of plant construction, was on hand. While local party officials took care of Maiorets, Marin and Maiorets's nuclear expert, Shasharin, immediately got into Kyzyma's Jeep-type vehicle and drove to the damaged reactor. There they got their first shock. The devastation was much greater than Briukhanov's reports had led them to believe. They could see the scope of destruction from the road. "Look at the mess we've landed ourselves in. That's just great. Now we're all lumped together with Briukhanov and Fomin," complained Marin. Kyzyma also blamed Briukhanov. Shasharin remembered Kyzyma saying that "he had never expected anything else from that marshmallow Briukhanov. In Kyzyma's opinion it was bound to happen sooner or later."[6]

Kyzyma stopped his car near a pile of rubble by the wall of Unit 4. They got out. "Without the slightest fear," remembered Shasharin, "Kyzyma was walking about, looking very much in charge, and lamenting the fact that after all the effort that had gone into building the place, now they were walking around among the wreckage of the fruits of their labor." Marin was enraged by what he saw. Swearing, he kicked a block of graphite. Not until later did Shasharin realize that the graphite was producing 2,000 roentgens of radiation per hour. The particles of uranium fuel on the site emanated 20,000 roentgens. "We had difficulty breathing," remembered Shasharin. "Our eyes smarted, we were coughing severely and, deep inside, we felt extremely worried and vaguely anxious to get out of there and go somewhere else." They drove to the plant's underground bunker, where they found the managers.[7]

The arrival of the experts from Moscow and then of Maiorets and his assistants came as a major relief to Briukhanov and his management team at the station. Feeling depressed and guilty, they were still confused about the causes of the accident. The arrival of the higher-ups removed primary responsibility for the future from their shoulders. They were now responsible only for the past. The visitors noticed that the managers of the power plant were performing any task given to them but making no decisions on their own. "The arrival of the top people had a great psychological effect. And they were all very serious, those in the top ranks. They inspired confidence. It seemed that people who knew and understood everything had arrived," recalled the party boss of the plant, Sergei Parashin. "We, the personnel, did something or other mechanically, like drowsy flies."[8]

No matter how angry Marin and Shasharin were with Briukhanov and Fomin for what had happened, they seemed to join the two locals in denying the severity and consequences of the accident. After experiencing difficulty breathing and kicking around pieces of graphite that could only have come from the active zone of the reactor, the two senior officials still refused to admit that the reactor had exploded. The myth of the reactor's safety was shared by everyone in the industry, from top to bottom. Besides, as Marin had observed earlier that day, they were now all together when it came to responsibility for the accident. Marin was the key figure in the Central Committee responsible for running the nuclear energy sector, and Shasharin, as deputy energy minister, bore the same responsibility on the government level. Minor accidents could be blamed on subordinates, but the one they had just seen was too big to be handled that way. And to assume that it was even bigger bordered on the impossible. It was easier to believe that the reactor was intact: in that case, they at least knew what could be done. If it had been destroyed, there was nothing in their training or experience to fall back on.

Meanwhile, Maiorets, the most senior man on the spot, convened the first meeting of the experts from Moscow and the local officials at the headquarters of Prypiat's Communist Party committee. He seemed unmoved by reports from local officials that radiation levels were high—information that he and the others had not received in Moscow. His task was still the same: to fix the problem in a few days, connect the reactor to the grid, and go back to Moscow—the May Day holidays were around the corner. Shasharin, who had just returned from his inspection of Unit 4 with Marin and Kyzyma, was the main speaker. He said that the situation was under control: cooling water was being pumped into the reactor, and boric acid was about to be delivered to stop the fire. Experts were taking a helicopter flight above the power plant to assess its current status.

Marin, upset by what he had seen at Unit 4, asked those present where the graphite had come from. Maiorets directed the question to Briukhanov, who looked extremely tired, "his face ashen, his eyelids puffed up," according to one eyewitness account. He found it hard to speak in public under normal circumstances. It was twice as difficult now. He rose slowly and took his time answering the minister's question. Eventually he spoke: "It's hard to imagine. The graphite we received for the new No. 5 unit is all there, intact. I originally thought it must be that graphite, but it's all there. In that case it might have been ejected from the reactor." In plain language, that meant the reactor had exploded. But the experts refused to grasp what had happened, or at least to admit it. When Maiorets asked Shasharin, who was complaining about the high radiation levels, what had caused the accident, his chief nuclear officer told him that it was not clear yet. "It seemed that all those responsible for the disaster were anxious to delay as long as possible the awful moment of reckoning, when the truth would be disclosed in every detail," concluded Vladimir Shishkin, a ministry official who was present at the meeting.[9]

The meeting went on, with local officials making their reports to the powerful minister from Moscow. The party boss of Prypiat reported that the situation in the city was calm; weddings were taking place. But the report of General Hennadii Berdov, a handsome, silver-haired Ukrainian deputy minister of the interior, was more worrisome. He had been at the scene since 5:00 a.m. and spoke of the sacrifice made by the police officers patrolling the contaminated areas—not one of them had deserted his post. But there were things Berdov wanted the commission to do. Trains had to be kept out of the area, as the railway line passed 500 meters away from the damaged reactor. Berdov also surprised the minister by telling him that the Ukrainian authorities had mobilized about 1,100 buses and were moving them to Prypiat in case there was a need to evacuate the city.

"What's all this talk of evacuation? Are you trying to start a panic?" asked Maiorets, who was still hoping to return to Moscow before May Day. Briukhanov joined the discussion, saying that he had suggested evacuation earlier that morning to an assistant to Boris Shcherbina, but had been told to wait until the arrival of the commission. Briukhanov's civil defense chief, Serafim Vorobev, reported to the commission that the radiation level around the reactor exceeded 250 roentgens per hour—the upper limit that could be detected with the equipment at hand. Vorobev, disheveled and distressed as always, demanded immediate evacuation. Briukhanov tried to calm him down.

The trip that had begun with the assumption that there was no significant radioactivity, and the meeting that had begun with assurances that the situation in Prypiat was under control, had suddenly taken a turn that Maiorets was desperately trying to avoid. He asked to see the operators of the reactor in order to question them about what had happened, but was told that they were all in the hospital. The medical officials present at the meeting told him that their skin had turned brown from radiation tan: they had received three to four times the lethal dose of radioactive exposure. A novice in the field of nuclear energy, Maiorets hoped that if they shut down the reactor, the ill effects would cease. But Shasharin told him that the operators had already shut down the reactor, which was now in an "iodine pit." This meant that the reactor was temporarily disabled because of a buildup of short-lived nuclear poisons, especially isotopes of iodine and xenon. When the reactor was in this "poisoned" state, the nuclear reaction usually slowed down. Although Shasharin did not say so, he and other nuclear experts were concerned that once the reactor emerged from the "poisoning," the reaction would pick up speed. That, in turn, could lead to another explosion, wiping the plant, the nearby city, and the members of the commission off the face of the earth.

Maiorets showed no visible signs of stress. As one of his subordinates noted, "He was his usual dapper self, his hair parted neatly on his pink skull, his round face as expressionless as ever. Perhaps he simply failed to understand." Maiorets seemed more concerned about the proposed evacuation of the city than about the possibility of a new explosion. The top officials wanted to avoid panic, but, even more, they were trying to avoid responsibility for ordering an evacuation. That would mean admitting that something terrible had happened. There had not been an evacuation for any reason since World War II. Even suggesting something of the sort could be detrimental to one's career. "What if you are mistaken?" Maiorets asked those advocating evacuation. He continued: "I am opposed to evacuation. The danger is clearly exaggerated."[10]

They decided to take a break. Maiorets was standing next to Shasharin, smoking in the corridor of the party headquarters, when two experts from Moscow, Boris Prushinsky and Konstantin Polushkin, showed up. Prushinsky was the chief engineer of the department of nuclear power stations at the Ministry of Energy and Electrification; Polushkin was a senior scholar at the research institute that had designed the RBMK reactor. They had more expertise on the reactor between them than anyone else on the high commission. They had taken the 9:00 a.m. plane from Moscow and arrived in Prypiat in the early afternoon. The first thing they did, after lunching in a restaurant that was also hosting a wedding party, was to find a helicopter. With a pilot and a photographer, they flew over the station to take a look at Unit 4 from above. What they saw left no doubt in their minds that the commission's working assumption was wrong. The reactor was not intact—it had exploded.

Prushinsky and Polushkin told Maiorets what they had seen. The central hall of the reactor building was gone, destroyed by the blast. So were the main circulation pump room and the steam separator room. An explosion of steam in the reserve water tank simply could not have caused so much damage. "The upper biological shield of the reactor was now a bright cherry-red from the extreme heat and was lying at an angle over the reactor vault," recalled Prushinsky. There were pieces of graphite all over the place. The reactor was no more. Its radioactive innards were scattered everywhere. "It was safe to say that the reactor was utterly finished." It was one thing for Prushinsky to say that, and quite another for Polushkin, a representative of the designer of the allegedly explosion-proof reactor. But he agreed with his colleague.

What should be done? asked the distressed Maiorets. "God knows," answered Prushinsky. "Right now I can't say. There is graphite burning in the reactor. That has to be extinguished before anything else. But how and with what? We've got to think."[11]

BORIS SHCHERBINA, deputy head of the Soviet government and chairman of the high commission, arrived in Prypiat after 8:00 p.m. A round-faced, slightly balding man of medium build, with dark hair and a decisive look in his eyes, he was calm and considerate, commanding the respect of his subordinates. A native of Ukraine who had begun his party career in the industrial city of Kharkiv, Shcherbina had spent most of his life as a regional secretary of the Communist Party in western Siberia, where he presided over the birth of the Tiumen oil and gas industry. In 1973 he was transferred to Moscow to become minister of construction of oil and gas enterprises and pipelines. Eleven years later, in 1984, he was promoted to deputy prime minister in charge of the energy sector of the government. Thus nuclear energy was becoming an ever larger part of his portfolio, even though oil and gas were remaining extremely important in the Soviet industrial complex, bringing in the lion's share of the government's hard-currency earnings.[12]

Shcherbina was a natural choice to lead an all-powerful commission on the Chernobyl accident. The only problem was that on the morning of April 26 he was away from Moscow, visiting the Siberian city of Barnaul. It was his routine to hop onto a plane on weekends and visit construction sites and enterprises throughout the Soviet Union. April 26 was a Saturday, and Shcherbina, faithful to his routine, was on the road. Summoned by his boss, Nikolai Ryzhkov, he returned to Moscow, where the two officials had a twenty-minute conversation about the accident, still believed not to be causing any rise in radiation levels. Then Shcherbina picked up a number of experts still in Moscow and flew with them to Kyiv.[13]

During the flight Shcherbina got a crash course in the history of nuclear disasters from Valerii Legasov. A bespectacled forty-nine-year-old of medium build, with a big nose and prominent lips, Legasov was the first deputy of Anatolii Aleksandrov, the director of the Kurchatov Institute of Atomic Energy. Instead of relaxing, Legasov had spent the entire Saturday at a meeting of party and managerial personnel at the Ministry of Medium Machine Building—a nuclear-power monster that all but owned the Kurchatov Institute.

"True to my character and to habit ingrained over many years, I called for a car and went to the party management meeting," recalled Legasov. He had first heard the news of the Chernobyl accident before the meeting, but it did not seem like a matter of major concern then. The main speaker at the event, the eighty-seven-year-old minister Yefim Slavsky, who had helped design and produce the RBMK reactor in the 1960s together with Legasov's boss, Aleksandrov, spoke for two hours but mentioned the accident only in passing.

"We had all become accustomed to hearing that aged but very demagogically active man speak energetically for an hour, in a loud and confident voice, about how excellent and wonderful everything was in our bailiwick," recalled Legasov. "This time, singing a paean to atomic energy as always and to the great successes achieved in its construction, he mentioned quickly that now, true, there had been some kind of accident at Chernobyl. The Chernobyl station belonged to the neighboring ministry, the Ministry of Energy and Electrification. So he noted quickly that they had made some kind of mess there, an accident of some sort, but it would not halt the development of atomic energy."[14]

During the break after Slavsky's report, Legasov learned that he had been appointed to the state commission headed by Shcherbina. He was told to be at the airport at 4:00 p.m. There went the weekend, thought Legasov. At the institute, he collected whatever technical information he could find on RBMK reactors and the Chernobyl nuclear plant, spoke with colleagues who knew the reactors inside and out, and headed for the airport to catch his flight.

On the plane, he shared his knowledge of the nuclear industry with Boris Shcherbina. Someone mentioned the Three Mile Island accident in the United States. Legasov stepped in. "I tried to tell Boris Yevdokimovich about the accident at the Three Mile Island station that had taken place in the United States in 1979," recalled Legasov. Caused by the malfunctioning of one of the plant's nonnuclear systems, the accident had led to the escape of nuclear reactor coolant and a rapid increase of radiation levels—partly the result of the operators' failure to identify the initial problem. It was rated a five on the seven-point scale of nuclear accidents, seven being the highest. One hundred forty thousand people had voluntarily left the twenty-mile zone around the nuclear plant. Legasov argued that "most probably the cause of that accident had no bearing on the events in Chernobyl because of the fundamental difference in the construction of the apparatus." He was right. The Americans had used a pressurized water reactor that was much safer than the RBMK. The American reactor also had a concrete containment structure, and the RBMK did not. The Soviet commission was about to deal with a much more dangerous situation.[15]

What struck Legasov once they landed in Kyiv was the "cavalcade of black government cars and the agitated crowd of Ukrainian leaders. . . . Their faces were all agitated; they had no precise information but said that the situation was bad there." Legasov and his colleagues got into the limousines waiting for them and drove to Chernobyl and then to Prypiat. The scenery on the way was beautiful, and those members of the commission who arrived during the day had a chance to fully enjoy it. "Spring was already in full flower, and the orchards were in bloom. There were geese on the river Uzh," remembered one of them. "The sensation of nature's bounty did not abandon us the whole way." Shcherbina and Legasov missed most of that scenery. The sun went down soon after 7:00 p.m., and they were confronted with a different picture. "As we were approaching the city of Prypiat, we were struck by the appearance of the sky some eight or ten kilometers from Prypiat. There was a deep red or, more precisely, a crimson glow above the station," recalled Legasov.[16]

At 8:20 p.m., Shcherbina and his entourage entered the building of the Prypiat party committee, where the rest of the members of the commission were meeting under the chairmanship of Anatolii Maiorets. Maiorets's deputy for nuclear plants, Gennadii Shasharin, who had just returned from his helicopter flight over the reactor, immediately approached him. Shasharin's claim that the reactor had been destroyed surprised Shcherbina. The temperature within the reactor was rising (Prushinsky had seen the graphite in the reactor heated to a cherry red, whereas Shasharin saw yellow). The radiation levels were high. Shasharin suggested the immediate evacuation of Prypiat. This came as a complete shock to Shcherbina, but he remained calm and collected. He told Shasharin that evacuation would cause a panic worse than the radiation itself.[17]

Shcherbina began a new meeting of the commission that turned into a brainstorming session. Everyone now accepted what had been unthinkable only hours, if not minutes, earlier—a meltdown had occurred, and the reactor's core was damaged, spreading radioactivity all over the place. The question was how to stop it from burning and producing ever more radioactivity. They bounced ideas off one another. Shcherbina wanted to use water, but they explained to him that water could make things even worse. It was one thing to pump water into the reactor to cool it off, and quite another to try dowsing a nuclear fire with water—it could make the fire more intense. Shcherbina did not give up on the idea but was willing to consider alternatives. Someone suggested using sand. But how would they bring it to the reactor? What about helicopters? Shcherbina had already called military helicopter and chemical units into the area. Their commanders were en route to Prypiat as they spoke.

It began to look as if the authorities finally knew what they were doing. Shcherbina's demeanor changed in a matter of minutes. He finally knew what had to be done. "Despite the late night (actually, the pre-dawn hours), [Shcherbina] did not appear to be tired," recalled his assistant, B. N. Motovilov. When General Nikolai Antoshkin, chief of staff of Kyiv's district air force command, crossed the threshold of the Prypiat party headquarters sometime after midnight, Shcherbina greeted him with the words: "Everything depends on you and your helicopter pilots now, general." He ordered Antoshkin to start dropping sandbags on the reactor right away. The air force commander replied that it was impossible: the helicopters were not there yet. Shcherbina agreed to wait, but only until daybreak—he had no other choice.[18]

Something else that he postponed until morning was the evacuation of Prypiat. What he had considered unacceptable spreading of panic only a few hours earlier had become an imperative by the late hours of April 26. Soon after 9:00 p.m., while the members of the commission were brainstorming about what to do with the reactor—which everyone now agreed was damaged—the reactor suddenly awakened. Three powerful explosions illuminated the dark red sky above Unit 4, sending red-hot pieces of fuel rods and graphite into the air. "It was a striking spectacle," remembered Leonid Khamianov, one of the experts from Moscow, who observed the scene from the third floor of the Prypiat party headquarters. "It's hard to say whether it was the result of the reawakening of the reactor or whether water happened to fall onto the burning graphite at that moment, causing an explosion of steam," he observed.[19]

It looked as if the worst-case scenario was now coming to pass. Earlier in the day, experts had predicted a possible chain reaction starting as soon as the reactor emerged from the iodine well. Some suggested 7:00 p.m., others 9:00 p.m., as the time when the reactor would reawaken. Now it seemed as if those predictions were on the mark. The explosions might be the first indications of a much bigger blast to come: they had no choice but to wait and see. But even without further explosions, those that occurred between 9:00 and 10:00 p.m. had put the population of the city in greater danger than it had been in during the day. The wind, which had previously been hardly noticeable, suddenly picked up, driving radioactive clouds northward from the damaged reactor to cover parts of the city. Radiation levels increased on the city plaza in front of party headquarters in downtown Prypiat, rising from 40 to 330 microroentgens per second, or 1.2 roentgens per hour.[20]

Late in the evening, Armen Abagian, the director of one of the Moscow nuclear power research institutes, who had been dispatched to Prypiat as a member of the government commission, approached Shcherbina and demanded the evacuation of the city. Abagian had just returned from the plant, where the explosions in the reactor had caught him unawares—he and his colleagues had had to seek shelter under a metal bridge. "I told him that children were running in the streets; people were hanging laundered linen out to dry. And the atmosphere was radioactive," remembered Abagian. "That made an impression on the normal human level." But, according to government regulations that had been adopted in the Soviet Union back in 1963, evacuation of the civil population was not necessary unless the radiation dose accumulated by individuals reached the 75-roentgen mark. Calculations had shown that with the existing level of radioactivity, the intake might be about 4.5 roentgens per day. The threshold of 75 roentgens had not yet been reached, and Yevgenii Vorobev, a deputy minister of health and the senior medical officer on the commission, was reluctant to take responsibility for ordering an evacuation.[21]

The debate continued. Some members of the commission later credited Legasov with making a key contribution to the debate. Ivan Pliushch, the head of the Kyiv regional administration, remembered: "Academician Legasov . . . placed his hand on his heart, as if taking an oath, and said: 'I am pleading with you to evacuate the people, because I don't know what will happen to the reactor tomorrow. It is ungovernable. We are pouring water, but we don't know where it's going. We are doing something that has never been done before. And as a scientist, I can't foretell today what the consequences may be. I therefore plead that you evacuate the people.'" According to Pliushch, Vorobev was opposed to evacuation. "People can be evacuated when radiation reaches 25 ber. That is our norm," he told the gathering, referring to the biological equivalents of roentgens. "But here, you see, people have been talking about milliroentgens and roentgens. We can't confirm the need for evacuation." According to Legasov, the nuclear scientists in the room, "having a presentiment that the dynamic would change in an unfavorable direction, insisted that a decision on evacuation be made without fail. . . . [T]he medical people naturally yielded to the physicists."[22]

Legasov, Abagian, and other scientists eventually convinced Shcherbina. "Sometime around 10:00 or 11:00 p.m. on April 26 [Shcherbina], having listened to our discussion, made the decision to evacuate without fail," remembered Legasov. But Shcherbina's decision was not sufficient. "They told one [party] secretary, and he said: 'I can't give you my agreement to this,'" recalled Pliushch. "They got through to another, who also expressed sympathy but said that he could not give his assent. And it was impossible to get through to a third." Eventually Shcherbina phoned his boss, Premier Nikolai Ryzhkov. "Shcherbina called me on Saturday evening," recalled Ryzhkov, "and reported on the situation. 'We've measured the radiation. . . . Prypiat has to be evacuated. Immediately. The station is close by, and it's emitting radioactive contagion. And people in the city are living it up full blast; weddings are going on . . .' I decided: 'Evacuation tomorrow. Prepare trains and buses today and tell the people to take only the bare necessities, money, documents, and nothing more. No furniture.'" By 1:00 a.m. on April 27, local officials in Prypiat had received an urgent order from Shcherbina to prepare lists of citizens for evacuation. They were given two hours to do the job. The evacuation was supposed to start early in the morning.[23]

It had been a long day for those assembled in the conference room of the Prypiat party headquarters. It had begun with the shocking news of the explosion and progressed through the stages of confusion, disbelief, and denial. But by now denial had given way to full cognizance of the disastrous situation, with the worst perhaps still to come. Had the chain reaction begun again? Would the explosions resume? And, if so, what would happen to the city of Prypiat and other population centers nearby? Kyiv, the capital of Ukraine, was only 130 kilometers away from the power plant. They had no answers to those questions, and there was little they could do during the night. They had to wait for dawn, hoping that it would not bring further trouble.

EXODUS

THE DECISION to evacuate Prypiat came as a relief to the seventy-year-old Ukrainian prime minister, Oleksandr Liashko. Under threat of being called "alarmist," he had begun preparations for possible evacuation soon after Nikolai Ryzhkov woke him up with his call about the accident at 2:40 a.m. on April 26. Liashko was never able to explain why he argued in favor of getting ready for that eventuality when everyone else in the top leadership of the Soviet Union and Ukraine seemed to be calm and confident that things at the Chernobyl plant were under control. As he observed, it was an intuition "that presented a clear foreboding of danger to people."[1]

In the Soviet political hierarchy, Liashko, the head of government of the Ukrainian Soviet Socialist Republic—the second-most-populous constituent of the Soviet Union after the Russian Federation—was an important but not all-powerful figure. Above him was the party boss of Ukraine, First Secretary Volodymyr Shcherbytsky, a close ally and protégé of the long-serving former Soviet leader Leonid Brezhnev. On the all-Union level, Liashko reported to the head of the Soviet government in Moscow, Nikolai Ryzhkov. Both Ryzhkov and Liashko came from the Donbas, the coal-mining region of eastern Ukraine, and they seemed to get along well. Indeed, Liashko seemed to get along with almost everyone. His no-nonsense, nonideological approach to running the second-largest Soviet economy had won him the respect of his colleagues and subordinates. But on April 26, at a 10:00 a.m. meeting of the supreme Ukrainian leadership, when Liashko began to talk about organizing Kyiv's city buses for the possible evacuation of Prypiat, many, including Shcherbytsky, were skeptical. Later, most of them would claim that at the time they knew only that there had been a fire at the Chernobyl nuclear plant and that it had been extinguished. The general feeling among the Kyivan officials was that they should avoid sounding the alarm and incurring the wrath of the bosses in Moscow.[2]

There was little surprise that the top leaders of Ukraine, such as Liashko, had learned of the accident in Prypiat, on their territory, from Moscow. The Chernobyl plant was under the control of all-Union officials in the Soviet capital. Although the Ukrainian party bosses and ministers could and did influence the appointment of top personnel at the station, including Viktor Briukhanov and Nikolai Fomin, they had no direct control over what was going on at the station itself. But the firefighters and police—the two rapid-response agencies that were on site immediately after the explosion—were under republican jurisdiction. Both reported to the Ukrainian minister of the interior. Under Kyiv control also were the party and city authorities of Prypiat and other towns and settlements in the vicinity of the plant. While radioactivity at the plant became not only the concern but also the responsibility of the all-Union authorities, the task of dealing with the consequences of the accident on the local level fell directly into the lap of the republican officials, who, as things progressed, increasingly got the feeling that they were being obliged to clean up the mess created by their Moscow bosses.

The Ukrainian leadership's main source of information on what had happened at the Chernobyl nuclear plant was the republic's Ministry of the Interior, which was under the joint jurisdiction of Kyiv and Moscow. The Ukrainian interior minister, General Ivan Hladush, was awakened at about 2:00 a.m. in a hotel room in the city of Kharkiv. The news of the accident that the duty officer was about to give him was so sensitive that the officer insisted the minister leave his hotel room and go to local party headquarters, where he could speak on a secure line. The Interior Ministry officials in Kyiv were the first to receive information about the fire, but they thought they could handle it by themselves without alerting Liashko. Lieutenant Pravyk radioed the level of alarm that mobilized all the fire brigades in the Kyiv region minutes after the explosion. Major Vasyl Kucherenko, the head of Prypiat's police department, managed to reach the plant ahead of Briukhanov; he called his superiors in Kyiv from there to report on the explosion and the fire. There was disbelief at the other end of the line. "Do you know what you're saying? Is there someone beside you?" asked a senior police official in Kyiv who was eager to verify Kucherenko's report.[3]

By 5:00 a.m., a deputy minister of the interior of Ukraine, General Hennadii Berdov, and a group of senior police officers were on their way from Kyiv to Prypiat. Berdov cut an impressive figure in the mayhem of the first days after the explosion. According to an eyewitness, he was dressed in "a brand-new uniform, with gold braid, a brightly colored array of ribbons, and a badge testifying to his record in the Ministry of Internal Affairs." Berdov was a calm but energetic officer who immediately took command of the more than four hundred policemen who were dispatched to Prypiat from neighboring towns and counties in the first hours after the explosion. They took control not only of the approaches to the plant but also of the nearby railway station of Yaniv. There were not only sergeants but also lieutenants, captains, majors, and colonels manning police patrols. Berdov was exposing himself and many of his subordinates to dangerous radiation, but in the first hours after the explosion few of them guessed that they were dealing with something more than a regular technological disaster, with fire constituting the main threat to human life and public order.[4]

The first deputy minister of the interior, Vasyl Durdynets, who took charge of the ministry's actions in Kyiv, later recalled that he knew nothing about heightened levels of radiation until midday on April 26. The civil defense authorities in Kyiv were silent. Earlier that morning, when Serafim Vorobev, the head of the civil defense department of the Chernobyl nuclear plant, finally reached those authorities on a civil defense line, the first question he heard was whether the fire had already been extinguished. "What do you mean, fire?" exclaimed Vorobev. "There's a general accident here!

Gen-er-al! The population has to be informed!" The Kyiv official shouted in reply: "Alarmist! Think of what you're saying! They'd have our heads for such a report." Vorobev kept insisting on the seriousness of the accident: "The DP-5 instrument [dosimeter] is off the scale! More than 200 roentgens per hour!" He was rebuffed. The record of incoming reports was later falsified to make it appear that Vorobev had called the Kyiv civil defense headquarters later than he actually did. Premier Oleksandr Liashko recalled that when a civil defense unit was dispatched to Prypiat, its members simply did not have the right equipment to measure the radiation levels.[5]

Around 9:00 a.m., Durdynets, who had spent most of the night in his office at the ministry, called the leaders of the republic to inform them of the previous day's developments. The report on the Chernobyl accident happened to be one of many items on his agenda. Valentyna Shevchenko, the chairwoman of the Ukrainian Supreme Soviet—the nascent republican parliament, which at that point served as a rubber stamp for the party authorities—and a member of the republic's supreme leadership, alleged later that Durdynets had mentioned Chernobyl at the very end of his brief summary. He told her that there had been a fire there, but it had been extinguished (on the following day, the report from the republic's Ministry of the Interior to the Central Committee suggested that the fire had been put out by 8:00 a.m.). "And what about the people?" Shevchenko had asked the deputy minister. "Nothing special," came the answer. "Some are celebrating weddings, others planting their gardens, still others fishing on the Prypiat."[6]

Shevchenko immediately called Prime Minister Liashko, who was in his office discussing the situation at the Chernobyl plant with one of his deputies. He had already ordered the mobilization of transport for the possible evacuation of Prypiat. The civil defense units were under the joint jurisdiction of Kyiv and Moscow, as was the republic's Ministry of the Interior, and Liashko used his power over both structures to start the mobilization. When the commander of the republican civil defense organization suggested that it would be difficult to get the bus drivers to come to work on Saturday morning, Liashko called the minister of the interior, Ivan Hladush, and ordered him to use police officers to inform the drivers relaxing at home about the mobilization. "Is it serious there?" Hladush asked the Ukrainian premier. "I don't know, but organize the mobilization and dispatch of transport to the area of the Chernobyl plant together with civil defense as if you were doing it under wartime conditions."[7]

Alerted by the news both from Moscow and from his Ukrainian subordinates, Ukrainian party head Volodymyr Shcherbytsky, the supreme leader of the republic, called a meeting of the Ukrainian Politburo for 10:00 a.m. The news received from Viktor Briukhanov in Prypiat via local party officials was not too disturbing: the fire had been extinguished, and radiation levels were high but within the norm. "Aren't you in too much of a hurry?" Shcherbytsky asked Liashko, who had already ordered the mobilization of transport facilities. "The commission will come and give a competent assessment. Take action then." Liashko refused to call off his order and asked for time until the situation became clearer. "Well, and if no evacuation is needed, how much will your action cost?" Shcherbytsky asked the premier.[8]

Liashko refused to succumb to pressure from the party boss. "I replied that if everything was all right there, we would charge all expenses to Civil Defense. Because we had done so in the past—we would charge several million to Civil Defense." Shcherbytsky finally gave his approval, but that was not enough. Around 11:00 a.m., Liashko called his all-Union boss, Premier Nikolai Ryzhkov, and got his permission as well. "By 1400 hours I had a report that 1,200 buses and 240 trucks were ready," recalled Liashko.

Shcherbytsky still had doubts. That day, the Kyiv and Ukrainian KGB headquarters had filed two reports on the situation at the Chernobyl power plant and the city of Prypiat. The first of them, which reflected the KGB's understanding of the situation in the late morning of April 26, already indicated heightened levels of radiation, but the levels reported were significantly lower than the actual ones recorded by Vorobev. According to the KGB report, the radiation level at the plant was estimated at 20–25 microroentgens per second, and in Prypiat it was 4–14 microroentgens per second. The rise in radiation levels was attributed to "contaminated water" that might have escaped from the cooling system of the reactor damaged by the explosion. The second KGB report, based on information available at

3:00 p.m. on April 26, estimated radiation levels near the reactor at 1,000 microroentgens per second, at the power plant at 100 microroentgens per second, and in Prypiat at 2–4 microroentgens per second. They were still relying on dosimeters with a scale of only 1,000 microroentgens per second.[9]

But Liashko already had all the authority he needed to start preparations for the evacuation, at this point as a precaution. Judging by reports submitted to the Ukrainian party's Central Committee, the order to begin the evacuation came at 8:00 p.m. on April 26. Columns of buses and trucks began to assemble along the rural routes of the Chernobyl district that evening. Two trains were ready to take passengers at the nearby Yaniv station. Bus drivers and train engineers would spend the whole night, sometimes in close proximity to the power plant, waiting for orders to move into the city. They did not know whether the orders would come or not.[10]

At the first session in Prypiat of the government commission summoned by Maiorets, on the evening of April 26, General Berdov had reported that buses were ready in case evacuation was called for, only to be rebuffed as alarmist. A few hours later, the members of the commission saw the reactor awakening and sending radioactive fireworks into the sky around Prypiat. The tone of the conversation changed, and sometime after 10:00 p.m., Ukrainian prime minister Liashko received word that the commission, now with Boris Shcherbina participating, had concluded that the reactor was damaged and that there was a threat of further radioactive contamination. It would take the all-Union authorities a while to give the final go-ahead for starting the evacuation. Liashko was on the phone with Prypiat throughout the morning of April 27 to make sure that the evacuation was actually taking place. By 2:00 p.m., he finally got word that the evacuation was about to begin. He was no longer an alarmist: he had turned out to be right. The evacuation was necessary, and now the government commission was ready to carry it out.[11]

As the party and city authorities debated the question of evacuation, residents spontaneously began to leave the city en masse. The first to leave on the morning of April 26 were the wives and children of the firefighters. Their husbands and fathers had paid the high and sometimes ultimate price for the knowledge that what had happened at the power plant was not just a fire. It was a nuclear explosion that had released invisible waves of radiation that incapacitated and killed without warning or mercy. Volodymyr Pravyk's parents could not refuse their son when he begged them from a hospital ward to take his wife and daughter out of Prypiat as soon as possible. From the Prypiat hospital they went directly to Pravyk's apartment, put his wife, Nadiika, and one-month-old daughter, Natalka, into a motorcycle sidecar, took them to the railway station, and got them on a train speeding away from Prypiat.[12]

The next to leave Prypiat were the firefighters themselves, especially those who had suffered the most from radiation poisoning. Leonid Shavrei, the firefighter who had reached the nuclear plant in the same truck as Lieutenant Volodymyr Pravyk, realized that there was something unusual about the fire he was trying to keep in check only after 5:00 a.m. on April 26, when he was allowed to leave the roof of the turbine hall, where he had spent most of the night. He lit a cigarette and suddenly felt a strange sweet taste in his mouth. "What sort of sweet cigarettes are these?" he asked his fellow firefighter. Pravyk had complained even earlier about the strange taste in his mouth. The medical personnel provided them with potassium iodide pills and water to wash them down. It was then that Shavrei began to vomit. "It was impossibly revolting," he recalled. "I wanted to drink, but drinking was impossible—I felt nauseous immediately." The medics wanted him to go to the medical center, but Shavrei refused. He went home instead, put his wife and young child in their car, and drove them away from Prypiat.[13]

In the late afternoon of April 26, when Leonid Shavrei returned to Prypiat after leaving his wife and child with relatives in a nearby village, he went straight to the hospital to check on his fellow firefighters. Shavrei's commander, Lieutenant Pravyk, was looking out of a window, his face swollen. "How are you?" asked Shavrei. "All right," came the not-very-assuring answer. "What about you?" Shavrei responded in the same tone: "All right too." In general, he found most of his colleagues in good spirits. Pravyk and others explained that they had been given intravenous medication and injections. They did not say much more. Shavrei left for home, hoping to see his friends again the next day, but when he showed up at the hospital, they were already gone.[14]

It took the doctors a while to figure out the most effective way to treat their patients, who were sick from irradiation. The first suggestion was to make them drink milk. "He needs milk, lots of milk," said a doctor to Liudmyla Ihnatenko, the wife of the firefighter Vasyl Ihnatenko, who was among the first to be admitted to the hospital. "But he doesn't like milk," responded Liudmyla. "He will drink it now," came the answer. Together with Tetiana Kibenok, the wife of Lieutenant Viktor Kibenok, who was in the same hospital room as Ihnatenko, Liudmyla drove to the countryside to get milk. They came back with enough three-liter jars to cure the whole ward, but Vasyl Ihnatenko and others began to vomit after taking their first sips. Only then did the doctors who were treating the firefighters for gas poisoning try intravenous injections, most likely of nitrites, which appeared to work. Later that evening, Liudmyla could already see her husband standing near a window.[15]

Viktor Smagin, one of the reactor operators, who had replaced Aleksandr Akimov and Leonid Toptunov on the night shift in the Unit 4 control room around 8:00 a.m., was brought to the medical center by ambulance six hours later, suffering from headache, dizziness, and uncontrollable vomiting. He began to feel much better after two hours of intravenous treatment. Smagin recalled that he had three bottles of liquid fed into him. Once off the intravenous feed, he wanted to go for a smoke. The patients had had to leave all their clothes and personal belongings behind when entering the medical center, so there were no cigarettes around. But friends who came to check on those hospitalized were happy to help. One of them attached a bunch of cigarettes to a string that Smagin lowered from his second-floor window. With the cigarettes in his pocket, Smagin went directly to the smoking room, where he found his fellow plant operators. They had been temporarily restored to a semi-normal state by the intravenous treatment. Aleksandr Akimov and Anatolii Diatlov were there, discussing the possible causes of the explosion. So were many other engineers and operators. But not everyone was able to walk, despite the intravenous treatment. Leonid Toptunov was among those who stayed in bed, his skin brown from the "radiation tan." His lips were swollen, making it difficult for him to talk.[16]

The firefighters and power plant operators who had been exposed to the highest doses of radiation were flown to Moscow for treatment in the late hours of April 26, even before they could bid farewell to their loved ones. Their evacuation was carried out in utmost haste, with the time of the departure kept secret even from family members. The decision to evacuate the firefighters and reactor operators most affected by radiation exposure was made by a group of doctors and scientists from Moscow who had flown into Prypiat on the afternoon of April 26. They came from the Institute of Biological Physics, which was attached to the Ministry of Health, and from Moscow's Hospital No. 6, which specialized in the treatment of radiation sickness and had the experts and equipment to deal with the immediate victims of the accident. The chief of the radioactive diseases ward, sixty-one-year-old Dr. Angelina Guskova, had been treating such patients since 1949, when, as a young doctor in the closed city of Cheliabinsk-40—the site of the first Soviet plutonium facility, which had been built to produce the atomic bomb—she had cared for the Gulag prisoners who were the guinea pigs of the early Soviet nuclear program. Now her doctors came to Chernobyl to treat the firefighters and nuclear plant operators.[17]

Guskova's doctors were looking for signs of acute radiation syndrome, which usually occurred in people exposed to more than 50 roentgens of radiation. For measuring acute radiation syndrome, the doctors used units called grays. Grays are defined in terms of the amount of energy absorbed by the human body in a dose of ionizing radiation, with 1 gray being equivalent to 1 joule per kilogram of irradiated material. How it relates to rem differs according to the type of ionizing radiation involved. In the case of beta radiation, 1 gray is equivalent to 100 rem. It corresponds to 1,000 rem in neutron radiation, and to 2,000 rem in alpha radiation.[18]

Depending on the units of measurement of radiation and its absorption and impact on the human body, signs of acute radiation syndrome start with a one-time exposure of 50 roentgen or more and absorption of 0.8 of a gray of ionized radiation. Symptoms include lack of appetite, nausea, and vomiting. After the absorption of more than 50 grays, symptoms begin almost immediately. They include nervousness, confusion, watery diarrhea, and sometimes loss of consciousness. For those absorbing a dose of between 10 and 50 grays, loss of appetite, nausea, and vomiting occur within hours after exposure; for those receiving 1 to 10 grays, the same symptoms set in anywhere between several hours and two days after exposure. In all cases, radiation kills stem cells in bone marrow, but radiation in the range of 10 to 50 grays also kills cells in the gastrointestinal tract. When radiation exceeds 50 grays, the cardiovascular and nervous systems are affected. All patients go through three main stages of the disease. In the first stage, symptoms of the syndrome appear; in the second, latent stage, patients may appear to be fine, and they feel well; in the final stage, the symptoms return with greater severity. Only patients in the first category, those who have absorbed less than 10 grays, have a chance of surviving, which is estimated at approximately 60 percent. Patients in the second and third categories die.[19]

The Chernobyl patients, 134 men and women diagnosed with acute radiation syndrome, were later divided into four categories. In the first, which included some twenty people with an absorption rate of between 6.5 and 16 grays, almost everyone would die. In the second category, which included almost the same number of people, who had absorbed between 4.2 and 6.4 grays, about one-third would die. For the patients who had absorbed between 2.2 and 4.1 grays, the statistics were much better, with only one person dying. There were no deaths recorded in the fourth category, between 0.8 and 2.1 grays. Altogether, twenty-eight people died of acute radiation syndrome within the first four months after the exposure. Twelve of the thirteen patients who received bone-marrow transplants did not survive the ordeal. The deaths of close to twenty patients in the second, third, and fourth categories that followed in the years after the accident were not attributed by the doctors to the radiation exposure.[20]

The doctors from Moscow who came to Prypiat on April 26 were highly experienced in dealing with patients affected by acute radiation syndrome, but they did not have the right equipment to measure how much radiation they had absorbed. They looked instead for the first symptoms and relied on the blood tests. One of Guskova's top experts, Georgii Selidovkin, had flown to Prypiat to assess patients on the spot and decide who should be brought to Moscow for immediate treatment. Many of those in the Prypiat hospital would remember Dr. Selidovkin because he had a beard, and that was uncommon in the Soviet Union of the 1980s, when it was taken to be a sign of either decadence or free thinking.

Selidovkin began examining patients sometime after 4:00 p.m. on April 26. Guskova, who stayed behind in Moscow, got on the phone with her colleagues in Ukraine. The doctors from Moscow examined close to 350 patients, looking at their skin, asking about when they had first started vomiting, and measuring the number of leukocytes in their blood. Leukocytes, or white blood cells, are produced in the bone marrow and have a rapid turnover rate; they are among the cells most easily damaged by radiation. A drop in the number of white blood cells is an easily recognizable effect of radiation exposure. By early evening, Dr. Selidovkin had twenty-eight patients whom he considered top-priority cases. Among them were Pravyk, Kibenok, Ihnatenko, and most of the other firefighters, as well as the reactor operators Akimov and Toptunov and the deputy chief engineer, Anatolii Diatlov. They had to be taken to Moscow without delay: time was of the essence and could mean the difference between life and death.[21]

The thirty-four-year-old deputy mayor of Prypiat, Aleksandr Yesaulov, was charged with making sure the twenty-eight people selected by Selidovkin and his colleagues were taken from Prypiat to the Boryspil airport in Kyiv and then to Moscow as soon as possible. A few days earlier, Yesaulov had reported in the local newspaper on the results of the Lenin *subbotnik*, the unpaid labor day organized by the party in a show of devotion to the founder of the Soviet Union and his communist ideology. Now all that seemed very far away. There were posters glorifying Lenin all over the city, but its inhabitants were now facing a new, post-ideological reality. Yesaulov's first task on the morning of April 26 had been to arrange for a vehicle to wash radioactive dust off the streets of the city. He then had to deal with the evacuation of those who had found themselves with too much of that same dust on their skin and in their intestinal tracts.[22]

The aircraft offered by the civil defense office was waiting at the Boryspil airport. Buses and ambulances were lined up as well. The main challenge turned out to be the documentation that had to accompany the patients to Moscow. They had had to surrender all their documents very early on the morning of April 26 upon entering the hospital in Prypiat, and there was no easy way to match individuals with documents and test results. No document in the Soviet Union was deemed legitimate without a stamp, but the requisite stamps had been left at the power plant. It was decided to forgo the stamps. With paperwork ready, Yesaulov faced a new challenge—dealing with the relatives of those to be evacuated to Moscow. Prohibited from entering the hospital, they besieged it from the outside, and once word spread that their loved ones were about to be taken to Moscow, they went into action.

"All the wives got together in one group. We decided we'd go with them," recalled Liudmyla Ihnatenko. Her husband, Vasyl, had just shouted to her from a window that they were about to be taken to Moscow. "Let us go with our husbands! You have no right! We punched and clawed," said Liudmyla, recalling her actions and those of the other wives. They were pushed back by soldiers, who had formed a cordon around the hospital. But they did not give up. Then a doctor showed up on the doorstep of the hospital and approached the wives. He told them that their husbands were indeed going to Moscow but needed a change of clothing—the outfits they were wearing when they arrived at the hospital were radioactive and had been burned. The wives rushed across the city to their apartments to fetch clean shirts, trousers, and underwear. It was late at night, public transportation was not running, and they had to walk. In fact, they ran. But when they came back, their husbands were already gone. "They tricked us," recalled Liudmyla Ihnatenko, "so that we wouldn't be there yelling and crying."[23]

Aleksandr Yesaulov left for Kyiv at the head of the two buses and two ambulances carrying the patients. The buses accommodated twenty-six patients who were still able to walk, while the ambulances took two reactor operators who could not—hot water and steam burns covered more than 30 percent of their bodies. On reaching Kyiv, they drove along its main boulevard, Khreshchatyk—an incongruous group of men in pajamas looking out the windows of buses that Kyivans, still unaware of what was going on in their backyard, largely ignored. The buses headed for Boryspil. They arrived soon after 3:00 a.m. on April 27. A few hours later, ambulances in Moscow, their interiors covered with plastic to prevent contamination, collected the Prypiat patients at the airport and delivered them to Guskova's Hospital No. 6. The doctor who had cut her teeth on prisoners of the Gulag was ready to start her job.

Yesaulov had carried out his task, starting the evacuation of his city. On the way back home to Prypiat, he passed hundreds of buses heading there as well. The evacuation of the entire city was about to begin.[24]

THE COLUMNS of buses, which had been waiting on the roads between Chernobyl and Prypiat for hours, absorbing high levels of radiation, began to move at 1:30 a.m. on the morning of April 27. Levels of radioactivity in the city were rising quickly. On April 26, radioactivity had registered in the range of 14 to 140 milliroentgens per hour, but by about 7:00 a.m. on April 27 it had risen to between 180 and 300 milliroentgens; in some areas close to the nuclear plant, it had jumped to as many as 600 milliroentgens per hour. The original plan had been to begin evacuation on the morning of April 27, but the decision was made too late to meet the deadline. It was decided instead to start the evacuation in the early afternoon.[25]

To many citizens of Prypiat the evacuation came not as a surprise, but as long-awaited relief. Intercity telephone networks had been cut, and the engineers and workers at the nuclear plant had been prohibited from sharing news of what had happened with their friends or relatives. But the family and informal networks that had always served Soviet citizens better than state-controlled media were active. Rumors about the accident at the power plant had begun to circulate in Prypiat a few hours after the explosion.

Lidia Romanchenko, an employee of a Chernobyl construction firm, recalled: "Sometime around eight in the morning [of April 26] a neighbor called me and said that her neighbor had not returned from the station; an accident had taken place there." That information was soon confirmed by another source: "Our dentist friend said that they had all been awakened at night because of an emergency and summoned to the clinic, to which people from the station were taken all night." A Good Samaritan, Romanchenko decided to share the news with her own friends and family. "I got in touch with my neighbors and close friends right away, but they had already 'packed their bags' that night: a close friend had called and told them about the accident."[26]

The city of Prypiat was slowly awakening to the reality of the disastrous accident in its midst. Liudmila Kharitonova, a senior engineer in a construction firm, was on her way to her country house—a dacha—in the neighborhood of Prypiat when she and her family were stopped by police. They had to turn back to the city, where they saw foam on the streets—roads were being treated with a special solution by water trucks. In the afternoon, military personnel carriers appeared on the streets, and military planes and helicopters filled the sky. The police and military were wearing respirators and gas masks. Children returned from school, where they had been given potassium iodide tablets, and were advised to stay indoors.

"We began to be more alarmed in the evening," remembered Kharitonova. "It's hard to say where the alarm came from, perhaps from inside ourselves, perhaps from the air, which by then was beginning to take on a metallic smell." A rumor began to circulate that those who wanted to leave could do so. Still, there was no official information on what had happened and what to expect. Kharitonova and her family went to the Yaniv railway station and got on a train to Moscow. "Soldiers were patrolling the Yaniv station," she recalled. "There were lots of women with small children. They all looked a bit confused, but they behaved calmly. . . . But I felt nonetheless that a new age had dawned. And when the train pulled in, it seemed to me so different, as if it had just come from the old, clean world we used to know, into our new poisoned age, the age of Chernobyl."[27]

Soon after 10:00 a.m. on April 27, the city authorities called in representatives of enterprises, schools, and institutions to draw up evacuation plans. Worried citizens rushed to the city party headquarters. Dressed in full uniform, General Hennadii Berdov, the deputy minister of the interior, appeared on the steps of the building and calmed the agitated crowd. His army of policemen was sent to apartment buildings to knock on doors and ask people to prepare for evacuation. The final decision was made by the government commission, with the blessing of the Moscow authorities, at noon, a mere two hours before the start of the evacuation.[28]

Deputy mayor Aleksandr Yesaulov, who had just returned from Boryspil, was given a new task—to evacuate the rest of the patients of the Prypiat hospital, altogether more than one hundred people suffering from radioactive poisoning. The party officials wanted him and his charges to be at Kyiv's Boryspil airport by noon. The rest of the patients were supposed to go to Moscow as well. It was 10:00 a.m., and the deadline was anything but realistic—the trip to Kyiv alone took close to two hours, and then there was all the paperwork to be done. They were finally ready to leave by noon. This time there was no hope of deceiving the wives and family members of those who were leaving, and Yesaulov did not try to do so. Amid the hugs, tears, and cries, he managed to assemble his group of patients and set out for Boryspil.

But no sooner had they departed than they had to stop—one of the patients needed urgent medical help. The buses stopped in the village of Zalissia, a few miles away from Chernobyl. The men, wearing pajamas, got out of the buses to stretch their legs and have a smoke. It was then that Yesaulov heard a woman's cry. It took him a while to understand that one of the young evacuees was from Zalissia. His mother had recognized him and could not control her feelings of shock, grief, and despair. "Just what we didn't need! Such a coincidence," recalled Yesaulov. "Where she came from, I had no idea." The words, "Mama, Mama," with which the young man tried to calm his mother, stuck in his mind. Eventually they left Zalissia. Yesaulov was so disoriented by the events of the past twenty-four hours that when, in the office of the director of the Boryspil airport, he was asked to pay for a few cups of coffee from a local coffee shop, he could not comprehend the question. "It was as if I had come from another world," he recalled.

The two worlds turned out to have borders and border controls. Before Yesaulov and his drivers were allowed to leave Boryspil, they had to wash their buses and take a shower. The high radiation levels caused by the accident were already an open secret. They left for home around 4:00 p.m. on April 27. As they approached Prypiat, they saw buses moving in the opposite direction—all 1,125 of them. The evacuation of the city had begun.[29]

PRYPIAT'S CITY radio station transmitted the announcement soon after 1:00 p.m. "Attention! Attention!" came the calm voice of a female announcer speaking Russian with a strong Ukrainian accent:

> In connection with the accident at the Chernobyl atomic power station, unfavorable radiation conditions are developing in the city of Prypiat. In order to ensure complete safety for residents, children first and foremost, it has become necessary to carry out a temporary evacuation of the city's residents to nearby settlements of Kyiv oblast. For that purpose, buses will be provided to every residence today, April 27, beginning at 1400 hours, under the supervision of police officers and representatives of the city executive committee. It is recommended that people take documents, absolutely necessary items, and food products to meet immediate needs. Comrades, on leaving your dwellings, please do not forget to close windows, switch off electrical and gas appliances, and turn off water taps. Please remain calm, organized, and orderly in carrying out the temporary evacuation.[30]

The radio repeated more or less the same announcement four times, but most people still did not understand the seriousness of the situation. "Just imagine," recalled Aneliia Perkovskaia, a city official, "it was only an hour and a half before the evacuation. Our children's cafeteria in a large shopping center was full of parents and children eating ice cream. It was a weekend day; everything was nice and quiet." For thirty-six hours after the explosion, people were given no reliable information about it and left virtually on their own. They were never provided with instructions on how to protect themselves and their children. Radiation levels that according to Soviet laws were supposed to trigger an automatic warning to the public about the dangers of radiation exposure had already been recorded in the early hours of April 26, but they had been ignored by one official after another. Finally, people were asked to gather their belongings and wait on the street a mere fifty minutes before the start of the evacuation. They were good citizens and did exactly what they were told to do.[31]

A film shot on April 26 and 27 by local filmmakers preserves images of a wedding taking place in the city attacked by radionuclides. It shows young men and women dressed in light summer clothes with their small children, walking the streets, playing soccer on sports fields, and eating ice cream in the open air. These scenes look surreal when juxtaposed with others shot by the same filmmakers: water trucks cleaning the streets, policemen and soldiers in protective gear atop troop carriers, patrolling the streets of Prypiat, and people waiting for buses that would take them away from their homes. One frame shows a doll on the window sill of an apartment building, seemingly waiting for the return of its owner. Sparks and white flashes in the film frames reveal the true meaning of what we see on the screen: these are scars left by radioactive particles attacking the film through the thick lenses of the camera.[32]

Liubov Kovalevskaia, the author of the recent article about quality control problems at the construction site of the Chernobyl nuclear plant, was among the thousands who boarded buses that afternoon, never to return to their homes. She had spent a good part of the previous night calming her elderly mother, who could not sleep after hearing rumors about imminent evacuation. Now their whole family—Kovalevskaia, her mother, her daughter, and a niece—were ready to leave. They were told that it would only be for three days. "There were already buses at every entrance," recalled Kovalevskaia. "Everyone was dressed as if to go camping, people were joking, and everything was rather quiet all around. There was a policeman beside every bus, checking residents according to a list, helping people bring in their belongings, and probably thinking of his family, whom he had not even managed to see in the course of those twenty-four hours. . . . The bus set off." The Chernobyl filmmakers kept shooting, now from the window of a departing bus. Theirs turned out to be the last images of the city still full of people.[33] By 4:30 p.m., the evacuation was all but complete.

The authorities were eager to report their first success to Moscow. Shcherbina called Premier Nikolai Ryzhkov. "Nikolai Nikolaevich, there are no people left in Prypiat. There are only dogs running around." People had not been allowed to take their pets— they were low on the scale of Soviet sensibilities, if they registered at all. A few days later, the police would create special squads to kill stray dogs. But dogs were not the only ones remaining in Prypiat. Close to 5,000 workers of the nuclear plant stayed to make sure the shutdown of the other reactors proceeded as planned. Young lovers took advantage of their parents' departure to have their apartments to themselves. Finally, there were the elderly who decided to stay behind. They could not understand why they had to leave when the evacuation was only supposed to be for three days.[34]

"We had to convince people," recalled the Ukrainian interior minister, Ivan Hladush. He was generally very proud of how well his people had carried out the evacuation. The next day, he reported to the Ukrainian Central Committee that altogether 44,460 individuals had been evacuated and moved to forty-three nearby settlements. "We took people to the villages," he remembered. "We took over clubs and schools and lodged people in apartments of local residents. Everyone had an understanding attitude." Valentina Briukhanova, whose husband, the Chernobyl plant director, had stayed in Prypiat, was evacuated like everyone else and ended up in one of the villages of the region. Newspaper reporters found her there a few days later, working on a cattle farm.[35]

The KGB officials informed Ukrainian party authorities that of the evacuees who had moved to the towns and villages of the nearby Chernihiv oblast on their own—close to a thousand people altogether—twenty-six had been admitted to hospitals with symptoms of radiation sickness. The KGB was busy curbing the "spread of panicky rumors and unreliable information," but it could do nothing about the diffusion of radiation. The evacuees brought not only their irradiated bodies to their temporary homes but also their contaminated clothing and personal belongings. With the evacuation of Prypiat and its nearby villages complete, the buses returned to Kyiv. They were assigned to their regular routes, where they spread high levels of radiation around a city of 2 million people.[36]

TAMING THE REACTOR

I N PRYPIAT on the morning of April 27, it was not only children playing in the sand who knew nothing about the radioactivity. The yellowish granular substance also seemed irresistibly attractive to the adults who knew the most about the accident and its consequences. About one-third of a mile from the Prypiat party headquarters where the government commission held its meetings, there was a large mound of sand intended for construction purposes. From the windows of nearby apartment buildings, one could see three men in their mid-forties, one dressed in a general's uniform and the others in expensive business suits, digging into the mound to fill the sacks they had brought with them. The three sand diggers sweated as they worked, their clothing becoming grimier and grimier. But the number of filled sacks grew as well.

The general was Nikolai Antoshkin, chief of staff of the Kyiv district air force command and commanding officer of the helicopter unit that had turned the plaza in front of party headquarters into its landing pad earlier that morning. The two men in business suits were Aleksandr Meshkov, the first deputy of the Ministry of Medium Machine Building, under the powerful Yefim Slavsky, and Gennadii Shasharin, the chief engineer of the nuclear power directorate of the Soviet Ministry of Energy and Electrification. They were loading sandbags to be dropped by Antoshkin's helicopters into the mouth of the damaged reactor in order to seal it off and stop further emissions of radiation.

The decision to begin the airdrops had been made the previous night, but its execution was postponed until daybreak of April 27. The time was needed to allow Antoshkin's helicopter pilots to get into the area, create a landing pad, and reconnoiter the approaches to the reactor. After the helicopters were in position, General Antoshkin had asked Boris Shcherbina, the head of the government commission, to provide people and equipment to fill the sandbags and load them onto the helicopters—his pilots were ready. Shcherbina did not conceal his irritation. He wanted Antoshkin's pilots to do the job; it was not the commission's responsibility. Antoshkin insisted that he needed extra help. Shcherbina told the general to take Meshkov and Shasharin and let them fill the sandbags. The senior officials followed orders. "Shcherbina was extremely impatient," recalled Shasharin. "With the helicopter engines roaring outside, he yelled at the top of his voice that we were lousy workers, that we were no good. He drove us like cattle—all of us, ministers, deputy ministers, generals, not to mention the others—telling us that we were very good at blowing up reactors but useless when it came to filling sandbags."

Shcherbina was running the show like an ancient despot. By the morning of April 27, he had already recovered from the earlier shock and uncertainty, when he could not figure out what had happened and had no plan of action. Now, with a clear goal before him, he switched to his hectoring mode of managing people and situations. That was the old, proven manner of dealing with subordinates; it had been adopted by Stalin's managers in the industrialization and forced collectivization drives of the 1930s. The protocol was to bully subordinates into submission and then demand the fulfillment of unrealistic production quotas. Unless Antoshkin, Shasharin, and Meshkov could organize others to do the job, they had to fill the sandbags themselves.[1]

After a bit they did round up more people and shovels. "With my own eyes I saw company commanders and junior officers filling sandbags, loading them onto helicopters, flying, reaching their target, dumping the bags, returning and repeating the operation," recalled Valerii Legasov. He was not impressed. "For about two days, April 27 and 28, both the Ministry of Energy and the local authorities were completely unable to organize the work, so urgent and clear-cut, of preparing the materials that had to be thrown into the well of the reactor," he remembered. "From about the 29th, order was established. Appropriate quarries were found, and lead was supplied. By that time people were in place, and things began to work properly. At about the same time the helicopter pilots worked out a highly effective way of carrying out their operations, establishing an observation post on the roof of the district party committee in Prypiat. From there they directed the teams above Unit 4 to their targets."[2]

The task of filling sandbags was eventually entrusted to the leaders of the Prypiat branch of the Young Communist League (Komsomol), who went to the workers' dormitories and asked people for help. The response was strong, but there was still a shortage of shovels, sandbags, and fabric to tie up the filled bags. At first, they began using the supplies of red fabric stored for the May Day parade. Once the city was evacuated, the workers were filling bags brought from outside, mainly from nearby villages. As in wartime, most of the workers were women. One of them, a Chernobyl resident named Valentyna Kovalenko, recalled: "They said that there was trouble and that we had to help by going to the sandpit and filling bags. . . . And there it was mainly we women who loaded that sand from morning to night."[3] The Ukrainian authorities provided the workforce, while Moscow officials offered financial incentives.

Once the Komsomol organizers and local officials took over the job of mobilizing workers to fill the sandbags, General Antoshkin returned to his primary duty—coordinating the efforts of helicopter pilots to dump the sandbags into the reactor in order to seal it off. Even before Shcherbina sent him to dig sand, Antoshkin had flown over the reactor with his officers and charted the approaches to the ruined Unit 4. There were two major challenges: first, pilots unfamiliar with the layout of the power plant initially found it difficult to identify the location of the reactor, which was spewing invisible clouds of radiation into the air but producing little smoke, and second, approaching the reactor by air was difficult because the tall exhaust pipe had survived the explosion. Antoshkin and his pilots dealt with both problems, charting a route that would be followed thousands of times in the coming days.

Soon, with a deafening roar, helicopter followed helicopter from the main square of Prypiat to the damaged reactor, less than 3 kilometers away. When a helicopter reached its destination with its load of sandbags and hovered over the reactor, the crew opened the hatch and dropped the bags by hand. The mission was all but impossible: the opening in the damaged reactor—the portion not covered by the "Elena" shield—was only 5 meters wide. Before they could close the hatch, the radioactive clouds produced by dropping the bags onto the reactor brought radioactive gases and particles into the helicopter cabin. The level of radioactivity would jump from 500 to as many as 1,800 roentgens per hour with every additional drop. By the end of the day, Antoshkin's men had made 110 sorties and dropped 150 tonnes of sand. That was a great achievement, but not great enough to satisfy Shcherbina.

On the evening of April 27, when the exhausted General Antoshkin reported the results to Shcherbina, the deputy prime minister did not respond with thanks. Instead, he shouted abuse at Antoshkin and his civilian counterpart, Shasharin, for what seemed an eternity. He removed Shasharin from the task of supervising the delivery of sandbags. Shcherbina demanded more sorties and more tonnes of sand. Ukrainian designers promptly created a hook that, attached to a helicopter, could carry dozens of bags wrapped in a parachute. The hook (they produced 30,000 of them) could be released from the helicopter cabin without the pilot opening the hatch and getting additional exposure to radioactive gases and particles. This, along with another innovation—the reinforcement of helicopter floors with lead plates—saved the lives of many pilots. But the task before them was almost impossible to achieve. Since the opening into the core of the reactor was so narrow, perhaps no more than 20 percent of the loads they dropped made it into the reactor.[4]

And then there was the radiation. At first the helicopter pilots did not know the potency of the radioactive fields, but that knowledge came quite soon, and only partly from the scales of dosimeters.

"The weather was wonderful, the sun was shining, everything was in bloom, coming back to life. But nearby there was a crow that could not take flight—it was too weak," remembered one helicopter pilot, Valerii Shmakov. "Then we understood that the situation was serious." Many of his colleagues were vomiting, while others got a radiation tan. Shmakov felt extremely tired the whole time—one of the symptoms of radioactive poisoning reported by many people who came to Prypiat immediately after the explosion. He and his fellow pilots began to believe that they were doomed. "Once we started making flights above the reactor and going through the disinfection process, when our apparatus and clothing were treated with a special substance, we discussed among ourselves that the flights were really dangerous, and that perhaps we should take the whole brunt of the blow, becoming condemned men, as it were, now that we had become involved," recalled Shmakov.[5]

The first pilots, including General Antoshkin himself, wore no protective gear as they hovered over the reactor. It was estimated that in order to drop the sandbags the pilots had to hover over the eye of the nuclear volcano for about four minutes. A single trip would thus expose the crew to between 20 and 80 roentgens, meaning that they should have been sent to the hospital immediately upon their return. In reality, for eight long days they would make flights day in and day out, accumulating extremely high, if not lethal, doses of radiation. Their doses were measured on the basis of the radioactivity of their clothes, not according to the level of radioactivity above the reactor. By the time the mission was completed in early May, none of the pilots who had made the first drops on the afternoon of April 27 were still on-site. Most of them were being treated for radiation sickness in the hospitals of Kyiv.[6]

BORIS SHCHERBINA wanted sand on the reactor, and he got it. But was he right to order the "bombing runs"? He followed the advice of his chief scientific expert, Valerii Legasov. But what if Legasov and his scientists were wrong? Some of Legasov's colleagues at the Institute of Nuclear Energy in Moscow believed that he was dead wrong. No one knew what had caused the original explosion or those that followed, or what was going on inside the reactor. Could it be that sealing the reactor with a pile of sand would cause a new explosion instead of extinguishing the fire caused by the first one?

On the night of April 27, Shasharin, recently released from filling sandbags, sat down with leading scientists, including Legasov, to assess the situation. "The first question that worried all of us was whether the reactor, or part of it, was working or not, that is, whether the production of short-lived radioactive isotopes was still going on," recalled Legasov. Measurements had to be taken on the spot. They used an armored vehicle equipped with instruments for measuring gamma and neutron activity, bringing it as close to the damaged reactor as they could. The results were devastating: the counters recorded high levels of neutron radiation. That could mean the reaction was still going on; the reactor was alive and could explode in another blast much more powerful than the one that had ruined Unit 4 and made the government commission order the evacuation of Prypiat. The next blast might not only destroy the entire power plant but also release radioactive clouds large enough to make a good part of Europe uninhabitable.

Legasov himself got into the armored vehicle and ordered that it be driven to the reactor. He soon realized that his measurements of neutrons—the ionizing subatomic particles that could reveal what was going on with the reactor—might be affected by powerful gamma fields, another component of the ionizing radiation. He therefore proposed a different method of measuring the radioactivity within the reactor. "We obtained the most reliable information on the status of the reactor by assessing the correlation of short- and relatively long-lived isotopes of iodine-134 and -131 and by means of radiochemical measurements," he recalled. They soon realized that "short-lived isotopes of iodine were not being produced; consequently, the reactor was not working and was in subcritical condition." That was a huge relief. They reported to Moscow that the measurements undertaken by the members of the state commission showed a value of 20 neutrons per square centimeter per second. "Admittedly," remembered Shasharin, "we were measuring from a distance and through concrete."[7]

Legasov's next major task was to keep the temperature of the burning graphite in the reactor at a level that would not allow the capsules of uranium fuel to release additional radiation. He proposed using boron, a rare chemical element that absorbed neutrons, to keep the reactor from "overheating." But first he had to convince his bosses in Moscow and Prypiat to accept that idea, and then get enough lead (there was enough boron in the plant's warehouse) to do the job. His superiors in Moscow, especially Anatolii Aleksandrov, the director of the Institute of Nuclear Energy and head of the Soviet Academy of Sciences, were reluctant to endorse the idea. Aleksandrov proposed using clay instead of boron. Shcherbina decided to use both clay and boron. He just needed to know how much lead, boron, and other substances were needed. When Legasov asked for 2,000 tonnes of lead but said that that would probably be insufficient, Shcherbina ordered 6,000 tonnes. It was a lot, but he did not care. The entire Soviet economy was at his disposal.[8]

The helicopter pilots would now drop not only sand but also clay, boron, and lead into the opening of the reactor. Sand was needed to extinguish the graphite fire, lead to lower the temperature of the burning graphite, and boron and clay to prevent a chain reaction. So went the argument adopted by Shcherbina and his commission. But some scientists tried to convince Legasov that a chain reaction was a theoretical impossibility, as the fuel channels had been destroyed, and that all that was going on inside the reactor was the burning of graphite—thus boron was not required. Valentyn Fedulenko, the Institute of Nuclear Energy's top expert on RBMK reactors, who arrived in Prypiat on the afternoon of April 27, was convinced that there was no need for sand, clay, or lead.

Anatolii Aleksandrov, one of the fathers of the RBMK reactor, had sent Fedulenko to Prypiat specifically to offer Legasov scientific advice, given that Legasov was a chemist, not a physicist by training or by academic interest, and was never involved in work on RBMK reactors. On the evening of April 27, the first night of his stay in Prypiat, Fedulenko visited Konstantin Polushkin, the first scientist to observe the reactor from above and conclude that it had been destroyed. Polushkin showed Fedulenko and some fellow scientists the film of the damaged reactor that he had shot from a helicopter that day. On the screen, Fedulenko could see that the massive "Elena" plate was still in place, covering most of the reactor and making it extremely difficult for the helicopter pilots to drop their loads into the reactor's half-open mouth.

Fedulenko believed that the strategy chosen on Legasov's advice was not only useless but also harmful. While most of the loads dropped by the helicopter pilots, at a huge cost to their health and safety, did not get into the reactor, every drop that did created a small radioactive explosion that increased radiation levels. "A second later, a black mushroom of fuel and graphite dust and ash (exactly like the mushroom of an atomic explosion, but miniature and very black) rose above the ruined unit," wrote Fedulenko, describing the scene that he witnessed after one of the airdrops. "In three or four seconds, the cap of the ominous mushroom rose to the height of about two-thirds of the ventilation pipe and slowly began to descend in heavy, shaggy black bands resembling rain from a cloud against the background of a gray sky. In ten to twelve seconds the mushroom disappeared, and the sky became clear. The wind blew away the mushroom cloud."[9]

Fedulenko argued against the continuing bombardment of the reactor, but Legasov would not listen. He told Fedulenko: "Yes, [radioactive] activity increased sharply after we began dumping sand and other substances. But that's probably temporary." He also added, after one of their conversations: "People won't understand if we do nothing." Indeed, the strategy of bombarding the reactor had been approved at the very top of the Soviet power pyramid. There, the main concern was not the spread of radiation because of the airdrops, but the possibility of a new explosion. That was the main question addressed to scientists on the evening of April 27 by Vladimir Dolgikh, the Central Committee secretary in charge of the energy sector.[10]

On the morning of April 28, Dolgikh informed Mikhail Gorbachev and the entire Soviet leadership in person that Unit 4 had been destroyed by the blast and had to be buried. The process had begun with airdrops of sand, lead, clay, and boron. Gorbachev asked him, "Are bags of sand and boron being dropped from the air?" "From helicopters," answered Dolgikh, giving a very low number. "Sixty bags have been dropped. Eighteen hundred are needed. But helicopter flights are not safe either." His data probably reflected activity as of the middle of the previous day. When Gorbachev asked

Marshal Sergei Akhromeev, the chief of the Soviet Army General Staff, what should be done under the circumstances, he responded: "The only realistic thing is to cover the reactor with bags of sand and boron."[11]

Boris Shcherbina did all he could to bury the reactor as soon as possible. Every day the amount of sand, lead, and boron dropped on the reactor increased: 300 tonnes on April 28, 750 tonnes on April 29, and 1,500 tonnes on April 30. The peak was reached on May 1, when 1,900 tonnes of sand were dropped. Further airdrops were canceled out of fear that the pile was becoming too big and heavy: its weight might bring the reactor crashing down into the underground structures of the plant, causing the radioactive poisoning of groundwater. Altogether, close to 5,000 tonnes of various substances, mostly sand, were dropped onto the reactor. It is a miracle that it did not collapse into the unit's lower levels.[12]

WAS THE sacrifice of the helicopter pilots justified? Legasov believed that it was. Levels of radioactivity around the reactor were falling. Later he and his fellow scientists would estimate that 25 percent of all the radioactivity released by the explosion had been emitted during the first twenty-four hours after the accident on April 26. By May 2, the emissions had fallen to one-sixth of what they had been on the first day. Legasov was convinced that the decline was a result of the commission's actions, especially the "bombardment" of the reactor from the air with sand and other substances.[13]

To many inhabitants of Ukraine, Belarus, and western Russia, this relative improvement seemed too little, too late. As the helicopters dropped their loads on the reactor, the wind, which had blown west and north during the first days after the accident, switched directions to the east and south, contaminating new territories. The chemical warfare detachments under General Vladimir Pikalov, who arrived immediately after the accident and was in charge of the measurement of radioactivity levels and the mapping of radiation zones, were spread thinly and could not cope with the rapidly deteriorating situation.[14]

On April 28, one day after the evacuation of Prypiat, Boris Shcherbina and his commission decided, with the approval of

Moscow, to create a 10-kilometer exclusion zone around the reactor and resettle the population from the entire zone. The commission also decided to abandon Prypiat, where the radioactivity level was increasing. The commission members and the operators of the power plant—altogether close to 5,000 people who stayed in the city to try to ensure the safe shutdown of the reactors, as well as police and military personnel brought to the city—were ill-equipped to deal with such high levels of radiation.

Respirators were not available in the quantities required, and dosimeters were in short supply: those available lacked batteries and were therefore useless. Potassium iodide pills were hard to get and were delivered too late, after residents' thyroid glands were already full of radioactive iodine from the atmosphere. Warning signs about high radiation fields were not posted, and the offices of high officials were among the most contaminated areas in the city—woolen carpets in offices and corridors became repositories of radioactive particles and absorbed radiation. At police headquarters, it took four days to figure out what was going on and remove the carpets.[15]

Most of the scientists who flew in from Moscow to deal with the consequences of the accident knew the kind of danger they faced but still often neglected to protect themselves. A macho "can do" attitude influenced the behavior of those who stayed in Prypiat. "Young guys (on the shift) were smoking and talking on-site," wrote Fedulenko, describing one of his trips to the environs of the damaged reactor. "A helicopter flew by. It was carrying its load in a net. The height was not great, and everything could be seen. It hovered above the ruined unit. It dropped its load. It flew off. The crowd on the open square was calm. They had happy faces, and no one was wearing a respirator. I felt for my respirator in my pocket, remembering that I had one. Somehow it felt embarrassing to put it on; everyone's face was open." Like the others, he remained unprotected. The only obstacle between Fedulenko and the reactor, only a few hundred meters away, was the concrete wall of a nearby building.[16]

In Prypiat, radiation levels fluctuated around 1 roentgen per hour. According to the norms later adopted for Soviet policemen in radioactive zones, an officer was allowed to spend no more than twenty hours in such conditions. The members of the government commission spent close to sixty hours in Prypiat. After that, along with everyone else who had stayed in the city after the first evacuation, they were finally told to leave for a safer place.[17]

Unlike the earlier evacuees, they knew they were probably leaving Prypiat not just for three days but forever. Valentyn Fedulenko later described a scene of the exodus that he found unforgettable: "Along the way we stopped at a place where people were filling paper bags with sand to be dropped into the well of the Unit 4 reactor," he recalled. "The supervisors were talking about something or other. I was struck by a scene that will long remain before my eyes. . . . [A]gainst the background of the shadowy bulk of the station, the little houses of a small village were visible about half a kilometer away. Behind the fence was a plowman with his plow, a horse in front of him. He was tilling his kitchen garden. A village idyll on a radioactive field."[18]

Before the last residents of Prypiat left their city, they performed their last duty to the first recorded victim of the accident, Volodymyr Shashenok. He had been taking readings of radiation levels at the time of the turbine test when the explosion destroyed pipes full of hot water and steam. Around 6:00 a.m. on April 26, Shashenok died of the severe burns to his body. When he was brought to the hospital, where his wife worked as a nurse, he could barely speak, but had begged people around him to keep their distance, saying that he was from the reactor unit. When the time came for his burial, the plant had no bus available to take his body to the cemetery. In one of his last acts as deputy mayor of Prypiat, Aleksandr Yesaulov requisitioned a bus passing through the city, making it possible to give Shashenok a proper burial.[19]

The new wave of Prypiat evacuees settled in a scout camp called "Skazochnyi"—a name meaning "Fairy-Tale"—35 kilometers south of the city. There was nothing fairy-taleish about the camp, though, or about the radioactive conditions facing its inhabitants. One day after the move, radiation levels began to increase there as well. But they did not exceed 1,300 microroentgens per hour. That was still a "good" level in comparison to Prypiat's 1 roentgen per hour.[20]

ALONG WITH the government commission, higher radiation levels were moving south, approaching the city of Kyiv.

On April 28, Valentyna Shevchenko, the fifty-one-year-old head of the Supreme Soviet of Ukraine, the anemic Soviet-era parliament, drove to those parts of the northern Kyiv region where inhabitants of Prypiat had been temporarily resettled. She did so on her own initiative, without consulting Volodymyr Shcherbytsky, the all-powerful boss of Ukraine. "I felt that a calamity had occurred but that no one understood the full extent of the danger in the first days. I was concerned about how such numbers of people from Prypiat could be accommodated, what they needed immediately, so I set out for Chernobyl early in the morning of April 28," she recalled. She found the evacuees housed in schools and public buildings, as well as in homes of collective farmers nearby. They were a diverse crowd. Many were city dwellers who felt disoriented in a village environment. But whatever their background, they were put to work. Among them was the former first lady of Prypiat, the wife of the director of the power plant, Valentina Briukhanova.[21]

Liubov Kovalevskaia, the Prypiat journalist who had written in March about construction problems at the Chernobyl plant, ended up in the village of Maksymovychi near the town of Poliske. It soon became clear that radiation levels there were also high. The evacuees were ordered to board buses again, pregnant women and children first. They were checked by dosimetrists and found to be already contaminated. "Imagine how a mother felt," remembered Kovalevskaia, "on coming up to a dosimetrist who measured [radiation on] her child's shoes: 'Dirty.' Pants: 'Dirty.' Hair: 'Dirty.'" The evacuation was done in a hurry and often resulted in families being divided: children might be taken to one village, parents and grandparents to another. Radio loudspeakers installed in the main squares of villages began transmitting news and announcements. The most common notices were about parents looking for their children.[22]

Valentyna Shevchenko observed that the evacuees in nearby villages seemed more bewildered than concerned. Many of them, comparing the new ordeal with the German occupation during World War II, found the current circumstances much less threatening. "I went around the dwellings and spoke with people, taking an interest in how the evacuees had been received," recalled Shevchenko. "People were calm, expecting that they would return to their homes very soon. Everyone told me: 'What sort of danger is this? When the Germans were here, that was real danger. But now? It's sunny and warm; we need to plant our gardens.'" Shevchenko, a high government official who had no reliable information about the situation, felt much the same way. She and her driver had lunch outdoors and tasted food offered them by the locals.

The danger posed by radiation came home to her late that night, when her car was stopped on the way back to Kyiv in the village of Vilche. A radiation control post that had not been there in the morning was now checking passersby, and her summer shoes were found to be highly radioactive. Shevchenko had to leave them at the post. The head of the Ukrainian parliament arrived home barefoot. On the following day, dosimetrists found radiation levels in Kyiv reaching 100 microroentgens per hour—five times the norm. That was just the beginning.[23]

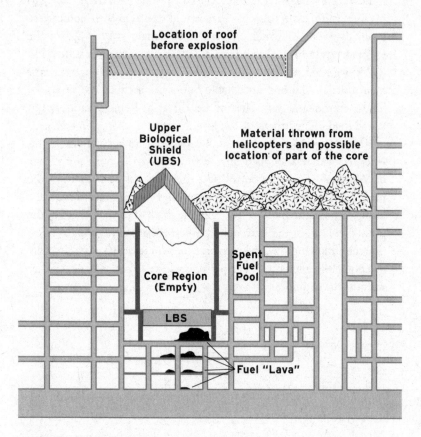

Unit No. 4 of the Chernobyl nuclear power plant in the first days and weeks after explosion.

IV

INVISIBLE ENEMY

DEADLY SILENCE

WHAT HAPPENED in Prypiat stayed in Prypiat. That had been the rule before the accident, and it remained in effect for days afterward. Even though tens of thousands of people were evacuated from Prypiat and nearby villages, the Soviet government refused to tell its citizens and the world at large what had happened. Television, radio, and newspapers, even local ones in Ukraine, remained silent about the accident.

The Kremlin had been quite successful in keeping previous large nuclear accidents under wraps, remaining silent about radioactive pollution and the dangers it posed to Soviet citizens and the rest of the world. That was the case with an accident that took place in 1957 in the closed city of Ozersk in the Urals, better known to the Soviet leadership under the code name Cheliabinsk-40—the home of the first Soviet nuclear fuel plant producing weapons-grade plutonium. On September 29 of that year an underground nuclear waste tank had exploded, blasting a concrete cap weighing 160 tonnes from the top of the steel and concrete container and releasing 20 million curies of radioactivity into the atmosphere. The authorities had had to resettle 12,000 people, mostly from the twenty-three villages of the region that were considered no longer fit for habitation. The houses and agricultural equipment used by the former inhabitants were buried, and an exclusion zone was established in the areas that had suffered the most.

The Soviet leaders refused to release information about the Ozersk explosion, thereby endangering the lives of hundreds of thousands of their own citizens, who went on with their daily routines not knowing how to minimize the risk caused by the accident. But although radioactive clouds never reached the borders of the USSR, it was still impossible to conceal such a massive release of radioactivity from the rest of the world. American military and civilian officials learned about the 1957 accident but decided not to use it in a propaganda campaign against their Cold War rival. Both sides had a stake in keeping it under wraps so as not to frighten their citizens and make them reject nuclear power as a source of cheap energy.[1]

The Ozersk accident led the Soviet authorities to develop a number of strategies that would be employed at Chernobyl thirty years later. The use of military conscripts to deal with the consequences of the nuclear explosion, decontamination techniques that included burying contaminated equipment and covering contaminated areas of the nuclear plant with concrete, mass resettlement of the population, the creation of an exclusion zone, and the handling of large numbers of patients with symptoms of acute radiation poisoning—all those strategies had first been employed in Ozersk. Yefim Slavsky, the Soviet nuclear industry tsar, and his subordinates had been in charge of dealing with the Ozersk disaster in 1957. They would be called back to deal with the Chernobyl explosion as well.

Original silence about the Chernobyl accident at home and abroad also followed the Ozersk pattern. In 1986, Mikhail Gorbachev, Nikolai Ryzhkov, and their subordinates in Moscow and Kyiv had a model they could follow not only for dealing with the nuclear disaster, but also for speaking—or, rather, remaining silent—about what had happened. The tradition of complete secrecy about the nuclear program, the regime's unwillingness to admit disaster after having prided itself on being the first to build a nuclear power plant and successfully managing the peaceful atom (incompetence in that regard had hitherto been attributed only to the United States and the capitalist world), and, finally, concern about unleashing panic and a resulting inability to mobilize the resources needed to fight the disaster—all that came together in the deafening official silence that prevailed for the first days after the explosion.

But this time it proved much more difficult to restrict the flow of information. Whereas the Ozersk accident had released 20 million curies of radiation into the atmosphere, the Chernobyl disaster released 50 million curies. Besides, Chernobyl was not in the Urals—the middle of the Soviet Empire—but closer to its western edge, and the winds there spread radiation, and thus awareness of the accident, to the countries of Northern and Central Europe. The architects who had prepared the general plan for the construction of Prypiat in the early 1970s had pointed out that the prevailing winds there blew toward the west and northwest. That was definitely the case in the first two days after the accident. On the night of April 26, the wind blew in a northwesterly direction, carrying the radioactive plume across the Ukrainian border to Belarus, then to Lithuania, and across the Baltic Sea to Sweden, Finland, and Denmark.[2]

The first to sound the alarm were the dosimetrists at the Forsmark Nuclear Power Plant near Uppsala, Sweden, 1,257 kilometers away from Chernobyl. Cliff Robinson, the radiation control officer at Forsmark, triggered the radiation alarm at about 7:00 a.m. on April 28, when passing from one zone of the plant to another. By midday, the Swedes had evacuated the workers at the plant and checked radiation levels at other plants—finding that they exceeded the norm everywhere. Swedish nuclear specialists soon figured out that the radiation they detected had been carried by winds blowing from the other side of the Baltic Sea. Swedish diplomats contacted three Soviet agencies whose activities involved nuclear power and asked for possible explanations. They got none. The Swedes were losing patience. The minister for the environment, Birgitta Dahl, made a statement suggesting that anyone withholding information on the release of radioactive materials was contravening international norms and agreements. This time the Soviet leaders could not conceal the facts, as they had done in 1957. Their secret was out in the open, and an international scandal was in the making.[3]

THE SOVIET media finally broke its silence on the Chernobyl accident at 9:00 p.m. on Monday, April 28, almost three days after the accident and more than twelve hours after high radioactivity levels had been detected in Sweden. In a dull voice, an announcer on the evening news program *Vremia* (Time) read a brief bulletin from the Soviet news agency TASS. Its text was as follows: "An accident has taken place at the Chernobyl atomic electricity station. One of the atomic reactors has been damaged. Measures are being taken to eliminate the consequences of the accident. Assistance is being given to the victims. A government commission has been struck to investigate what happened." That was it. Everything was supposedly under control, and nothing was said about rising radiation levels or the evacuation of Prypiat. No Soviet newspaper published the news, either that day or the next.[4]

But even that brief announcement was no easy step for the Kremlin leadership, which was still deeply grounded in the Soviet tradition of secrecy and neglect for the immediate well-being of the people while allegedly staying focused on the greater good and a better future. The decision to release the limited information about the accident was made only after prolonged debate in the Soviet Politburo at its first meeting after the accident.

On the morning of April 28, a few hours after high radiation levels were detected in Sweden, Vladimir Dolgikh, the Central Committee secretary in charge of the energy sector, reported to Gorbachev and his Politburo colleagues. He stated that Unit 4 had to be buried, and that sandbags were already being dropped on the reactor. The Politburo members reluctantly accepted the conclusion that sandbagging was the only feasible measure to stop the spread of radiation. The reasons for the accident were still unknown. The working hypothesis was a hydrogen explosion, but no one knew for certain. "That is guesswork for the time being," said Gorbachev. Dolgikh reported that 130 people had been hospitalized, and the evacuees were being accommodated and given jobs. That was regarded as the task of the Ukrainian authorities, who were working with the refugees. "There is nothing alarming as far as we are concerned. The population is calm," reported the head of the KGB, Viktor Chebrikov. "But it has to be taken into account that so far the accident is known only to a small circle."

Then Gorbachev asked a key question: "What will we do about information?" The first to respond was Dolgikh. "We have to finish localizing the extent of the radiation," he told the gathering. But Gorbachev was not prepared to wait. "We have to make an announcement quickly. We can't delay. We have to say that there has been an explosion and that necessary measures are being taken to localize its consequences." Yegor Ligachev, who was then Gorbachev's right-hand man but would later lead a conservative opposition to him, was the first to express approval. Gorbachev's liberal supporter Aleksandr Yakovlev added, "The sooner we make an announcement about it, the better." The others went along. Anatolii Dobrynin, the Central Committee secretary in charge of international relations, and previously a longtime ambassador to the United States, had his own reasons for supporting Gorbachev's proposal. "The Americans will discover the fact of the explosion and the spread of the radioactive cloud in any case," he told his colleagues.

Andrei Gromyko, Dobrynin's former boss, a former Soviet foreign minister, and now the holder of the largely ceremonial post of chairman of the Supreme Soviet, the Soviet parliament, did not object to the proposal either, but he urged his colleagues to be cautious. "The announcement should be formulated in such a way as to avoid arousing excessive alarm and panic," he said. A few minutes later, he made another suggestion: "Perhaps we should inform our friends in particular about what has happened. After all, they buy atomic-energy technology from us." This suggestion was a throwback to the old Soviet practice of parceling out information: restricting the whole truth to members of the Politburo and releasing it selectively, with different emphases, first to "friends" in the socialist camp, then to "enemies" in the West, and last of all to their own people. But Gorbachev wanted none of that. "We have to inform our own public first," he told Gromyko.

When the Politburo members considered in the late morning of April 28 whether or not to make the accident public, they believed that the nuclear contamination was limited to the territory of the Soviet Union and that it was entirely a Soviet matter. Dolgikh asserted that the radioactive "stain" was only 60 kilometers in diameter. Prime Minister Nikolai Ryzhkov remarked that radioactivity had reached Vilnius in the Baltics, and the chief of the Soviet general staff, Marshal Akhromeev, estimated the diameter of the radioactive cloud at 600 kilometers. Unbeknownst to the members of the Politburo, contaminated clouds had already crossed Soviet borders and were beginning to create havoc abroad.[5]

By the time the Soviet television anchor read the terse announcement late in the evening of April 28, the international scandal was already brewing. The Soviet authorities were now barely admitting something already known to Western governments, and, through word of mouth, to large numbers of Soviet citizens. Gorbachev later defended his record, maintaining that he and the Politburo simply did not have all the requisite information at the time. "I absolutely reject the accusations that the Soviet leadership intentionally held back the truth about Chernobyl," he wrote in his memoirs. "We simply did not know the whole truth yet." In fact, they knew much more than they admitted. Even those prepared to break with the Soviet tradition of secrecy, such as Gorbachev and Ryzhkov, were afraid of losing control over the situation. "And what should I have told the country?" mused Ryzhkov later. "People, a reactor has exploded, the radiation is off the scale, save yourselves as best you can?"[6]

ON APRIL 29, the day after the Soviet announcement, a Pan American flight carrying the US ambassador to the Soviet Union landed in Moscow, and a Soviet Aeroflot flight made a landing at Dulles International Airport in Washington, DC. In both cases it was a special occasion. Direct flights between the two countries had been canceled in December 1981 at the initiative of the Reagan administration to protest the imposition of martial law in Poland. Now they were finally being resumed as a result of the first meeting between President Ronald Reagan and Mikhail Gorbachev in Geneva in December 1985. There the two leaders had sized each other up and decided, despite major ideological and philosophical differences, that they could work together.[7]

The Soviet delegation on the Aeroflot flight was led by the deputy minister of civil aviation, who made a statement to American journalists expressing his hope that the resumption of direct flights would promote increased cooperation between the two countries. He was ready to discuss the future, but the reporters seemed more interested in Chernobyl, asking him about the scope of the catastrophe and the number of victims. The high official was at a loss. He knew less about the accident than the journalists asking the questions did.[8]

On the same day, the American public received its first substantial information on what was going on in and around Chernobyl. "The severity of the accident, which spread discernible radioactive material over Scandinavia, was not immediately clear," wrote Serge Schmemann, the Associated Press reporter in Moscow, in the *New York Times*. He continued: "But the terse statement distributed by the TASS news agency and read on the evening television news suggested a major accident. The phrasing also suggested that the problem had not been brought under full control at the nuclear plant." Schmemann reported on the deep concern in Scandinavia about the rising levels of radiation, noting that the Soviet announcement had come "hours after Sweden, Finland and Denmark reported abnormally high radioactivity levels in their skies."[9]

Ronald Reagan was first informed about the Chernobyl accident by his foreign policy and national security team on the same day, April 29. He was on board Air Force One, traveling from Guam to Bali to attend a South Asian summit. The administration immediately set up a task force to monitor the accident and its possible impact on the United States. It was overseen by Vice President George H. W. Bush and headed by Lee Thomas, the administrator of the Environmental Protection Agency.

The first intelligence memo on Chernobyl prepared by the CIA is also dated April 29. The CIA experts called the Chernobyl accident the worst nuclear disaster in world history and reported on rumors suggesting that hundreds, if not thousands, of people were dying of the explosion. They also admitted an intelligence failure—their inability to gather any information on the disaster prior to the Swedish detection of high radiation levels and the Soviet announcement of the accident. "We are disturbed by the lack of intelligence evidence, prior to the Swedish disclosures and the near simultaneous official Soviet announcement, that this disaster was unfolding over some three days. We are especially disturbed because it occurred right in the heart of NATO's 'warning of war' domain. We need to find out why," read the CIA report.[10]

On the basis of satellite surveillance data, the authors of the CIA report speculated that the graphite in the reactor destroyed by the original blast was still burning and releasing additional radioactivity into the atmosphere. CIA experts predicted "the disruption of large areas downwind of the radiation source due to evacuations, shutdown of plants and facilities, and decontamination activities." They also foresaw "the impact of radiation on agriculture, especially the dairy industry," and the "pollution of water supplies, especially downstream on the Dnieper toward Kiev." The CIA made the safe bet that the Soviets' withholding of information about the accident would hurt them in the international arena: "Governments and publics in both East and West Europe could be alienated in lasting ways by Soviet failure to provide an early warning, when something like three days were available. This will noticeably dull Soviet persuasiveness on all manner of negotiations, from arms control to trade, and their related propaganda efforts."[11]

The Reagan administration's first public reaction to the news from Chernobyl was an offer of assistance. It was presented to a Soviet diplomat who came to the State Department to discuss nuclear arms control on April 29. He was asked for additional information but had none to offer. It was then that the State Department voiced its first criticism of the Soviets for not being forthcoming with details about the accident. "The delay in saying anything about the disaster, the fragmentary information that has been issued, all are typical of the Soviets," a State Department official told the press. Expectations that the Soviet Union would actually accept US help were quite low, especially if, as was wrongly suspected, the Chernobyl nuclear plant was part of a military-run program.[12]

On the next day, April 30, Soviet diplomats delivered a message from Mikhail Gorbachev to Ronald Reagan. The message confirmed that the accident had taken place. "They say that a leak of radioactive material has required the partial evacuation of the populations in regions immediately adjacent to the accident," the White House deputy press secretary, Larry Speakes, told the media. "They indicate that the radiation situation has been stabilized; and, finally, they note that the dissemination of radioactive contamination in the western, northern, and southern sections has been

detected. The message further states that these levels of contamination are somewhat above permissible norms but are not in the extent
which would require special measures to protect the population."
Speakes said that the US government was pushing the Soviets for
more information.[13]

Reliable information on what the US intelligence community
and media called the world's largest nuclear disaster was soon sub-
stituted with rumors. On April 29, Luther Whitington, a correspondent for United Press International in Moscow, reported that
80 people had died on the spot as a result of the explosion, and
that 2,000 more were on the way to hospital. He cited a telephone
interview with an unnamed source in Kyiv. The same source allegedly claimed that between 10,000 and 15,000 people had been
evacuated from Prypiat. In reality, the former numbers were gross
overestimates, while the latter were significantly below the actual
numbers of people evacuated from Prypiat and its environs. Nonetheless, many Western newspapers picked up and disseminated the
United Press International report. The Chicago Board of Trade lost
no time in betting on mass contamination of Soviet crops and a
dramatic increase in American agricultural exports to replace potential losses—not only in the USSR, but also in the countries of
Northern and Eastern Europe affected by the disaster and the radioactive fallout.[14]

The Soviets were furious. Vladimir Fronin, a Soviet reporter
who accompanied the Soviet delegation that landed at Dulles on
April 29, was upset but also probably frightened when he began to
watch coverage of the Chernobyl disaster by the leading American
television networks. Fronin concluded that the accident was being
used to undermine emerging Soviet-US cooperation. In the report
that he later published in the Soviet media, Fronin ridiculed claims
that the Soviet authorities were hiding the truth from their own citizens. He compared what he called the hostile attitude of the American media toward the USSR with the Soviet reaction to the death of
seven American astronauts as a result of the space shuttle explosion
in January 1986. "If we remember the 'Challenger,'" wrote Fronin,
"when there was nothing but sympathy for the Americans both in
our press and in our hearts, then the soul grows bitter."[15]

Fronin and the regime he represented were clearly on the defensive. Fronin forgot to mention the American offer of assistance. All but silent about Chernobyl, the Soviet propaganda chiefs flooded their media with information about nuclear disasters abroad. "The terse Soviet announcement of the Chernobyl accident," wrote Serge Schmemann in the *New York Times,* "was followed by a TASS dispatch noting that there had been many mishaps in the United States, ranging from Three Mile Island outside Harrisburg, PA, to the Ginna plant near Rochester, NY. TASS said an American antinuclear group registered 2,300 accidents, breakdowns and other faults in 1979."[16]

ON APRIL 30, the Soviet Union's leading newspaper, *Pravda* (Truth), finally broke the printed media silence on the accident by publishing a short announcement at the very bottom of the second of its six pages. It restated the information released by TASS on April 28 and added some new details: the name of Boris Shcherbina as head of the government commission was mentioned; the evacuation of the city of Prypiat was acknowledged, although it was called not a city but a settlement; and the public was assured that radiation levels were being closely monitored. As Mikhail Gorbachev, Nikolai Ryzhkov, and their Politburo colleagues released limited information on the disaster, their main challenge was to balance the desire to maintain a degree of credibility as responsible world citizens with concern about losing control on the ground and creating panic in the areas most affected by the disaster.[17]

The Ukrainian KGB reported to the party bosses in Kyiv on April 28 about growing concern among the population of the areas close to the nuclear plant regarding "the spread, as they believe, of radiation." The situation was becoming especially tense in Kyiv, only 130 kilometers away from the damaged reactor. The mobilization of buses, nuclear specialists, and police to deal with the consequences of the disaster caused rumors to spread among the worried inhabitants of Ukraine's capital. So did the arrival of patients with radiation sickness in Kyiv hospitals. And yet the government was saying nothing about the scope of the catastrophe or the need for citizens to protect themselves. "The whole city is alarmed, and they say that the hospitals are overcrowded," noted Ukraine's leading writer, Oles Honchar, in his diary. "But there is not a word of information on the radio, just upbeat music and cheerful songs."[18]

On April 30, the Ukrainian Politburo met to discuss the situation. The main issue on the agenda was the May 1 parade scheduled to take place in downtown Kyiv the following day. May 1 was one of the two most important events on the political calendar of the Soviet Union, the other being November 7, the day of the Bolshevik takeover in 1917. May 1, officially known as the Day of International Worker Solidarity, traced its origins to the events on Haymarket Square in Chicago on May 1, 1886, when the police had opened fire on striking workers. It was a living reminder of the international origins and ambitions of Russian communism. On May 1, the Soviet authorities always organized mass demonstrations. Many Soviet citizens regarded them in a nonpolitical way as a spring holiday, an opportunity to get together with friends and coworkers and take part in the only form of public gathering allowed by the regime.

On April 29, the day before the Ukrainian Politburo meeting, the head of the Ukrainian KGB, Stepan Mukha, sent a memo to the party boss of Ukraine, Volodymyr Shcherbytsky, summarizing the work of the secret police in preparation for the all-important public holiday. The main concerns were about any possible manifestations of disloyalty toward the regime, especially the dissemination of leaflets critical of the government. Starting on April 21, reported Mukha, KGB officers had met more frequently with their informers. Surveillance was increased on 4 Western diplomats stationed in or visiting Kyiv, 6,060 foreign students, of whom 38 were suspected of espionage and 22 of belonging to radical Muslim organizations. Among the Soviet citizens who received special attention before the holidays were 89 individuals suspected of espionage, 54 labeled as Ukrainian nationalists, 24 Zionists, 17 religious dissenters (mostly Protestants), 6 Crimean Tatar activists, and 223 mentally ill individuals. The KGB was also investigating the Chernobyl accident as a possible act of sabotage and increasing its surveillance over the rest of the nuclear power plants and areas adjacent to Chernobyl, where the main concern was "the spread of alarmist rumors and tendentious information."[19]

On the political front, everything seemed to be under control. But that was not all. There was also the spread of radiation, against which the KGB was powerless. The main question before the Ukrainian leadership was whether to even hold the parade, as the winds had changed direction and the radiation front was approaching Kyiv. On April 28, the KGB reported that radiation levels in the city were within the norm, or less than 20 microroentgens per hour, but by the next day they had already increased to 100 microroentgens. Ukrainian leaders were confused, as they could not properly understand the significance of the radioactivity data. Shcherbytsky kept writing on KGB reports about the radiation level: "What does this mean?"[20]

The party leaders in Moscow believed that it did not mean much, or, rather, that the radiation levels were basically within the norm. Valentyna Shevchenko, the head of the Ukrainian parliament, remembered that before the Politburo meeting the Ukrainian leaders received instructions from Moscow to go ahead with the rally. The May Day parade in Kyiv had to send a signal to the world that things were normal, that people were safe and felt secure, and the Western media was engaging in a propaganda war by spreading false news about huge destruction caused by the explosion and about thousands of victims. Television images and newspaper photos of happy Kyivans marching in the center of their city were supposed to send a message to both domestic and foreign audiences that everything was fine, and that the party was in control.[21]

After some deliberation and consultations with experts, the Ukrainian leaders decided to proceed with the rally, but to reduce the time allotted for the parade and the number of participants. Traditionally, each of Kyiv's ten districts sent between 4,000 and 4,500 participants to the rally. This time the quota was set at 2,000, and those called upon to march in the parade were to be younger people. Shcherbytsky also wanted Politburo members and city leaders to participate with members of their families, including children and grandchildren, to show the citizens of Kyiv that the situation was safe.[22]

On the morning of May 1, *Pravda* carried its usual front-page slogan: "Long live May 1, the Day of International Worker Solidarity. Proletarians of all countries, unite!" Buried at the very bottom of the second page was the official update on the situation in and around Chernobyl. The statement said that it was improving. It also attacked those in the West who were allegedly trying to arouse panic in the Soviet Union. "Some Western agencies are spreading rumors to the effect that thousands of people died in the accident at the atomic energy station. As already reported, two people actually died, and 197 were hospitalized in all, of whom 49 left the hospital after observation. The work of enterprises, state farms, and institutions is proceeding normally," wrote *Pravda*.[23]

Technically, *Pravda*'s figures were correct. The firefighters and plant operators most affected by the explosion and release of radio-activity were still hanging on to what remained of their lives at the hospitals in Moscow and Kyiv, and the medium- and long-term consequences of the disaster were yet to be determined. The regime had clearly decided to take a more proactive stance in the information battle over the disaster without releasing much more information. On May 1, the Operations Group of the Politburo, led by Premier Ryzhkov, adopted a resolution: "To send a group of Soviet correspondents to areas adjacent to the zone of the Chernobyl atomic energy station in order to prepare material for the press and television attesting to normal everyday activity in those areas."[24]

Meanwhile, radiation levels in Kyiv had become more dangerous. According to data collected by experts of the Ukrainian Institute for Nuclear Research, gamma radiation had begun to climb rapidly in the Ukrainian capital on the morning of April 30. By noon it had reached 1,700 microroentgens per hour, but the levels then subsided. By the time the Ukrainian Politburo meeting ended at 6:00 p.m., the gamma levels had dropped to 500 microroentgens per hour. It was a promising sign. The levels held steady throughout the night, but a rapid new increase was registered at 8:00 a.m. on May 1, just as people started to gather for the rally in downtown Kyiv. It was a disaster in the making.

The increase in radiation levels was especially rapid on Kyiv's main street, Khreshchatyk, the site of the May Day parade. It was located on lower ground between two of the city's hills. "The threat to all participants in the rally was very serious," recalled the city's mayor, Valentyn Zgursky. "Radioactive currents in the air went directly to Khreshchatyk from the Dnieper side."[25] Sometime after 9:00 a.m., Politburo members and city fathers gathered near the monument to Vladimir Lenin on the city's main square in the middle of Khreshchatyk. They were waiting for the leader of the republic's party, Volodymyr Shcherbytsky, to arrive for the opening of the rally—only he had the right to begin the ceremony, according to the event's unwritten protocol. By that time, radiation levels had reached 2,500 microroentgens per hour, the highest level that would be recorded that day. But Shcherbytsky was nowhere in sight.

Then, just before 10:00 a.m., those gathered near the Lenin monument saw Shcherbytsky's car speeding toward the square. The limousine stopped near the stage constructed for the parade. Shcherbytsky emerged from the car, swearing angrily. Zgursky heard him say: "I told him that the parade could not be held on Khreshchatyk. It's not Red Square but a valley; radiation concentrates here! And he said to me: 'Just try not holding the parade! I'll leave you to rot!'" According to the version of events given later by Shcherbytsky's wife, Rada, Gorbachev had threatened to expel her husband from the party: "If you don't hold the demonstration, you can say goodbye to the party." Speaking to his comrades, Shcherbytsky never called his recent interlocutor by name, but few doubted that he was talking about Gorbachev. "To hell with him; let's go and start the parade," he said.[26]

They all climbed the stairs to the podium, and Shcherbytsky took his place in the center of the leadership line, with Valentyna Shevchenko, the head of the Ukrainian parliament, to his left, and Oleksandr Liashko, the head of the republic's government, to his right. "Everyone was bareheaded," recalled Liashko. "My grandchildren were in the column of marchers; my wife was on the guest platform along with the wives of the other leaders. After all, none of us was fully informed at the time. On the contrary, people were trying to minimize the danger." No one was trying harder than the political and industry bosses in faraway Moscow. "I got a call from Slavsky, the all-Union minister of medium machine building, whose department dealt with atomic energy," recalled Liashko. "'Why are you raising such a fuss? I'll come and shut down your reactor with my ass alone.'"[27]

The photos taken that day in downtown Kyiv show Shcherbytsky, Liashko, Shevchenko, and other Ukrainian leaders waving to the crowds of Kyivans marching on Khreshchatyk just as the level of radioactivity in the city reached its peak. Unlike those on the podium, the marchers knew nothing about the danger. The photos show members of folk ensembles dressed in traditional Ukrainian costumes and young people carrying portraits of Marx, Engels, and Lenin, as well as photos of Politburo members, starting with Mikhail Gorbachev. The marchers are lightly dressed: it is a warm, sunny morning. Many came with small children; some fathers are seen bearing them on their shoulders. Children were also marching in their own formations. "Our future successors—a large group of children—brought up the rear of the column," remembered one of the participants. "They were catching up with us, laughing and dancing."[28]

One of the marchers, Natalia Petrivna, describing the event later, said that at first, everything seemed the same as in previous rallies. But when she and her fellow marchers entered the main square, and she looked at the podium—normally occupied by representatives of various branches of the Soviet economy, top managers, and "shock workers," that is, workers who had been honored for their exemplary discipline and productivity—she could not believe her eyes: it was half-empty. "I asked in wonder: 'Where are the energy people?'" she recalled, referring to employees of the nuclear energy sector whom she knew from previous parades. She was then approached by a man in plain clothes, whose behavior suggested that he was from the KGB. Natalia Petrivna remembered that the man "whispered . . . 'Go away, go away.' He took me by the arm and led me after the departing column of marchers." The KGB did not want anybody disrupting the orderly procession of people in front of the government tribune, and was not going to tolerate questions that could cause panic in the city. [29]

After the rally ended, Natalia Petrivna recalled, "I sat down on a bench to rest. I felt weak and dizzy. My throat was dry, with a strong tickle." Those were clear signs of exposure to radiation, but Natalia Petrivna suffered no long-term consequences. Another woman, Natalia Morozova, who came to Kyiv from Odesa for the May Day holidays, was not so lucky. "Damn them all," she wrote later to the special commission appointed by the Ukrainian parliament to investigate the handling of the Chernobyl accident. Her curse was addressed to the Ukrainian leaders who greeted people from the podium. "I was pregnant and went to Kyiv on [April] 24 to visit a female relative. I went to the rally and did some boating on the Dnieper. I managed to leave Kyiv only on May 12, and in July my child was stillborn."[30]

On May 1, Ukraine's leading writer, Oles Honchar, noted in his diary: "A parade on Khreshchatyk, people cheerfully shouting slogans, everyone pretending as if nothing had happened." The Soviet leaders in the Kremlin may have prevented panic, but the unintended consequence of the "radioactive" parade was the loss of legitimacy of the regime it was supposed to enhance. "My government deceived and betrayed me. When the Chernobyl disaster took place, I learned of it not from my government but from a foreign one abroad," wrote the Kyivan worker Heorhii Ral to the Ukrainian parliamentary commission. Indeed, the Western media, alerted to the danger by rising radiation levels in Sweden, had been the first to break the news, and Russian- and Ukrainian-language broadcasts on Voice of America and Radio Liberty had taken the lead in advising Soviet citizens about precautionary measures against radiation, but the Soviets had jammed the broadcasts. Meanwhile, KGB agents were busy confiscating leaflets that, in their opinion, contained "tendentious fabrications about the consequences of the accident at the Chernobyl atomic energy station."[31]

Mikhail Gorbachev never took responsibility for what happened in Kyiv that day, but he did later admit that holding the parade was a mistake. "Rallies were not canceled because on May 1 there was as yet no full picture of what had happened," he said in an interview given in 2006, twenty years after the disaster. "Indeed, we were afraid of panic—you yourself can imagine the possible consequences of mass panic in a city of many millions! It's clear today that that was a mistake."[32]

EXCLUSION ZONE

A YOUNG KGB officer named Anatolii Shumak was lucky to get hold of the steering wheel of the van when he did. Otherwise, he and nine other KGB officers would have ended up in a ditch and might have died. It was late on the night of May 1, and the commander of the group refused to disclose their final destination, so secret was the mission for which Shumak had been chosen. As the van moved through the dark streets of Kyiv, the commander directed the driver: right, left, straight, then right again. At some point, frustrated by the last-minute commands, the driver lost his nerve and grabbed his head with his hands, leaving the wheel unattended. Shumak jumped and took control of the vehicle, preventing a crash.

Only then did the commander tell the driver that they were going to the Boryspil airport near Kyiv. Their task was to receive a special radiation-proof vehicle designed for use by the top Soviet leaders in case of nuclear attack and escort it to Chernobyl. The commander told his subordinates that it was the only vehicle of that type in the Soviet Union, and it would be their responsibility to transport and protect it. They followed orders, and by midday on May 2 the vehicle was at the Chernobyl nuclear plant. Rumor had it that Mikhail Gorbachev himself was going to visit the damaged power plant.[1]

Gorbachev never came—his first visit to Chernobyl would take place almost two years after the accident, in February 1988—but on the day Shumak and his fellow officers delivered their secret cargo, the site was visited by two of Gorbachev's closest assistants, Premier Nikolai Ryzhkov and Gorbachev's second-in-command at the Central Committee, Yegor Ligachev. They flew to Kyiv from Moscow on the morning of May 2. From there, in the company of the Ukrainian leaders Volodymyr Shcherbytsky and Oleksandr Liashko, they took a helicopter to the power plant. Some of those involved after the accident, including the chief scientific adviser of the state commission, Valerii Legasov, believed that the visit was a response to reports from Kyiv and other major Ukrainian centers about rising levels of radiation.[2]

After having ordered that the Kyiv rally proceed as planned, the leaders from Moscow now came to the area to assess the situation for themselves. They brought along their personal dosimeters but had little understanding of the danger posed by the damaged reactor. As the helicopter approached the nuclear plant, Ryzhkov ordered the pilot to descend and fly over the reactor. "The ever more frequent beeping of the instruments turned into a frenzied, continuous wail; the numbers ran up the scale at a furious pace," remembered Liashko, who was on the flight with Ryzhkov and Ligachev. He recalled that the helicopter had no protection whatsoever against radiation; Ryzhkov believed that there was a lead plate on the bottom, but nothing else. Looking at the reactor, they were able to comprehend for the first time the scope of the damage the explosion had caused. But they were still far from fully understanding the consequences of the disaster.[3]

In Chernobyl, the plenipotentiaries of the Kremlin presided over a meeting of the state commission, only slowly gaining a better appreciation of the enormous problems facing them. Anatolii Maiorets, the all-Union minister of energy, who was one of the main speakers at the meeting, exuded optimism about the future of the plant. Trying to anticipate the expectations of the senior officials, Maiorets concluded his presentation by stating: "We will take all necessary measures and have Unit 4 in working order by October, and Unit 5 by December!" The director of the plant, Viktor Briukhanov, who was no longer in any position to make decisions but still present at the meeting, was shocked. Later, he recalled his thoughts on listening to Maiorets: "And nobody said to him, 'Why are you talking nonsense? The unit can't be restored!' The atomic experts remained silent. And I couldn't say a word so as not to be expelled from the meeting." One participant who did not remain silent but also did not speak out was Liashko, the head of the Ukrainian government. "What is he talking about?" he quietly asked Ryzhkov. "How can any units be brought onstream when a ten-kilometer zone is contaminated with radiation above the normal limit?" Ryzhkov did not respond. The meeting went on.[4]

It was a hot day in May. The windows were open, and next to one of them sat Volodymyr Shcherbytsky, chain-smoking cigarettes. He wiped tears from his eyes with a handkerchief. He was probably suffering from a spring allergy, but the situation was grim enough to justify real tears. Not all the speakers were as optimistic as Maiorets. General Vladimir Pikalov, the commander of chemical units in the region, reported on radiation levels, and leading scientists left no doubt that radioactivity was high and increasing every day. The head of the all-Union meteorological agency, Yurii Izrael, produced a map of the contaminated areas around the power plant. They extended up to 30 kilometers, with the locations of radioactive "tongues" and "dirty" spots depending on the direction and strength of winds at the time of the explosion and in the following days. Many members of the state commission believed that the 10-kilometer exclusion zone on which they had agreed earlier would have to be extended.[5]

Ryzhkov, who was slowly grasping the extent of the problem caused by the spreading radiation, asked how large the new exclusion zone should be. Those present suggested a radius of 30 kilometers, although there were some "clean" spots within that area. Ryzhkov later recalled, "We had several sources of information: ecologists, geologists, meteorologists, the military, and civil defense. We compared all those maps, analyzing why some of the data was inconsistent. We placed all the maps one on top of another and got a 'blot' encompassing the contaminated areas of Ukraine, Belarus, and Russia. . . . I sat thinking for a long time: a decision had to be made." After some hesitation, Ryzhkov went along with the recommendation to create a 30-kilometer zone. The zone would cover more than 2,000 square kilometers of territory, include more than

80 settlements, and result in the additional evacuation of more than 40,000 people.[6]

Liashko recalled the moment of decision somewhat differently. After the meeting in Chernobyl, the senior officials drove to nearby villages to check on the evacuees from Prypiat. Liashko was in the middle of a conversation with one of the women evacuated from the city when the commander of the civil defense units approached him with the map of contaminated areas produced by General Pikalov and his officers. Liashko looked at the map and realized that the village they were visiting, to which Prypiat citizens had been evacuated a few days earlier, was itself in the radioactive zone. It was located 20 kilometers from the nuclear plant. Liashko showed the map to Ryzhkov, who made the final decision on the evacuation of all settlements within the 30-kilometer zone.

Liashko then returned to his conversations with evacuees. The woman with whom he had just spoken complained that the physics teacher in the local school, in whose home she had been billeted with her family, had asked them to move to the summer shack on the premises because they were "bearers of radiation." "And I had a fleeting thought," recalled Liashko later. "What would that teacher, who had treated the family of evacuees in unfriendly fashion, say if he were ordered to evacuate his dwelling the next day?"[7]

The members of the state commission were pleased with the visit of senior Soviet officials to Chernobyl and the discussions they had there. "That was an important meeting," recalled Valerii Legasov. "First of all, they understood from our reports, and it fell to me to be one of the reporters, they understood the situation, grasping that this was not just a local accident but one of great significance that would have long-term consequences, and that huge efforts would be required to continue localizing [the consequences of] the damaged unit; that preparations had to be made for large-scale deactivation measures; that a cover for the damaged Unit 4 would have to be designed and built." There was no more talk of restoring Unit 4 to working order or bringing it back onstream by the end of the year. The highest officials in Moscow were beginning to comprehend the consequences of the disaster.[8]

Ukrainian officials had mixed feelings about the visit of their superiors from Moscow. Ryzhkov, who took political responsibility for the evacuation of Prypiat and then for extending the exclusion zone, was not above showing the locals who was boss. Considering the medical response to the crisis tardy and inadequate, he was quick to reprimand local officials. After listening to the Ukrainian deputy minister of health complain that he did not have enough ambulances at his disposal, Ryzhkov all but fired him. "That makes you a nobody today," he told the frightened Ukrainian official. "Because we have enough of everything in the Union." That was the Soviet style of management, deeply rooted in the Stalinist tradition of terrorizing subordinates into meeting their production quotas. The local officials, who felt they were being punished for the shortcomings of the all-Union government, which ran the plant, were resentful.[9]

The party ideologue Yegor Ligachev also rubbed the locals the wrong way. "Like a commissar, he appealed for responsibility to the party, for consolidating communist ranks," remembered Vasyl Synko, the head of the Kyiv region's agricultural administration. Vitalii Masol, the head of the Ukrainian planning agency and a member of the state commission investigating the accident, felt that Ligachev had little grasp of the realities of everyday life in the disaster zone. When one of the workers approached Ligachev with the question of whether vodka would be available to those working in the contaminated areas—rumor had it that alcohol helped cleanse the body of radioactive particles—Ligachev responded: "No vodka; we shall observe the resolution of the CC CPSU Politburo." He had in mind the launch a few months earlier of an all-Union anti-alcohol campaign of which he was a sponsor. "At that point I became impatient and intervened," recalled Masol. "'Don't worry,' I told the worker, 'you'll get 100–200 grams of vodka in the dining hall.'"

Vasyl Synko was struck by what he considered Ligachev's lack of empathy for the locals. He remembered Ligachev saying, "Well, it's certainly a great misfortune, but we'll learn from your experience." "Those words simply finished me," recalled Synko. "So the point of the whole disaster was to let the USSR learn from our misfortune?" He suspected that Moscow regarded the Ukrainians as guinea pigs.

"So that was why there was no evacuation order for such a long time!" wrote Synko, recalling his thoughts many years later. "Obviously, the Moscow bosses wanted to see how radiation would affect the health of Ukrainians."[10]

The Chernobyl disaster was slowly driving a wedge between party officials in Ukraine and their bosses in Moscow. The former had to deal with the consequences of the disaster, although they had no control over the nuclear plant, while the latter visited, reprimanded, and fired those whom they held responsible for failing to keep the situation under control. Even the KGB operatives were unhappy. Scores of them absorbed high doses of radiation while sitting for hours in the bushes in contaminated areas around the power plant. They had been posted there to protect members of the leadership from possible attack or assassination.[11]

ON MAY 3, the day after the departure of Kremlin representatives from Ukraine, Oleksandr Liashko called a meeting of a special commission of the Ukrainian Politburo dealing with the consequences of the Chernobyl disaster. With the all-Union authorities and the state commission trying to stop radioactive emissions from the reactor, the Liashko commission focused on the refugee crisis and protecting the civilian population from radiation. It was a difficult task, partly because the authority of the Ukrainian commission was limited by policies made in Moscow regarding the release of information. The Liashko commission had to protect people without telling them what was going on.

The rising radiation levels were the primary concern of Liashko and his colleagues. Since the first days of the disaster, the Ukrainian KGB and the Ministry of Health had been reporting daily to party headquarters in Kyiv on the number of people affected by radiation sickness. It was growing among adults and children alike. According to KGB reports, 54 people had been admitted to Ukrainian hospitals with symptoms of radiation poisoning on April 28. That day, the Politburo in Moscow learned of 170 cases throughout the USSR. On May 1, *Pravda* wrote of 197 people admitted to hospitals. By the morning of May 3, there were 911 such patients in Ukraine alone, and by May 4, 1,345 patients. The number of children with symptoms of radiation poisoning grew as well, from 142 on May 3 to 330 on May 4. With the radiation sickness wards created in Kyiv's hospitals full or approaching capacity, the medical authorities prepared hospitals outside Kyiv and were ready to take in an additional 1,680 patients.[12]

Anatolii Romanenko, the Ukrainian minister of health, urged members of the Liashko commission who had been traveling to Chernobyl to have checkups at government clinics. He reported that 230 medical brigades were being sent to the Chernobyl area to deal with the growing radiation crisis. Given the shortage of medical personnel, many of those sent to the area were students of medical schools in Kyiv. With the KGB reporting that parents were protesting the dispatch of young female students of nursing schools to the exclusion zone, mobilization was limited to the male students of the medical schools in the six- and seven-year programs. The announcement to prepare for a trip to the zone had been made in the middle of classes on the morning of May 4, recalled one of the mobilized students, Maksym Drach, the son of the prominent Ukrainian poet Ivan Drach. "We got into the bus feeling happy and joking," he remembered. Maksym and his friends were assigned to dosimeter control posts on the borders of the exclusion zone. It was a heartbreaking scene. "Those who came were mainly old, hunchbacked men and women and small children," recalled Maksym. He and his friends had no protective gear, and after spending a few days patrolling the zone they ended up in hospitals themselves.[13]

Ukrainian officials were struggling with the question of what could and should be said about the danger posed by the reactor to the general public. The continuing information blockade, orchestrated from Moscow, was nowhere as obvious and scandalous as in the exclusion zone itself. There, the Prypiat newspaper ceased publication after the accident, but the Chernobyl newspaper, *Prapor peremohy* (Flag of Victory), continued to appear, although it was forbidden to publish anything about the accident or its consequences. As people in the region developed radiation sickness and packed their bags in response to evacuation orders, the paper published not a word about the accident, either in its issue of April 29—the first to appear after the explosion—or in the next issue, dated May 1. On the front page of the latter was an image of Lenin and a selection of party slogans, one of which, taken from Gorbachev's report to the Twenty-seventh Party Congress, asserted: "The Soviet people may be assured that the party is deeply aware of its responsibility for the country's future."[14]

At the meeting of the Liashko commission, Anatolii Romanenko pushed for more openness. "Our way is to tell people the truth, but we are avoiding it," he told the gathering. But Liashko, mindful of the monopoly on information exercised by Moscow, was reluctant to make any public statement on the accident. He shot down a proposal to produce a television report on the damaged power plant and proposed to postpone discussion of the issue until the next day, when they would have a better understanding of the situation. Stepan Mukha, the head of the Ukrainian KGB, did not object. His only concern was the coordination of information policy with Moscow. "Moscow issues data without coordinating it with us," he said. "They write that 17 people are gravely ill; we write 30." They decided to do nothing. At the next meeting of the commission, on May 4, Liashko finally told Romanenko to prepare information for the general public on how to protect themselves from radiation. He added: "Tomorrow we'll present the text to the [Ukrainian] Politburo, and, if permission is granted, we'll issue it later in the day."[15]

Subservience to Moscow was part and parcel of the historical DNA of the Ukrainian political elite. Ukrainian officials postponed public announcements they considered essential for the protection of public health and downplayed the severity of the ecological impact of radiation because they feared being reprimanded by Communist Party bosses. They were determined to fulfill government quotas for agricultural production in the severely contaminated areas. The Soviet government had trouble feeding its citizens, importing 45 million tonnes of grain and close to 1 million tonnes of meat in 1985 alone. Moscow relied on a steady supply of agricultural products from Ukraine, the breadbasket of the Soviet Union.

The Liashko commission ordered the police and military to patrol the exclusion zone. It was less clear what to do with all the crops in the fields. Peasants leaving the zone had to abandon not only their houses but also 10,000 hectares of winter crops and 13,000 hectares of spring ones, as well as 45,000 hectares of planted potatoes. Not fully understanding the effects of radiation, the authorities tried to save what they could of the affected crop. "Radioactive contamination presents almost no danger to the growth of crops," Oleksandr Tkachenko, the overoptimistic Ukrainian minister of agriculture, told the Liashko commission on May 4. "There is a loss of 25–30 percent of winter crops at 80 roentgens, and complete loss at 330 roentgens." The numbers they received were much lower. Whatever radiation did not kill immediately was good enough to be consumed, according to the leaders of the Ukrainian collective farming sector.

It was only after Moscow officials refused to take Ukrainian vegetables for sale in their city's stores and markets that the Ukrainian authorities began to question their own policies. For the time being, however, they allowed the production of butter from the contaminated milk of cows eating radioactive grass. In a country with chronic food shortages, they believed that they could not refuse any agricultural products, including those from the exclusion zone.[16]

As LIASHKO and his colleagues debated what to do and how much to say about the disaster, the radiation situation in and beyond the newly created 30-kilometer zone was rapidly deteriorating. Radiation levels that had been declining after the explosion began to rise again. On April 27, the reactor released close to 4 million curies of radioactive particles; by May 1, that amount had dropped by half, but it jumped back to 4 million curies on May 2, the day when Ryzhkov and Ligachev toured the area. On May 3, the radioactivity release reached 5 million curies, and on May 4 it climbed to 7 million.

What had caused that dramatic increase of radioactivity, and where would it lead? Nuclear scientists were at a loss. One theory was that the 5,000 tonnes of sand, lead, clay, and boron dropped onto the reactor were preventing heat exchange between the core of the reactor and the outside air, while still allowing oxygen to get into the reactor and accelerate the burning of graphite in its core. It was decided to stop the airdrops, but the situation continued to deteriorate. There was concern that the overheated reactor, weighed down by the heavy load, would burn its way through the concrete foundations to the basement of Unit 4, which was full of water pumped there during the first hours after the accident. That could lead to another explosion, potentially much more powerful than the one of April 26. Some argued that up to 10 percent of the reactor's core had been blasted into the atmosphere. If that was true, then the remaining 90 percent might be hurled thousands of kilometers from Chernobyl, making the first explosion an overture to a global disaster.[17]

While the scientists were confused and unsure about what to expect, the local authorities were terrified by prospects of a new explosion. Preoccupied with the logistics of uprooting tens of thousands of people, the managers did not lose sight of the state of the reactor, which they knew to be worsening. There was one major indicator to watch—the temperature within the reactor. "They could find no way of arresting the temperature of the reactor, which kept rising by 100 degrees every 24 hours," remembered Vasyl Synko, the agricultural chief of the Kyiv region, who attended the meetings of the state commission and understood the full extent of the danger. "At the time of the accident, the temperature was 1,200 degrees. A value of 2,200 degrees was defined as a critical threshold that the reactor could not be permitted to attain, as it might result in a new explosion hundreds of times greater than the previous one. In that case, Ukraine and Europe as a whole would become a desolate wasteland."[18]

Synko learned of the new danger emanating from the reactor at a meeting of the state commission that he attended late on the night of May 2. "And so, at 3:00 a.m. on May 3 I was walking to my workplace after that meeting," he recalled. "Buses full of people passed, and trucks with livestock, all bellowing, squealing, mooing. . . . And I counted on my fingers: April 26, 27, 28, 29, 30, May 1, 2, 3, in which time the temperature of the reactor had already risen by 800 degrees, attaining 2,000 degrees Celsius on that symbolic thermometer. So the terrible, irreparable event could happen at any moment, although theoretically two days remained until the critical point was reached."[19]

That night Synko managed to save thousands of people from continuing exposure to radiation. In the previous days, the military had built a pontoon bridge across the Desna River that would make it possible to evacuate the population and domestic animals from areas north of the river. The bridge helped with the evacuation but disrupted the normal navigation of vessels. By May 3 there were dozens of ships near the bridge that could not proceed to deliver their cargos to ports in Belarus. The head of the state commission, Boris Shcherbina, ordered that the pontoon bridge be temporarily dismantled so the ships could get through. Synko was terrified when he learned of Shcherbina's order. Following it would mean stopping buses loaded with passengers and trucks full of animals from leaving the left bank of the Desna and finding safety south of the river. "If they were unable to cross, the people would spend the night on buses, absorbing radiation. And what would the hungry livestock do? It was frightening even to think about that," recalled Synko. Along with the regional party boss, he managed to convince Shcherbina to let the people and animals pass before the bridge was disassembled to let the ships proceed. Synko breathed a sigh of relief.[20]

Among the Chernobyl-area evacuees forced to abandon their homes during the first days of May were the parishioners of an Orthodox church from the village of Krasno near Chernobyl. The orders to evacuate the village came as a complete surprise to the parish priest, Father Leonid, who believed not only in God but also in the power of Soviet science. "We now have powerful science, so they'll fix all the problems," he told his wife soon after the explosion. Father Leonid's belief in the power of science came crashing down on May 2, which happened to be Good Friday. Around 2:00 p.m., as he was celebrating the liturgy in his village church, parishioners came to tell him that there were visitors from the Communist Party committee in Chernobyl. They were calling a meeting of the entire adult population of the village. The service had to be stopped.

The party officials told the peasants that they had four hours to get ready for evacuation—the power plant was too close for them to stay any longer. Indeed, from their church on the hill, Father Leonid and his parishioners could see helicopters dropping their loads on the reactor. As in Prypiat, the people of Krasno were told that they would be leaving for three days and should take only essentials with them. But the evacuation of villages presented problems that urban evacuations did not: there were cows, pigs, geese, and rabbits whose owners insisted on taking them along. Trucks dispatched to pick up cattle finally arrived at 2:00 a.m. on May 3. "You should have seen what went on!" recalled Father Leonid. "They wrote down data—who was delivering how many head of livestock, their weight—and they loaded the animals onto the trucks and took them to the designated places."

It was only when the sun had risen and the cattle were gone that buses arrived to pick up the villagers. "I was among the people," remembered Father Leonid, "visiting the sick, giving communion to those who required communion. There were old men and women there, sick people who had lain ill for years." Some refused to be evacuated. "We are staying here, Father. We're not going anywhere. We're going to die in any case," they told the priest. Father Leonid did his best to persuade the elderly to leave with their families. Grudgingly they followed his advice. The priest closed the church, leaving all his garments and icons, in the hope of a speedy return, gave the keys to the head of the parish, and boarded a bus. Finally the buses started to move. "As we departed Krasno," recalled Father Leonid, "everyone was grieving on leaving home; people wept. They all blessed themselves, making the sign of the cross in hopes of returning."

In the atmosphere of rising tension and complete uncertainty, even communist members of the village administration began to pay attention to the priest and the church he represented. One of the officials offered to have a drink with Father Leonid when they returned. When the priest responded that he did not drink, the official told him: "No problem. You're a priest, and we're communists: when we come back and there is nothing to fear, we'll drink 100 grams each for such delight. As long as we come back." They never did. Having had his service interrupted in his own church, Father Leonid celebrated the Easter liturgy in Chernobyl. The night vigil ended with the blessing of Easter bread at 3:00 a.m. on Easter Sunday, May 4. At 9:00 a.m., Father Leonid, his son, and the remaining inhabitants of Chernobyl boarded buses and left the city. The Easter exodus from Chernobyl had begun.[21]

The head of the Kyiv region's agricultural administration, Vasyl Synko, his communist credentials aside, celebrated Easter in the old Orthodox fashion as well. He had returned to his staff after the meeting of the state commission on the night of May 3 depressed, but also defiant. Facing what might be a global catastrophe, he refused to follow what he considered ridiculous orders coming from the top. After telling his subordinates to stop compiling data for the civil defense authorities, who were trying to keep track of the number of animals evacuated from the exclusion zone, Synko told the colonel in charge to get lost. The colonel threatened to report Synko's insubordination to his authorities, but Synko did not care. He believed that he was responsible to a higher authority. The Orthodox Easter was coming.

"We brought potatoes and marinated mushrooms from the cellar, placed a bottle of alcohol on the table, and had a festive supper or, more properly, breakfast," said Synko, recalling his Easter meal with staff members. "For it was Easter morning. Although we considered ourselves atheists then, after the disaster we recalled the prophecies of our wise grandparents about the end of the world and began to gaze at the heavens in supplication. Someone is there after all—someone much stronger and more influential than the CC CPSU. So everything is in his hands." In the face of disaster, atheists were turning into believers, and the power of communist ideology— like that of Moscow—was dissolving under the influence of the radiation in Chernobyl.[22]

Easter Sunday turned out to be the worst day on record for radioactive emissions from the reactor. They increased from 5 million curies the previous day to 7 million. According to KGB reports, radiation levels near the reactor that amounted to 60 to 80 roentgens per hour on May 1 rose to 210 roentgens per hour on May 4. Moreover, scientists recorded an increase of ruthenium-103 in radioactivity from the reactor. Since ruthenium melts at temperatures above 1,250 degrees Celsius, this was evidence of mounting heat within the reactor. Then came Monday, May 5, bringing another major increase in radiation levels. That day the reactor produced a record amount of radiation, estimated at anywhere from 8 million to 12 million curies.[23]

Having no influence over what was going on at the power plant, the Ukrainian authorities did their best to speed up evacuation from the exclusion zone. "On May 3 the evacuation of inhabitants (9,864 people) from the 10-kilometer zone to Borodianka county was completed; 12,180 head of large horned cattle were also evacuated," read a KGB report for that day. "Complete evacuation from the 30-kilometer zone is scheduled for May 4 and 5." It was easier said than done. About 30,000 more people were subject to new evacuation orders. "The evacuation was carried out with great difficulty," recalled Vasyl Synko. "The impression was that a war was going on. There were instances of panic, confusion, and ill-considered decisions; nevertheless, what prevailed in the actions of those in charge of taking people out of the zone was stamina, courage, and self-sacrifice. Everyone knew that every day and hour spent in the zone would have a deadly effect on the organism of every individual."[24]

Many of the locals, both officials and rank-and-file workers and peasants who could remember the days of World War II, compared the evacuation with their wartime experiences, when people were either shipped east by the Soviet authorities or had to flee to the forest to avoid German retribution for partisan activity. But this was different. Back then, they knew who the enemy was, and the forest provided them with cover. Now danger was everywhere, and the forest, which had helped them survive during the Great Ukrainian Famine of 1932–1933 and sheltered them against German reprisals, was the most dangerous place to be. Leaves on the trees and grass accumulated radioactivity at an extremely high rate. One of the forested areas near Prypiat soon became known as the Reddish Forest: the Scotch pines turned red as a result of heavy radioactive contamination. The name of one of the nearby villages, Strakholissia, or Fearsome Forest, suddenly took on a new and disturbing meaning.

CHINA SYNDROME

MIKHAIL GORBACHEV and his colleagues had a full agenda for their Politburo meeting on May 5, 1986. But apart from reports on May Day parades that had taken place throughout the huge country, including in Kyiv, there was little to lift the spirits of the Soviet leadership. The view from the Kremlin was that of a country besieged at home and abroad.

Abroad, the continuing war in Afghanistan weighed heavily on Soviet finances and damaged the image of the USSR in the international arena. "It is clear that a military victory is impossible, no matter how many troops we have there," Gorbachev told his colleagues that morning. "It is clear that we have not carried out the social revolution for the Afghans—the intention was doomed to failure from the very outset and 'by definition.' It is also clear that instead of building up our 'southern underbelly,' we have gotten ourselves a zone of instability and internal conflict. We have provoked a much larger U.S. presence in the region than before. We have incited the Muslim world against ourselves and pushed Pakistan into open hostility. In a word, this is a total defeat."[1]

At home, the Chernobyl explosion had punched another big hole in the Soviet budget and further damaged the international prestige of the country and its new leader. Even more disturbing, despite all the propaganda efforts to suggest that things were improving, no one could get a handle on the situation, which was becoming worse with every passing day. On May 5, Premier Nikolai Ryzhkov got an update on the health impact of the catastrophe. "The total number of those hospitalized has reached 2,757, of whom 569 are children," read a report. "Of these, 914 show symptoms of radiation sickness, and 18 of them are in very serious condition." The previous day, the number of hospital admissions had been reported as 1,882, so the new data represented a daily increase of more than 45 percent. The patients showing symptoms of radiation sickness were no longer only firefighters, reactor operators, and local civilians. Also affected were members of the country's establishment—leaders and members of the state commission who had been sent from Moscow to fix the problem. Now they were back, with the problem growing worse and their own health in jeopardy.[2]

On the previous evening, the plane carrying Boris Shcherbina, the head of the state commission, and his numerous deputies and assistants, including government ministers and heads of departments, had landed at the Vnukovo airport in Moscow. Shcherbina and his colleagues boarded a waiting bus that took them directly to Hospital No. 6, where victims of the Chernobyl disaster were being treated for radiation sickness. They were examined by dosimetrists, undressed, washed, and shaved bald. Most of them remained in the hospital for further examination and treatment. The decision to bring Shcherbina and his team back to Moscow for health reasons had been made by Premier Ryzhkov during his visit to the area two days earlier.[3]

But before Shcherbina was allowed to take care of his health, he was asked to report to the Politburo on the results of his work. What he told the party bosses was anything but uplifting. The situation was grim. The emission of radioactive particles from the reactor, declining before May 2, had now shot back up and was climbing ever higher. The authorities no longer knew what to say, and May 3 had passed without a press release. Bombing the reactor with sand and lead now seemed ineffective, and the helicopter runs had been reduced in number; on that day they were canceled entirely.

Many scientists were afraid of a new explosion. There was also another danger: it was termed the China syndrome, after the title of a movie produced by Michael Douglas and featuring Jack Lemmon and Jane Fonda along with Douglas. The film was released in the United States in March 1979 and shown to Soviet viewers in 1981. Its title came from an inside joke among nuclear scientists about nuclear reactor meltdown. The joke imagined that radioactive fuel from a damaged reactor could burn through the earth's crust to its core and then reappear on the other side of the globe in China. The greatest real concern, as explained in the film, was that the radioactive fuel would reach the water table. In Chernobyl, some scientists were worried about the radioactivity released by the damaged reactor contaminating the Dnieper River Basin and, eventually, the world's oceans. It would be a catastrophe affecting not just the region, or even all of Europe, but of global proportions.[4]

Apart from Shcherbina, Gorbachev summoned the chief scientific adviser to the state commission, Valerii Legasov, to the Politburo meeting. Legasov flew to Moscow early on the morning of May 5. Before going to the Kremlin, he managed to get to his institute, where he was cleansed of Chernobyl radiation. He also managed to spend a few minutes at home, where he tried to calm his wife as best he could: she had been worried about him, as no one, including Legasov, was allowed to place private calls from Chernobyl. The KGB-imposed security measures, mostly intended to prevent the spread of information about the situation at the Chernobyl power plant and its environs, did not distinguish between highly placed academics and simple workers and engineers.

Legasov arrived at the Kremlin at 10:00 a.m. and was brought to the Walnut Room, where Politburo meetings had taken place since Stalin's time. The key question on everyone's mind was what to do next. Legasov later recalled that Gorbachev "said right away that for now he was not interested in the problem of blame and reasons for the accident. He was interested in the state of affairs and in the necessary measures, and even additional ones, that the state would have to take in order to cope more quickly with the situation that had arisen." The discussion led to the adoption of a Politburo resolution stating that "the number and pace of measures to eliminate the consequences of the accident at the Chernobyl atomic energy station are not yet commensurate with the scale and complexity of the problems posed by that extraordinary event." Gorbachev ordered everyone to go back to work. But there was little understanding of what had to be done to change the situation decisively.[5]

GORBACHEV AND his advisers were at a loss. The measures already taken were no longer working. Water pumped into the reactor during the first hours after the accident had flooded its lower levels, increasing the danger of a much more powerful explosion if the reactor disintegrated completely or burned down to its foundations. Many now believed that attempting to seal the reactor with sand and lead had contributed to a rise in temperature within the reactor, which also increased the possibility of an explosion. Shcherbina had been trying one method after another, hoping that something would work, but so far nothing had proved effective. He had exhausted not only all options at his disposal but also the amount of time he could stay in the Chernobyl zone without jeopardizing his health even further. He was ordered to stay in Moscow and undergo medical treatment.

The torch of chief troubleshooter at Chernobyl was passed from Shcherbina to Ivan Silaev, the gray-haired fifty-five-year-old deputy head of the Soviet government and future leader of the Russian government who, together with Boris Yeltsin, would organize the defense of the Russian parliament against coup leaders in August 1991. Calm, confident, and efficient, Silaev replaced Shcherbina as interim head of the government commission on May 4. But if top managers such as Boris Shcherbina were replaceable, scientists were not. Silaev was badly in need of scientific advice and wanted Legasov back at his side. Legasov was asked to return to Chernobyl after spending only a few hours in the capital.[6]

Silaev wanted to reduce the probability of a new explosion by removing radioactive water from beneath Unit 4. It was a gruesome reminder of the first day of the catastrophe, when the operators had poured water on the reactor in the hope of preventing its destruction, which had already occurred. Once again, firefighters were assigned to the front line, this time with the task of using their equipment to drain water from the basement. Altogether there were close to 20,000 tonnes of radioactive water underneath the reactor, which was rapidly heating up. The scientists endorsed this measure.

"We were afraid that some of the melted fuel would get in there and produce so much steam as to cause additional radioactivity outside," recalled Valerii Legasov.[7]

The only way to accomplish the task was to penetrate the flooded underground corridors and open the valves. "It was quite difficult to approach those bubbler pools because the corridors alongside had been filled with water from the moment they tried to cool the reactor with water," remembered Legasov. "The water level and its radioactivity were high: at particular times and in particular places the level of radioactivity in the water reached one curie per liter." It was a suicide mission, and Silaev fully understood that. It was then that he decided to save the situation by providing capitalist incentives. With the approval of the political leadership in Moscow, he offered financial bonuses to the divers who agreed to do the dangerous work, including free cars and apartments for them or their families.[8]

The operators of the power plant, who knew its layout well, went first. A crew of three engineers in diving suits reached the submerged gates of the pools and opened them to release the contaminated water into adjusting chambers, from which it could be pumped by the firefighters. Legasov later recalled the mixed feelings of a participant in the operation when he was presented with a monetary award during a public gathering. "I saw the face of a man who was, on the one hand, proud of having accomplished that difficult task under difficult circumstances," remembered Legasov. "But on the other hand, one could see him crumpling that packet of money, not as an award: in sum, he found it inconvenient to refuse the money, but at the same time the monetary form of the award itself gave him little satisfaction because in fact, and particularly at that time, people were struggling to deal with the accident, to bend all their efforts and do all they could, not thinking of any incentives, whether material or moral." All three engineers-turned-divers would miraculously survive the ordeal.[9]

NOW THAT water was being removed from underneath the reactor, the possibility of a steam explosion in Unit 4 was significantly diminished. But the threat of the China syndrome—the radioactive poisoning of underground waters—was not eliminated. Radioactive water could flow into the Dnieper River Basin, then into the Black Sea, the Mediterranean, and finally the Atlantic and other oceans. It would be a realization of the biblical prediction of the star named Wormwood falling into one-third of the world's rivers and making them bitter—that is, poisoned. The scientists wanted to keep radioactivity away from the water table but disagreed on the means of doing so.

Valerii Legasov believed that the main threat was the radioactive water pumped from underneath the reactor, which could get back into the ground if not taken care of. He insisted on installing filters to purify it, which was done. On May 4, the military had begun elevating the banks of the Prypiat and other rivers and covering them with a chemical solution that would prevent radioactive particles washed by rain from penetrating the river system. It was the beginning of a long and laborious process.

One of Legasov's colleagues, Academician Yevgenii Velikhov, who arrived in Chernobyl in early May, had a different view, arguing that the main threat was that of the overheated reactor burning its way down to the water table. He proposed freezing the earth under the reactor in order to cool it down, and then building a concrete platform beneath the reactor's foundation to prevent the contamination of underground waters. Legasov was skeptical at best. Since Legasov and Velikhov were at odds, the new head of the government commission, Silaev, was at a loss, not knowing which project of the two academics to implement.

In Ivankiv, a town 50 kilometers south of Chernobyl and the new headquarters of the state commission, Legasov and Velikhov shared the same room but little else—they were not just colleagues but rivals. As first deputy director of the institute, Legasov, a chemist by training, was one rung above Velikhov, who was a physicist specializing in thermonuclear reactors and one of many "regular" deputy directors of the institute. But Velikhov, who was only one year older than Legasov, had become a full member of the Soviet Academy of Sciences seven years earlier than his competitor and was already vice president of the academy, a position that he had assumed in 1978. In 1979, an asteroid had been named after him,

and in 1985 he was given the highest Soviet award—Hero of Socialist Labor. Although Legasov had been overlooked for that award, it
now appeared that the authorities could not manage without him.
Indeed, Legasov believed that Ivan Silaev wanted him back to coun-
terbalance the influence of Velikhov.[10]

According to Velikhov, he was sent to Chernobyl almost by accident on the fourth day after the disaster. He had attended one of the
first meetings of the Politburo group headed by Nikolai Ryzhkov,
which was meeting in Moscow almost daily to coordinate the activities of the Soviet ministries and agencies dealing with the Chernobyl
catastrophe. Velikhov had gone to one of the first meetings of Ryzhkov's group to share advice from one of his American friends, Frank
von Hippel, a professor of physics at Princeton University and chair
of the Federation of American Scientists, an organization created in
1945 by participants in the Manhattan Project to promote peace and
security through science. After hearing of the Chernobyl accident,
von Hippel had telegrammed Velikhov suggesting that children be
given potassium iodide tablets. Velikhov brought von Hippel's telegram to the meeting. "Ryzhkov told me that everyone who had been
there [in Chernobyl]—Shcherbina, Legasov—had received a dose,
so they had to be relieved [from duty]," recalled Velikhov. Off he
went to Chernobyl to help manage the situation while the others
were getting ready to take a break.[11]

Once in the Chernobyl zone, Velikhov played it safe, making no
decisions on matters in which he did not consider himself an expert.
As a physicist, however, he became increasingly concerned about
the possibility of the active zone of the reactor burning through its
foundations and reaching the water table under the power station.
Legasov disagreed, saying Velikhov was unduly concerned about
it—perhaps he had watched the American movie too many times.
Legasov did not believe it was possible for the reactor to burn its
way down into the lower levels of the unit. "The possibility of such
an event was extraordinarily slight," he recalled. "Nevertheless, Yevgenii Pavlovich [Velikhov] insisted that a [concrete] platform be
built beneath the slab at the foundation of the reactor." Velikhov
never publicly questioned any of Legasov's proposals, but others did.
Yulii Andreev, who worked on the deactivation of the nuclear plant,

later remembered that Legasov "was impetuously active in installing absorption filters to eliminate radionuclides from water, but it did not occur to him to check how much of the emission was soluble in water. It was wasted effort—the result of illiteracy in engineering. While Velikhov may have known something about reactors, Legasov's knowledge of the subject was quite insignificant."[12]

Caught between the two quarreling scientists, Ivan Silaev decided to follow the advice of both. Legasov was allowed to install his water filters, while Velikhov could go ahead with freezing the ground under the reactor and constructing a concrete platform there. The latter was an enormous undertaking. They began drilling around the reactor and pumping liquid nitrogen at a temperature of 100 degrees Celsius below zero into the tunnels. It was calculated that 25 tonnes of liquid nitrogen per day would be required to keep the earth around the reactor frozen and the reactor cool. It was anybody's guess whether that would work. The reactor kept heating up and erupting with clouds of radioactive dust, releasing close to 7 million curies into the atmosphere on May 4. Those clouds carried the radiation all over Europe.[13]

ON THE evening of May 6, Anatolii Romanenko, the Ukrainian minister of health, finally received approval to make a televised address to the population of Kyiv and the surrounding region about the dangers presented by the high levels of radiation. He assured the citizens of Kyiv that the radiation levels in the city were too low to cause any harm, but he also announced that "wind direction and strength has changed recently, causing some increase of background radiation in the city and province." He then proceeded to advise on dealing with "some increase" of radiation: "The republican Ministry of Health considers it advisable to inform inhabitants of the city and province of Kyiv of recommendations to be observed in order to reduce substantially the degree of possible effect of radioactive substances on the organism." Romanenko continued: "As much as possible, the amount of time spent by children and pregnant women in open areas should be restricted. Above all, it should be taken into account that radioactive substances spread mainly in the form of aerosols; hence it is advisable to close windows and ventilation panes in dwellings and to prevent drafts."[14]

The traditional phobia of drafts, common to people in the Soviet republics, now took on new significance in Ukraine. The health ministry's recommendations were finally given to the population more than ten days after the explosion and more than a week after contaminated clouds had reached the city. The population did not trust the authorities, and the official admission of what rumors had claimed long before that was interpreted as a worsening of the situation. The void created by the lack of reliable information was filled by rumors, and the news circulating among the citizens of Kyiv was disturbing indeed.

Vitalii Masol, the head of the Ukrainian planning committee and a member of the state commission, later recalled that what was at stake in early May was not just the future of the 30-kilometer exclusion zone but also that of the city of Kyiv. "At the very first meeting, which took place on May 2 at the Chernobyl atomic energy station itself, it was said that there might be another explosion affecting a zone with a radius of 500 kilometers, and in the 'dead zone' (with a radius of 30 kilometers) nothing would be left alive," remembered Masol. "To be perfectly honest, we were already preparing on the quiet for an evacuation of Kyiv." The Ukrainian authorities dreaded such a possibility. "I'm not even speaking of panic," recalled Masol. "What pillaging there would have been: all the stores would have been looted, private apartments, museums. . . . Hundreds of people would have been crushed to death in train stations and airports."[15]

Kyiv, a city of more than 2.5 million citizens, was located 130 kilometers south of the epicenter—not in the 30-kilometer exclusion zone, but well within the 500-kilometer zone that would be devastated by another nuclear explosion. News of a possible new explosion and the potential evacuation of the city spread like wildfire. It came from scientists, engineers, managers, and bureaucrats who had access to the information available in Chernobyl. A secretary of the Ukrainian Central Committee, Borys Kachura, recalled that a group of experts "sat in a room in the Central Committee building working out an evacuation plan. . . . A great many people directly involved—scientists, doctors, and others—obviously knew about this, and it produced, well, a corresponding response on the part of the population."[16]

The KGB reported that a scientist in the Ukrainian Academy of Sciences had made his own prediction about a new explosion in the event of a complete meltdown of the reactor, followed by the spillage of radioactivity into underground waters. Kyiv was awaiting a new catastrophe, one of truly biblical proportions. People rushed to the airport and railroad ticket offices, but soon no tickets remained for planes, trains, or buses leaving the capital of Ukraine. People began storming ticket offices.[17]

On May 6, the Chernobyl commission of the Ukrainian Politburo, led by Premier Oleksandr Liashko, heard a report indicating that more than 55,000 people had left Kyiv the previous day, double the average number of railroad passengers leaving the city or passing through it. Close to 20,000 had left in buses and cars, and 9,000 by air. Kyivans were trying to protect themselves as best they could under the circumstances. The main concern was the health of their children. City authorities reported to the party bosses that on May 4, some 33,000 students, or 11 percent of the Kyivan student body, did not attend classes. On May 6, that number grew to 55,000 students, or close to 17 percent of the student body, and on May 7, 83,000 students, or 28 percent of the student body, failed to show up. On that day, 62 percent of the students were absent from schools in the Lenin district of the city, where most of the party and managerial elite—the people with the best access to information— resided. They were leaving before anyone else.[18]

Disturbed by the growing exodus from the city, Volodymyr Shcherbytsky, the party boss of Ukraine, called a meeting of the Ukrainian Politburo. The main question on the agenda was the evacuation of schoolchildren from the panic-stricken city of Kyiv. At the beginning of the meeting, Shcherbytsky welcomed two guests from Moscow, Academician Leonid Ilin, the fifty-eight-year-old director of the Institute of Biophysics and the Soviet representative to the United Nations' Committee on the Effects of Nuclear Radiation, and Yurii Izrael, the fifty-six-year-old director of the State Committee on Hydrology and Meteorology. They had been sent to Kyiv by Boris Shcherbina at the request of the Ukrainian government.

According to the memoirs of Borys Kachura, who was present at the Politburo meeting, Shcherbytsky opened it with the following question, which was directed to Ilin and Izrael: "We have been advised that we are receiving incomplete information, and in that connection we would like to obtain a clear answer from you: Under what conditions and from what zones do we need to evacuate people? Where are they in danger?" Kachura recalled that the Ukrainian leader received a lukewarm answer from the Moscow scientists: "We are not authorized; we can't; the situation is dynamic; it's changing." Shcherbytsky refused to take no for an answer. "I have . . . permission . . . from Mikhail Sergeevich Gorbachev not to let you out of here until you give us such recommendations," he told his guests, turning them into hostages.

Shcherbytsky later admitted to his aides that he had been bluffing: Gorbachev knew nothing about his meeting with the scientists. But Ilin and Izrael took his statement at face value. The problem was that they had no immediate answer to Shcherbytsky's question and needed time to consider it. "They were given appropriate conditions, a room was assigned to them, all necessary arrangements were made, and they sat down with our minister of health, Romanenko, and began to prepare this document," recalled Kachura. Eventually they signed a statement indicating, according to Kachura, that "there was no danger to the population in Kyiv and other cities of Ukraine beyond the 30-kilometer zone, and that there was no need to evacuate anyone from anywhere; all that was required was to monitor strictly where milk was coming from, since there were very highly radioactive radionuclides in milk."[19]

Valentyna Shevchenko, the chairwoman of the Ukrainian parliament and a Politburo member, was not satisfied with that answer. "When the issue was raised of taking children out of the city, they shook their heads: that was out of the question; there was no need," she recalled. "I burst into tears and asked, 'If your children and grandchildren were in Kyiv, would you take them out?' They remained silent. For us, that was a signal that the children had to be taken out." Premier Liashko suspected that Ilin and Izrael "were reluctant to take responsibility for measures as radical as those we were proposing, obviously because they understood that such measures would require huge expenditures." They agreed only to the evacuation of children from the vicinity of the Chernobyl power plant, who were being evacuated anyway.

As the bureaucratic practice of shifting responsibility continued, the Moscow scientists wore down the Ukrainian party bosses. "We will have to make the decision ourselves," Shcherbytsky told Liashko. "Just make sure that you don't allow panic to spread." They decided to end the school year for children younger than fifteen earlier than usual, in the second half of May instead of late June, and to send those children to Pioneer camps in the southern parts of the Soviet Union. Shevchenko got on the phone with her counterparts in other republics, asking them to take in children from Kyiv. They agreed. But when government officials began to ask Union authorities for additional trains to transport the children, they got a negative response from Moscow. Boris Shcherbina personally sent an angry telegram asking the Kyiv authorities to stop spreading panic and cancel preparations for the partial evacuation.[20]

In Moscow, Mikhail Gorbachev felt increasingly concerned about the international fallout resulting from the Soviet mishandling of the Chernobyl crisis. "What's going on there? I'm very anxious about that problem. The name of Gorbachev is starting to get a thrashing throughout the world in connection with that accident, so a kind of mass psychosis is developing in the world. What is the real situation there?" he asked Legasov when he called him a few days after his return to the Chernobyl Zone. Legasov told the general secretary that in his view the worst was over: "The basic emissions from the destroyed unit have essentially been halted, and the situation is now under control. On the whole, we now more or less grasp the extent of pollution in the zone around the Chernobyl station and in the world at large." Indeed, on May 5, radioactive emissions from the reactor had begun to subside as suddenly as they had risen a few days earlier. On May 6, the emissions were estimated at 150,000 curies, which was 100 times lower than they had been on May 5. Gorbachev was satisfied with Legasov's answer.[21]

On May 9, Soviet Victory in Europe Day, a newly confident Gorbachev called Shcherbytsky in Kyiv to ask what was going on in the city. He was concerned about plans for evacuation. Shcherbytsky decided to play it safe and told Gorbachev: "It was Valentyna Semenivna [Shevchenko] who raised a panic, and we all succumbed to it." It was a sexist defense, but it worked in the macho culture of the Soviet leadership. When Shcherbytsky informed Shevchenko about Gorbachev's call, she asked, with tears in her eyes, what they were going to do now. "We will take them out," responded her boss. "No one is going to punish us for children." The government's plans called for the evacuation of 986,000 children from Kyiv and the vicinity surrounding it before the end of May. The authorities wanted to put an immediate stop to the chaotic exodus of children and their parents from the city. That day, Shcherbytsky, together with his young grandson, was seen at the memorial events in downtown Kyiv on the occasion of V-E Day. Despite their own doubts, the authorities were determined to send reassuring messages to the citizens of Kyiv: if the first person in the republic kept his grandson in Kyiv, it was apparently fine to keep their children there as well.[22]

That evening, at the Ivankiv headquarters of the state commission, Valerii Legasov was in a good mood as he prepared to celebrate V-E Day. The previous day had brought more good news: close to 20,000 tonnes of water had been pumped from beneath the reactor. Coupled with the news about falling levels of radiation, it seemed reason enough to look much more confidently into the future. Legasov and others were getting ready to take a few hours off to celebrate the holiday with a festive dinner. Legasov was spending the evening with Vitalii Masol, the representative of the Ukrainian government in Chernobyl, who later remembered that "suddenly a flame leaped above the reactor, and then a pink glow appeared. We couldn't understand what had happened." Legasov did not hide his disappointment, writing, in his own memoirs, "We were embittered, of course. The May 9 holiday was ruined."[23]

Memories of the World War II victory aside, it was too early to celebrate success at Chernobyl. The reactor was by no means under control. No one knew what had caused the drop in radioactive emissions. The explanations provided later included three possibilities:

The XXVII Congress of the Comunist Party of the Soviet Union, which set new targets for the development of the nuclear industry. On the podium: Mikhail Gorbachev, Nikolai Ryzhkov, Yegor Ligachev, and Volodymyr Shcherbytsky, who would play an important role in dealing with the consequences of the Chernobyl accident and in the cover-up of its true impact on people and the environment. Moscow, February 1986 (Yurii Abramochkin, Sputnik Images).

The control room of a nuclear reactor at the Chernobyl power plant before the explosion. The desk of the shift leader is in the center of the hall; operators are seated at the control panels. November 1985 (RIA Novosti, Sputnik Images).

A helicopter flight over the damaged reactor (visible in the background). During the first days after the explosion, helicopter pilots dropped thousands of tons of sand, clay, boron, and lead into the opening created in the roof of the reactor by the explosion, exposing themselves to extremely high levels of radiation. May 1986 (Igor Kostin, Sputnik Images).

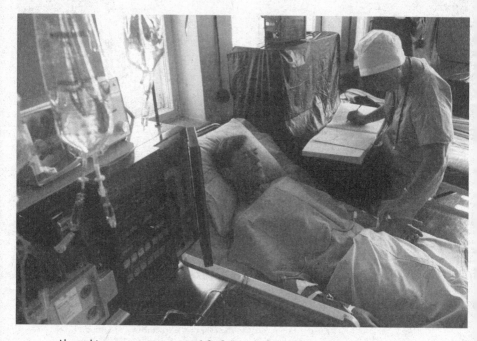

About thirty reactor operators and firefighters with symptoms of acute radiation syndrome were flown to Moscow on the first day of the accident. Only a few of them survived. The photo shows one of the first victims of the Chernobyl accident being treated at Moscow Hospital No. 6. August 1986 (Vladimir Vyatkin, Sputnik Images).

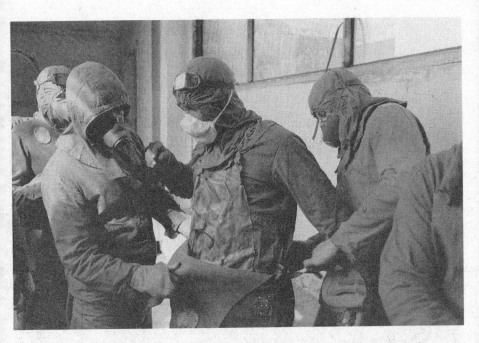

Clean-up workers, many of them reservists, known as "biorobots," preparing to climb onto the roof of the damaged reactor to remove radioactive debris with shovels and handbarrows. October 1986 (Igor Kostin, Sputnik Images).

Abandoned radioactively contaminated vehicles in the Chernobyl Exclusion Zone. Thousands of firefighters' vehicles, helicopters, and pieces of heavy machinery used for clean-up operations were ruled to be beyond salvaging and were never allowed to leave the Chernobyl Exclusion Zone (Liukov, iStock, Getty Images).

The guilty innocents. The trial of the managers of the Chernobyl nuclear plant. Left to right: plant director Viktor Briukhanov, deputy chief engineer Anatolii Diatlov, and chief engineer Nikolai Fomin. Chernobyl, July 1987 (Igor Kostin, Sputnik Images).

Children were affected the most by Chernobyl radioactive fallout, with 3,000 cases of thyroid cancer registered in the 1990s in Belarus, Russia, and Ukraine in the population under 14 years of age. Here are the photos of the children born to the families of the liquidators and resettlers from the Chernobyl ExclusionZone on display in Kyiv Chernobyl museum, May 2013 (Oktay Ortakcioglu, iStock, Getty Images).

Ukrainians protesting the cover-up of the consequences of the Chernobyl accident and demanding independence for their country. Kyiv, April 1990 (Igor Kostin, Sputnik Images).

Nuclear Pompeii—the abandoned city of Prypiat, which was home to 50,000 people before the accident, April 1990 (STF/AFP/Getty Images).

COUNTING LIVES

ON MAY 9, Valerii Legasov and Yevgenii Velikhov were in the middle of a discussion with the energy minister, Anatolii Maiorets, in the Ivankiv headquarters of the state commission when they were joined by a new arrival from Moscow, Grigorii Medvedev. A nuclear specialist and ministry official who had been a deputy chief engineer of the Chernobyl plant in the early 1970s, Medvedev had visited it again a few weeks before the explosion.

Now, Medvedev had just arrived to help with the task of bringing the damaged reactor under control. There were a number of things that struck him. The first was the subject of conversation: the scientists were doing their best to persuade Maiorets to take charge of what was going on at the plant, which belonged to his ministry. Two weeks after the explosion, the organization of the whole effort to deal with the disaster was anything but clear. In fact, there was no organization: the state commission was serving as a fire brigade running from one emergency to another. "There are dozens of ministries working here at present," said Maiorets to Medvedev. "The Ministry of Energy is unable to coordinate them all." Velikhov believed that it could, and that it was Maiorets's responsibility to coordinate the effort of other ministries. "The Chernobyl Nuclear Power Station is yours, so you must organize everything, . . ." argued Velikhov. "Today, Anatolii Ivanovich, you have to count people and lives."

Only later did Medvedev realize what "counting people and lives" meant under the circumstances. "At the evening and morning meetings of the government commission," recalled Medvedev, "whenever they discussed ways of performing a particular task, such as collecting the fuel and graphite ejected by the explosion, entering the highly radioactive zone, or opening or closing some gate valve or other, Silaev, the new chairman of the government commission, would say, 'To do that, we are going to have to count two or three lives. And for that task, one life.'" Two weeks after the explosion, everyone was becoming accustomed to a grim reality: to do the job, one had to count not only minutes and hours spent by personnel in the highly radioactive zones, but also lives that might be lost in carrying out the task. That included their own lives. Legasov had returned to Chernobyl instead of undergoing treatment for radiation poisoning, and Medvedev noted that Velikhov, who had arrived a few days after Legasov, looked not only tired but pale—he had already sustained 50 roentgens of radiation, twice the permissible dose. They believed that they had no other choice. Sacrificing themselves and others was the only way to bring the nuclear monster under control. That task came first. Counting people and lives was secondary.[1]

IN THE Soviet vocabulary of the time, they became known as "liquidators" (*likvidatory*)—hundreds of thousands of people, mostly men, summoned by the state in the aftermath of the explosion to "liquidate [eliminate] the consequences of the Chernobyl nuclear plant accident." Some of them became known as "biorobots," their task being to remove radioactive debris from the premises. Close to 600,000 men and women were mobilized through the party apparatus, through government ministries and institutions, and, most of all, through the Soviet army reserve system to go to Chernobyl for clean-up tasks. Inept at ensuring the safety of the nuclear power industry, the authoritarian Soviet regime proved exceptionally good at mobilizing resources to deal with the consequences of the disaster.

"Everything is proceeding in organized fashion: just one phone call, and a decision is made," Velikhov told *Pravda* correspondents. "Earlier, months were required to reach agreement, but now one night suffices to solve practically any problem. There is no one at all who would refuse to work. Everyone is working selflessly." Legasov went on all-Union television praising the new rules of the game, which no longer restricted scientists and engineers with red tape and the need for endless bureaucratic approvals. Decisions had to be made and implemented quickly to head off another, potentially much more deadly, disaster. Everyone understood that. "There was never such precise work as then," recalled Borys Kachura, secretary of the Ukrainian Central Committee in charge of the Chernobyl disaster efforts.[2]

People and resources were mobilized from all over the Soviet Union. While the logistical center of the liquidation efforts was in Ukraine, the centralization of the command economy in Moscow made it possible to bring in resources from all over the country. Newspapers wrote about the heroism of firefighters and successes in extinguishing the source of the radiation. They also celebrated the vaunted Soviet "friendship of peoples." "We have a sacred principle," wrote *Pravda*, quoting one of the early liquidators, Dmitrii Zhuravlev, who helped install a pontoon bridge over the Prypiat River. "The sacred principle of brotherhood. Specialists who came here from Belarus are working side by side with us. I have met people from Moscow, Leningrad, and other cities of our Motherland in the town center. They all want to help Ukraine overcome the disaster that occurred here as quickly as possible."[3]

Although the scientists did not always know what to do, there seemed to be no shortage of people who could be dispatched to the most dangerous place on earth. Something that the Soviet Union could mobilize in almost unlimited quantities was human resources. The first to be sent into the eye of the nuclear hurricane were members of the military. The firefighters who fought the radioactive fire were on active duty and belonged to the interior troops. The helicopter pilots were members of the Soviet air force, and the officers and soldiers of the chemical troops belonged to the Soviet ground forces. In the last category, the absolute majority were draftees, boys aged eighteen to twenty.

THE PRACTICE of using the military—draftees, in particular—to carry out dangerous jobs in the nuclear industry began at the very start of the Soviet nuclear program. The military would be sent in to perform the most dangerous tasks after scientists, engineers, and skilled workers had sustained the maximum allowable doses of radiation. The latter were needed to run the program, while the former were deemed expendable. Since the regime did not trust Gulag inmates, soldiers were used as the first "biorobots." Draftees from the Central Asian republics, who had limited knowledge of the Russian language and lacked understanding of the dangers involved in the nuclear industry, were particularly liable to such exploitation. Whereas the maximum dosage of radiation for industry personnel was 25 rem, soldiers were kept at their tasks until they had absorbed 45 rem, almost twice the otherwise permissible dosage.[4]

In the fall of 1957, when a tank containing radioactive waste exploded at the Maiak complex in the closed city of Ozersk in the Urals, soldiers whose camp was located near the tank found themselves beneath a nuclear cloud. They were among the first victims and liquidators of the first Soviet nuclear catastrophe of mass proportions. Military men were dispatched to the most dangerous spots to remove radioactive debris. Some refused to follow orders, but most did as they were told. It was in Ozersk that the Soviets acquired their first experience not only of dealing with nuclear disasters but also of using the military to conduct decontamination efforts.[5]

The Chernobyl accident surpassed anything the Soviets had already dealt with or expected to happen in the domain of nuclear energy. The army soon ran out of chemical corps troops trained to perform decontamination tasks. The authorities then took the unprecedented step of calling up reservists—the only way to mobilize not only large numbers of men, but also qualified cadres that were not otherwise available. In late May, the all-Union government decreed that, "considering the large scope of decontamination tasks, the deployment of appropriate units and subunits is to be accelerated, and the requisite number of those liable for military service is to be called up for special instruction meetings for a period up to six months." Nobody knew what the "requisite number" might be. Altogether, close to 340,000 members of the military, most of them reservists, took part in the liquidation of the consequences of the

Chernobyl accident between 1986 and 1989. Every second liquidator was a military serviceman or a reservist.

When the first reservists began to arrive in the Chernobyl zone in May 1986, there was no existing law permitting the use of military reservists for dealing with the consequences of technological disasters, and the one adopted in the following year allowed mobilization for only up to two months. The Soviet Union, however, was anything but a law-based state. People were summoned from the reserves and sent to Chernobyl on orders of military commissars, and some of the commissars promised the draftees salaries five times higher than their normal compensation, as well as all sorts of privileges for them and their families—none of that sanctioned by the government. The commissars had quotas to fulfill. Some draftees were mobilized at their places of work, with no opportunity to see their families and bid farewell to loved ones. It was equivalent to a wartime situation, and, although some did all they could to avoid mobilization, others joined the ranks of the liquidators out of civic duty.[6]

The decontamination effort at Chernobyl became the task and responsibility of the military, and it did its best to carry it out. "At the station itself and within a 30-kilometer zone, the work of decontaminating villages, hamlets, and roads was shouldered by the army. They did a colossal job," recalled Legasov. Helicopter pilots spread a liquid substance called "water soup" to make radioactive dust adhere to surfaces, and chemical troops on the ground cleansed the surfaces of buildings as well as soil and vegetation with a special decontaminating solution. Engineering troops destroyed and buried buildings, structures, and machines that were beyond salvaging by decontamination. Among the best-known objects buried by the military was the so-called Red Forest, a stand of pine trees 10 square kilometers in area that had turned red after absorbing radiation released at the time of the accident. The hardship was not only physical but also emotional as the young soldiers bulldozed and buried entire villages abandoned by their residents in the 30-kilometer exclusion zone around the power plant.

But the best-known cleanup operation conducted by the military took place on the roof of the third reactor of the power plant.

Altogether 3,000 officers, reservists, and cadets, under the command of General Nikolai Tarakanov, did the job machines could not do—picking up radioactive pieces of graphite from the roof of the reactor adjoining the damaged Unit 4. Dressed in self-made leaden protective gear, including lead aprons and "swimsuits," as well as pieces of lead placed to cover the genitalia, they were allowed to stay on the roof for only a few minutes, sometimes even seconds. The task was to get there, grab a piece of radioactive debris with a shovel, run to the edge of the roof, drop off the debris, and run back into the relative safety of the concrete building that housed the reactor. That was done in order to reduce the radiation level on the roof of Unit 3 and make it operational once again.[7]

General Tarakanov and his soldiers were following orders, but the experts were often divided on whether what the troops did was worth the sacrifice they were asked to make. Some claimed that cleaning up the roof did not do the trick—the radiation levels remained extremely high. Others were more concerned with the count of lives that the whole operation called for. "I was tremendously moved by our young soldiers, who collected shards of the reactor and fuel from the roof of the unit practically with their bare hands," wrote the energy minister of Ukraine, Vitalii Skliarov, sometime later. "The radiation levels there were inconceivable! Who sent them? Who gave the order? How can that insanity and crime be represented as heroism? And the whole country saw that horror; it was televised on the first program." General Tarakanov's "biorobots" were neither the first nor the last to risk their lives in order to carry out a plan that did not really work. No one could tell at the time what would or would not work.[8]

THE MILITARY could do a great deal to mobilize human resources, talents, and skills. But it could not do everything. The nuclear and energy industries conducted their own mobilizations, sending more people into the zone. So did managers in the construction, coal, and oil industries and in the water management departments. As the state commission switched from one way of bringing the reactor under control to another, it asked for more experts, more equipment, and ever more people.

The unexpected "awakening" of the reactor that ruined V-E Day celebrations for Valerii Legasov and his colleagues on the evening of May 9 reminded everyone that despite the fall in the intensity of radioactive emissions that began on May 5, the reactor was still dangerously alive. On the next day, May 10, Legasov got into a helicopter to assess the situation from above. "It was hard to establish," he recalled, "whether it was the parachutes used to drop lead and other materials that were burning. In my view, it did not look like that by any means. Most probably it was a red-hot mass, as I understood much later—a red-hot mass of sand, clay, and everything else that had been dropped." That day the Ukrainian prime minister, Oleksandr Liashko, told the members of the Ukrainian Chernobyl commission that the explosion had been caused by the collapse of the crust built on top of the reactor by an accumulation of substances dropped on the unit in the previous days—the crust had melted from the heat. The good news was that after a temporary spike, radioactive emissions from the reactor continued to diminish.[9]

Even so, no one could tell what the reactor would do the following day. Fears of the China syndrome were far from over, the main concern being that the burning reactor and radioactive fuel would sooner or later bulldoze their way down to the water table. At first the state commission had decided to pump liquid hydrogen underneath the reactor to cool it down. Construction workers, drilling crews, and equipment were brought into the area, but it soon became clear that the idea had to be scrapped. The construction crews could not get close to Unit 4 because of the high levels of radiation, and horizontal drilling undertaken from a safer place behind Unit 3 turned out to be extremely difficult. With the drilling equipment available they could create frozen pockets beneath the damaged reactor, but not the solid frozen platform that was needed to cool the reactor from below.

Despite Legasov's skepticism, the new head of the state commission, Ivan Silaev, gave Velikhov his approval for the construction of a concrete platform that would be cooled by pipes conducting ammonium hydrate. To build such a platform, it was necessary to dig a tunnel and then a chamber beneath the reactor to house the freezing equipment. Only then could pipes be brought in and concrete poured under the reactor. Those who supported Velikhov's idea argued that a concrete platform would have to be built under the reactor one way or another. They explained to officials of the International Atomic Energy Agency, who were monitoring the situation at the Chernobyl nuclear plant from their headquarters in Vienna, that the platform would also serve as a foundation for the structure that would have to be built to enclose the reactor and make it safe.[10]

Drilling equipment was removed from the site and miners were brought in. The first of them came to the construction site from the Donbas mining region in eastern Ukraine, but soon other crews began to arrive from mining towns in Russia and other parts of the Soviet Union. Altogether there were more than 230 miners from the Donbas and more than 150 from the Tula region (approximately 240 kilometers south of Moscow) who took part in the project. Communist Party committees at mines all over the USSR were given the task of selecting the very best workers. Vladimir Naumov, a thirty-year-old miner from the Tula region of Russia, who came to Prypiat on May 14, recalled that every miner sent to Chernobyl was approved by the committees.

The scientists were worried that the use of heavy machinery could make the reactor building's foundation shift and possibly crack, releasing radioactive substances from the reactor's core into the earth and affecting groundwaters. The miners were therefore forbidden to use any heavy equipment. They had to dig virtually with their bare hands and push carts full of soil out of the tunnel, also by hand. And dig and push they did. "When that chamber was being dug, the soil was taken out in a wagonette (they installed it in the tunnel immediately)," recalled Naumov. "It was a half-tonne wagonette that we used. Just imagine that every shift brought out 90 wagonettes, and once there was even a record of 96! And now, do the math: three hours are 180 minutes. That means two minutes for every trip. To fill the half-tonne wagonette, push it a distance of 150 meters along the track, empty and return it. And it was pushed by two men, while five or six filled it with their hands or with shovels."

The miners worked in three-hour shifts and, according to Naumov, they were eager to do the job. They would later be paid handsomely by Soviet standards, but at the time, said Naumov, they knew nothing about the bonuses for the liquidators decreed by the government on May 7. "People seized shovels from one another!" he recalled. "A shift would show up, but the previous one would tell them that it was too early. The latter would say in turn: 'No, our shift began two minutes ago!' There was no shortage of enthusiasm. After all, the ideology at the time was Soviet, a different upbringing. There was a well-known motto: 'Who if not we?'"[11]

Khem Salhanyk, a veteran of World War II and Ukraine's leading documentary filmmaker, who led one of the first film crews that came to the Chernobyl power plant after the accident, recalled his days in the exclusion zone with nostalgia. "All in all—perhaps it's a sin to say so—it was a wonderful time! I remembered the war and my comrades in arms. I didn't want to leave that place—that was the way we related to one another. And everyone concentrated on the job and nothing else." Understandably, those in the exclusion zone compared their battle against radiation to a war, the major difference being that they could not see the enemy. As in wartime, there were those who abandoned the front without leave, and others who volunteered to serve, sacrificing themselves in the process.[12]

THOSE WHO refused to duck in the face of danger often died first. For more than two weeks after the explosion, the count of lives lost and sacrificed as a result of the disaster remained at two, but on May 7 it began to rise. That day saw the death in Kyiv of Oleksandr Lelychenko, a forty-seven-year-old deputy chief of the electrical unit, who was hospitalized in the Prypiat hosital but left it without permission to join his colleagues at the plant, and received a deadly dose of radiation.

Lelychenko's death was just the beginning. Two twenty-three-year-old lieutenants, Volodymyr Pravyk and Viktor Kibenok, commanders of the firefighting units at Chernobyl, passed away three days later, on May 11. Before Pravyk died, his mother asked God to take her instead, recalled Liudmyla Ihnatenko, the wife of Sergeant Vasyl Ihnatenko, who was in the same ward. Pregnant with their first child, Liudmyla sacrificed her own health and that of the unborn baby to stay with her husband, whose body turned into a source of radioactivity. Although she became ill, she lived to tell her story and recount the last days of those who had fought and extinguished the radioactive fire at the price of their lives.

Liudmyla arrived in Moscow almost immediately after Vasyl was evacuated there on the night of April 26. She found the hospital by asking a policeman at the Moscow railway station, then bribed her way into the radiation unit by giving a few rubles to an old woman on duty. She begged the personnel to show her the ward where her husband was being treated and eventually found herself in the office of the chief radiologist, Angelina Guskova, who did not seem very receptive. Guskova asked Liudmyla whether she had any children. The only thing on Liudmyla's mind at the time was to get Guskova to let her see Vasyl. She assumed that claiming to have children would strengthen her case and said that she had two, a boy and girl. In reality, she had none. Liudmyla decided not to mention that she was six months pregnant. Guskova expressed satisfaction on hearing Liudmyla's story because, given Vasyl's condition, it was unlikely that the couple could have any children in the future. Liudmyla thought to herself that if Vasyl's health had been undermined by the accident, she could accept that: the most important thing was to see him. Guskova gave her permission.[13]

Vasyl was in better shape than he had been in Prypiat, when his face was swollen. Now it looked normal. He was in the same ward as the rest of the firefighters hospitalized a few hours after the explosion, including the lieutenants Pravyk and Kibenok. On seeing Liudmyla, Vasyl joked that there was no escaping her. The others were also happy to see her and asked about the latest news from Prypiat. They were still trying to figure out what had caused the accident. Most believed that it had been an act of terrorism or sabotage: someone had blown up the reactor on purpose.

Liudmyla became a caretaker not only for Vasyl but also for other firefighters in his ward whose family members were not yet in Moscow. She would cook for them at her friend's apartment, where she was staying, and when they were no longer able to chew, she would grind the food for them. At some point she moved into the hospital building to be closer to Vasyl and his comrades. Dressed in hospital attire, she was often mistaken for a medical aide. She was also learning the basics of radiation medicine. As Dr. Guskova explained, acute radiation illness proceeded in stages. The last days of April were relatively good for the Chernobyl firefighters and reactor operators in Moscow.[14]

Arkadii Uskov, one of the reactor operators, who had arrived at Unit 4 a few hours after the explosion and taken control of the reactor from Aleksandr Akimov and his crew, noted in his diary: "I feel normal." Uskov ended up in Hospital No. 6 with the first group of patients evacuated from Prypiat on the night of April 26. His only complaint was that he was constantly thirsty; his main concern that day was the discomfort caused by a blood test. "Blood from a finger is nothing, but blood from a vein is hardly a pleasure," wrote the thirty-two-year-old reactor operator. On May 2, he made another note in his diary: "I feel fine. I could eat a horse."

At that point, most of the patients were concerned not about their health but about the cause of the explosion and further developments at the power plant. Many saw it as their responsibility to go back and help remedy the situation. "We often think of our own workplace. Our guys," wrote Uskov on May 4. "What a bad time to end up here. At this time, our place is there." Two days earlier, concerning his conversation with Anatolii Diatlov, a key figure in the events of the fateful morning of April 26, he had written: "All we talked about were the reasons for the accident."

It was on May 6 that Uskov first noted a change in the condition of his fellow patients—a doctor told him that the "invisible" period of radiation sickness was coming to an end. Diatlov developed visible burns on his face and legs and suffered from a major burn on his right hand. Viktor Proskuriakov, one of the two interns Diatlov had sent to the reactor hall a few minutes after the explosion to find out what had happened, was in particularly bad condition. People began to die. "It's a shame to die young, at the height of one's powers," recorded Uskov on May 9, V-E Day. "In the evening we watched the holiday salute," he noted, "but we took little pleasure in it." On May 11, Uskov recorded the appearance of radioactive burns on his own body, beginning with one of his fingers. He was among the few lucky ones who survived the ordeal and eventually returned to the Chernobyl power plant. Most of his fellow patients did not.[15]

On May 9, Vasyl Ihnatenko presented Liudmyla with flowers for the last time. She was in his hospital room when he opened his eyes and asked her: "Is it day or night?" She told him that it was around nine in the evening. He asked her to open the window. There were fireworks outside. "I told you I'd show you Moscow," said Vasyl. He had been stationed in Moscow during his service in the Soviet Army and had always promised his wife that one day he would bring her to the capital of the Soviet Union. "And I told you that I'd always give you flowers on holidays," Vasyl said to Liudmyla as he took three carnations from under his pillow. Running to him, she hugged and kissed him. He protested, well aware that his own body was a source of radioactivity and that Liudmyla was pregnant. She stayed with him all night. Vasyl was scheduled for surgery in a few days to replace his bone marrow. It had been damaged by the radiation and had stopped producing white blood cells, which are essential to survival. His twenty-eight-year-old sister donated her bone marrow to save her brother; she was therefore also put under general anesthesia during the procedure while some of her bone marrow was removed and transferred to her brother. His wife, Liudmyla, was hoping for a miracle.

Vasyl Ihnatenko died on May 13, the same day that the bodies of Lieutenants Pravyk and Kibenok were interred in sealed zinc caskets at the Mitino Cemetery in Moscow. Their corpses were wrapped in plastic bags before being placed in coffins. The coffins themselves were then wrapped in plastic and fitted into larger caskets made of zinc. Only then were the double coffins lowered into the graves, which were covered with cement tiles. The families were informed that the bodies were too radioactive to be released to them or to be buried anywhere else or in any other way. Family members signed the papers they were given, indicating their consent. They were told that their sons were heroes. But they no longer belonged to their families: their earthly remains now belonged to the state, which decided what to do with them and how to honor them.

Vasyl Ihnatenko's funeral took place under conditions of utmost secrecy. The concern was no longer the spread of radiation but the dissemination of news about people dying of radiation sickness. A colonel of the Soviet Army who was in charge of the burial ceremony drove the bus with Ihnatenko's coffin and members of his family around Moscow for hours before finally approaching the cemetery. He told the members of the funeral party: "We are not allowing anyone into the cemetery. The cemetery's being attacked by foreign correspondents. Wait some more." No longer capable of controlling her emotions, Liudmyla exploded: "Why are they hiding my husband? He's—what? A murderer? A criminal? Whom are we burying?" The colonel gave up. "Let's enter the cemetery. The wife is getting hysterical," he told his subordinates. Once inside, the small funeral party was completely surrounded by soldiers. "No one was allowed in," recalled Liudmyla. "It was just us. They covered him with earth in a minute. 'Faster! Faster!' the officer was yelling. They did not even let me hug the coffin. And—onto the bus. Everything on the sly."[16]

The bad news about the firefighters dying was not supposed to reach foreign correspondents, or, through them, the Soviet population at large. The death count was just beginning. Within a few months, twenty-eight people had died of acute radiation syndrome. And in the months and years to come, many more would die of complications caused by exposure to high levels of radiation. The average dose sustained by the almost 600,000 liquidators in the exclusion zone was 12 rem, or 120 times the yearly dosage considered safe by the International Commission on Radiological Protection. In subsequent years and decades, death and disability rates among the liquidators would significantly exceed those in the general population.[17]

V

RECKONING

WAR OF WORDS

MIKHAIL GORBACHEV finally broke his public silence on May 14, eighteen days after the disaster. "Good evening, comrades!" he said, beginning his televised address to the country. "You all know that misfortune has recently befallen us." He did not address his fellow citizens as "brothers and sisters," as Joseph Stalin had when Germany had invaded the Soviet Union in 1941, but by speaking of "us," he tried to invoke a spirit of trust and solidarity between the rulers and the ruled that, if it had ever existed in Soviet times, had been shattered by his government's handling of information on the Chernobyl disaster.

Gorbachev clearly did not believe that telling the truth was the best way of governing. "For the first time, we actually encountered a force as menacing as nuclear energy out of control," he said, maintaining the official silence on the 1957 catastrophe in Ozersk. Gorbachev was, however, completely honest in assuring viewers that the authorities were doing everything in their power to deal with the disaster and that work was going on around the clock. He also provided the most accurate figures he had at the time on the number of people directly affected by the accident—299 men and women diagnosed with radiation sickness, with the death toll rising from two to seven. He mentioned the names of the first two, who had died on the first day, but not the names of those who had died in the Moscow and Kyiv hospitals in the first weeks of May.

Gorbachev claimed that every effort had been made to evacuate people from affected areas as quickly as possible. "As soon as we received reliable information from primary sources, it was made available to the Soviet people and sent to governments of foreign countries through diplomatic channels," he asserted. That would become his and his government's defense from then on. What might be considered "reliable" information was, of course, a matter of opinion. There was clear disagreement about it between Gorbachev, on the one hand, and the citizens of Prypiat and Kyiv and foreign governments, on the other.

More than half of Gorbachev's first address to the country on the Chernobyl disaster was dedicated to polemics with and attacks on the West. "The ruling circles of the USA and their most zealous allies—among them, I would mention the Federal Republic of Germany in particular—perceived in the event nothing but a further opportunity to raise additional barriers to the development and deepening of the dialogue between East and West, which was already proceeding with difficulty, and to justify the nuclear arms race," complained Gorbachev. "As if that were not enough, an effort was made to show the world that negotiations, to say nothing of agreements, with the USSR are impossible in general, thereby giving the green light to further preparations for war."[1]

Gorbachev was reacting to the wave of indignation and criticism that had rocked Central and Western European countries and eventually reached the United States as a result of the initial Soviet refusal and subsequent reluctance to share information on the occurrence and consequences of the disaster. As news of the radioactive cloud moving beyond the Soviet Union reached the European public, politicians and ordinary citizens alike began sounding the alarm about immediate and long-term consequences.

The reaction was strongest in West Germany, where the foreign minister, Hans-Dietrich Genscher, demanded the closure of all Soviet nuclear reactors. The Italians refused to receive Soviet ships in their ports if they were carrying any cargo originating in Ukraine. But a country's politics and the importance of nuclear power in its economy influenced the reaction. In France, where most of the electricity was produced by nuclear plants, the government was in denial, refusing to acknowledge that the radioactive Chernobyl cloud had ever entered French air space. In Britain, where the cloud went after floating through France, there was no attempt to deny or hide its presence. In the communist countries of Eastern Europe, government officials were silent, but the people were not. "That the Soviets said nothing and let our children suffer exposure to this cloud for days is unforgivable," *Time* magazine quoted a Polish citizen as saying.[2]

In the United States, which was not directly affected by the Chernobyl disaster but had the greatest stake in maintaining the international order and information exchange on nuclear energy accidents, President Ronald Reagan, then in the second term of his presidency and at the peak of his popularity, expressed his sympathies to those affected by the disaster in a May 4 radio address to the nation. "We stand ready, as do many nations, to assist in any way we can," said Reagan, who then proceeded to attack the Soviets for their "secrecy and stubborn refusal to inform the international community of the common danger from this disaster." He continued: "The Soviets' handling of this incident manifests a disregard for the legitimate concerns of people everywhere. A nuclear accident that results in contaminating a number of countries with radioactive material is not simply an internal matter. The Soviets owe the world an explanation. A full accounting of what happened at Chernobyl and what is happening now is the least the world community has a right to expect."[3]

That was the first time President Reagan, or any other Western leader, for that matter, expressed criticism of the Soviet handling of the Chernobyl disaster. When asked by journalists soon after his radio address about his critical remarks, Reagan responded: "Hasn't that been rather their way about many things in their own country? They are a little mistrustful of all of us." For Reagan, a seasoned Cold War warrior, his remarks on the Soviet system were mild indeed, but they came only a few months after his promising first meeting with Gorbachev in Geneva in December 1985. The two had decided there to meet again the following year, and the media was actively discussing the possible timing of the future summit and its agenda. Gorbachev, addressing the Communist Party Congress in February 1986, had spoken not only of American imperialism but also of the new interdependence of the great powers. And now Chernobyl, or, more precisely, the Soviet handling of the disaster and the American reaction to it, were threatening to derail the normalization of relations between the two superpowers.[4]

On May 5, the leaders of the Group of Seven (G-7), the most advanced democratic economies, including Canada, France, Germany, Italy, Japan, the United Kingdom, and the United States, meeting in Tokyo, issued a joint statement on the Chernobyl accident that largely followed the line taken by Reagan the previous day. They expressed sympathy for those who had died or been affected by the disaster, but they also noted their responsibility as nuclear states to inform their neighbors about nuclear accidents, especially those with transborder consequences, and made a similar request of the USSR. They welcomed the news that the Soviet government had begun to cooperate with the Vienna-based International Atomic Energy Agency, which was charged with promoting cooperation in the peaceful uses of nuclear energy. But they also demanded more openness and cooperation. "We urge the Government of the Soviet Union, which did not do so in the case of Chernobyl, to provide urgently such information, as our and other countries have requested," read the statement.[5]

The outside world was eager to get as much information as possible. Between April 27 and May 16, there were twenty-two visits of foreign diplomats to Kyiv—unprecedented attention to a city that had only a few consulates, and those only of Eastern European communist countries. The KGB worked hard to prevent foreign diplomats and journalists from acquiring any nonofficial information about the accident. Foreign correspondents' phone calls were monitored, and reporters stationed in Moscow began to experience technical problems in wiring their stories from the Soviet capital. Soviet officials blamed what they called an anti-Soviet campaign on warmongers in Western governments and Ukrainian nationalists abroad, who were allegedly lobbying the US Congress for increased pressure on the Soviet government to release more information to its own people and the world.[6]

On April 30, foreign ambassadors had been summoned for a briefing at the Soviet Ministry of Foreign Affairs in Moscow.

Deputy Foreign Minister Anatolii Kovalev provided casualty figures but generally downplayed the danger presented by the release of radioactivity. It was a hard sell. The meeting lasted until 2:30 a.m. on May 1. Kovalev then issued instructions to the foreign ministers of the Soviet republics on how to handle the growing crisis. The local authorities were to explain to concerned foreigners that the accident posed no danger to their health, but if the foreigners wanted to leave anyway, they should be allowed to do so. Their requests for medical checkups were to be met immediately, but anyone diagnosed with symptoms of radiation sickness would have to stay in the country. As the Ukrainian foreign minister summarized Kovalev's instructions, "the task is to prevent people who have fallen ill from leaving the country so as not to allow our enemies to exploit chance incidents for anti-Soviet purposes."[7]

Kovalev's assurances to the foreign ambassadors did not have the desired effect. The British withdrew one hundred of their students from Kyiv and Minsk, the capital of neighboring Belarus. The Finns evacuated their students from Kyiv. A total of eighty-seven language students from the United States and Britain left Kyiv, as did sixteen Canadians, whom the KGB failed to persuade that the Soviet government was not hiding the truth from them. Students from "developing countries," alerted by the exodus of their wealthier classmates, claimed that they were being discriminated against and demanded that their embassies evacuate them as well. Students from Nigeria, India, Egypt, Iraq, and other countries took a vote and decided to leave Kyiv before the end of the academic year. The KGB reported that some of them simply wanted a free ticket home, longer vacations, and lenient treatment in early exams. One way or another, they were leaving.[8]

In late April, a group of American tourists in Kyiv, alarmed by news of the accident, tried to get airplane tickets to Leningrad in order to leave Ukraine as soon as possible. The head of the Ukrainian KGB reported to the Ukrainian party boss, Volodymyr Shcherbytsky, that his agents had managed to "normalize the situation," meaning to postpone the group's departure. A group of fourteen Canadians insisted on leaving the country immediately, claiming that the Soviet media was "concealing the actual situation."

The KGB was working on that group as well, the goal being to convince the foreigners—and, through them, governments and public opinion in the West—that nothing extraordinary had happened in the Soviet Union.[9]

While foreign students and visitors already in Kyiv began to leave, others who had planned to visit the city refused to come. Tourism companies were canceling trips to Kyiv. Whereas in May 1985 the city had hosted up to 1,000 tourists per day from "capitalist countries," in the first weeks of May 1986 the KGB counted no more than 150. Bicyclists from the United States, Britain, Norway, and most other Western countries refused to come to Kyiv for the May 6 start of an international bicycling competition. In an attempt to convince viewers that the situation was normal and under control, Soviet television showed Soviet bicyclists and their colleagues from communist countries riding through the streets of Kyiv. The same footage clearly showed, however, that the streets of Kyiv, which should have been full of people welcoming the athletes and cheering for their team, were empty.[10]

MIKHAIL GORBACHEV, always sensitive about his image abroad, was watching the rising wave of Western criticism of him and his government with great concern. He even complained to Academician Valerii Legasov that his name was being taken in vain in the West. Something had to be done, and done quickly.[11]

On May 6, as radioactivity from the Chernobyl reactor began to decline after the unexpected increase of the previous few days, the Soviet Ministry of Foreign Affairs hosted a press conference to discuss the accident. The deputy foreign minister, Kovalev, who had met with foreign ambassadors a few days earlier, was in charge. He pushed the traditional Cold War line, attacking the United States for organizing a "campaign of hysteria." But there was also a new voice at the press conference—that of the first head of the Chernobyl state commission, Boris Shcherbina, freshly arrived from Prypiat. He admitted that the radiation levels had previously been underestimated and that the evacuation of civilians had been delayed. Soviet journalists and their colleagues from the socialist camp were allowed to ask questions on the spot, while Western reporters had to submit theirs in advance. The Westerners were disappointed, but the Soviet government had finally begun to speak truthfully to its people and the world.[12]

On the same day, *Pravda* published an article explaining that the explosion at the Chernobyl nuclear plant on April 26 had caused a large fire. The paper's journalists described the heroic struggle of the firefighters who had extinguished the blaze. In a measured attempt to provide more information on the consequences of the disaster, the Soviet news agency TASS reported on the spread of radioactivity beyond the exclusion zone into Ukraine and Belarus and the threat that it might contaminate the Dnieper River. But the Soviet media did not only inform: it also attacked, its target being the West, whose indignation had forced the Soviet government to break its silence on the rising radiation level.

"It is to be regretted, however, that against the broad background of sympathy and understanding, particular circles are trying to exploit what has happened for unseemly political purposes," read the TASS statement. "Rumors and speculations that run counter to elementary moral standards are being circulated as propaganda. For example, nonsensical exaggerations are being spread about thousands of dead and panic among the population." The Soviets were referring to unverified Western reports on the number of victims published in the first days after the accident, their goal being to discredit the requests of Western governments and journalists for more information. The Soviets were talking, but they were also trying to save face.[13]

Around the same time, after days of stonewalling, the Ministry of Foreign Affairs in Moscow finally allowed a select group of foreign journalists, including Westerners, to visit Kyiv and then the site of the Chernobyl disaster. The Ukrainian Chernobyl commission discussed the coming visit at its meeting of May 5. Orders were given to prepare the venues and instruct those people who would be speaking with foreign journalists. The latter task was entrusted to the head of the party propaganda department and the future president of Ukraine, Leonid Kravchuk. It was assumed that the journalists would want to visit hospitals and areas undergoing decontamination. Apart from the regime of secrecy, an issue of pride was involved for those charged with hosting the foreigners. "We have to start with hospital wear, which is unsightly, starting with bed linen," remarked the party boss of the Kyiv region, Hryhorii Revenko. The deputy head of the Ukrainian government, Yevhen Kachalovsky, assured those attending the preparatory meeting that he could "sign the order to provide additional supplies of new sheets and bed linen." The authorities were not eager to show visitors the actual state of Soviet hospitals and the standard of living in the communist superpower.

The hospitals received new linens and were ready by May 8, but the need for window dressing did not end there. The KGB was especially concerned about long lines at railroad ticket offices—Kyivans were leaving the city en masse. "They [foreign journalists] will go to the ticket offices first and foremost and release information that we can do without," said the head of the Ukrainian KGB, Stepan Mukha, to the commission. He informed his colleagues that of twenty reporters descending on the city that day, half were from "capitalist countries." Ukrainian officials promised to shorten lines by opening new ticket offices, a move that helped confuse reporters. A *New York Times* article of May 9 claimed that Kyivans were leaving their city by the hundreds: in fact, they were leaving by the tens of thousands.[14]

Reporters visiting Kyiv on the evening of May 8 noticed a heavy police presence on the streets, but otherwise no signs of panic—Kyivans were still strolling about, and some were even fishing in the Dnieper. This was a far cry from the unverified reports about thousands of people killed or severely injured by the explosion. Premier Oleksandr Liashko, who addressed the journalists, could not resist scoring a point in the propaganda war. He asked the reporter who had written a panicky story about the consequences of the accident to rise. He probably had in mind Luther Whitington, a United Press International reporter who had obtained erroneous information from a female resident of Kyiv whom he had met a few weeks earlier in Moscow. Some of his colleagues believed that Whitington, whose Russian was shaky, at best, never understood his source properly and got her story wrong. One way or another, however, Whitington was not in Kyiv. "He spread dirt and went into hiding," commented

Liashko. He read an excerpt from the erroneous report. "The correspondents were considerably embarrassed, as was apparent from the murmur that swept through the room," recalled Liashko, feeling good about the whole experience.[15]

On May 8, the day the foreign journalists arrived in Kyiv, Hans Blix, the former Swedish foreign minister and now director general of the Vienna-based International Atomic Energy Agency, visited the Chernobyl nuclear plant—a sign of the new openness the Soviets were demonstrating. In the company of Academician Yevgenii Velikhov, Blix and his American nuclear security adviser, Morris Rosen, took a helicopter flight from Kyiv to Chernobyl, observing the area around the damaged reactor.[16]

Blix's visit to Chernobyl presented the Soviets, who had sent him an official invitation on May 4, with a number of challenges. The Soviets wanted Blix in Chernobyl to calm Western public opinion by showing that the original reports and fears had been exaggerated, and that every effort was being made to contain the damage caused by the explosion. But how could that be accomplished when the top Soviet experts themselves still did not know what had caused the explosion or what was to be expected of the unruly reactor, which kept heating up or cooling down in seemingly arbitrary fashion?

When Blix's visit was being planned, Velikhov had suggested that bringing him to Chernobyl by car was a bad idea, as toilet facilities at the power plant (and, apparently, along the road, although he did not mention that) were in disrepair, which would reflect badly on the hosts. In actual fact, Velikhov was afraid that on the road to Chernobyl Blix and his party would meet with clouds of radioactive dust, which would show up on his Geiger counter and defeat the whole purpose of the visit. He suggested a helicopter flight, but that raised an additional set of problems.

A few miles from the Chernobyl power plant, the Soviets had built a huge radar system, called Duga (Arch), in the mid-1970s. It was one of two installations that made up the core of the Soviet antiballistic-missile early-warning system. The radar required a large supply of electricity to operate and was linked by a secret electric cable to the Chernobyl nuclear plant. The radar had been built to detect NATO missile launches. Another, similar installation was located near Komsomolsk on the Amur River in the Soviet Far East to monitor the American West Coast. The Chernobyl accident rendered the first installation inoperable, because the military unit that operated the Duga, in the compound called Chernobyl-2, shut it down as levels of radioactivity began to rise during the first hours after the explosion. But the huge radar apparatus, which American experts called "Steel Works," remained plainly visible from the air. There was no way to observe the nuclear plant from a helicopter without seeing the top-secret radar system.

The Soviets had to choose whether to show Blix the toilet facilities and hide the super-secret radar or vice versa. According to Velikhov, Gorbachev personally ruled on the issue, allowing the helicopter inspection. This decision eliminated the possibility that radioactive dust absorbed by Blix on the way to Chernobyl and measurements taken at the nuclear plant would allow him and his experts to comprehend the true scope of the disaster. That would contradict what the Soviets were telling the world: that the radioactive emissions from the reactor had stopped, and that the remaining radiation levels were being caused by the debris that had been dispersed by the original explosion. As the Soviet authorities were well aware, this half-truth amounted to a lie.[17]

It is not clear whether Blix and his companions noticed the Duga radar system when they took their helicopter flight. But Blix did see smoke coming from the reactor—an indication that graphite elements were still burning. Velikhov recalled that Blix's nuclear security expert, Rosen, lacked the equipment required to measure high radiation levels and, when asked whether he wanted to get closer to the reactor, responded in the negative. Blix reported that inside the helicopter cabin at a height of 400 meters and a distance of 800 meters from the reactor, his dosimeters showed a radiation level of 350 milliroentgens per hour. The team members did not measure radiation levels outside the relatively safe haven of the helicopter; nor did they visit the power plant itself. Instead, they landed in the relatively safe part of Chernobyl, as opposed to the heavily contaminated Prypiat, and then flew back to Kyiv. Except for the smoke, the situation looked quite good from the helicopter. "In general, the station was in one piece, somebody was stirring down below, and there was no trace of any tens of thousands of corpses," recalled Velikhov.

At the subsequent press conference in Moscow, Blix was, if anything, optimistic about the future of the area affected by the disaster. "We were able to see people working in the fields, livestock in the pastures, and cars driving in the streets," he told the journalists. "The Russians are confident that they will be able to clean up the area. It will be available for agriculture once again." Blix proposed to convene an international conference in Vienna that would look into the causes of the disaster and ways of avoiding them in the future. He assured the audience that the China syndrome, or the poisoning of underground waters and the world's oceans, was not a threat, and Rosen concluded that there had been no meltdown of the reactor. In an interview with a Soviet reporter in Vienna, Rosen would later estimate the radiation level that he measured during the helicopter flight at 10 millirem. "That is no great amount of radioactivity," he allegedly said. "It is equivalent, for example, to the dose of radioactivity received by an airline passenger in the course of two trips from Europe to the United States."[18]

Blix's visit to Chernobyl gave the Soviets their first victory in the propaganda war with the West. On May 9, one day after Blix's visit to Chernobyl, *Pravda* published an article by the leading Soviet expert on international affairs, the director of the Institute of the USA and Canada, Georgii Arbatov, who stated that the West was not united in its criticism of the Soviet Union. There were good and bad guys. The good ones, who were truly compassionate and wanted to help, included Dr. Robert Peter Gale, an American bone-marrow-transplant expert who had flown to Moscow on May 2 to operate on accident victims, and his colleague Dr. Paul Terasaki. The opposite camp was represented by unnamed practitioners of psychological warfare against the Soviet Union who were allegedly afraid of Soviet peace initiatives and now claimed that, because the USSR had hidden information about the accident, its proposals could not be trusted. "In order to make a propagandistic stir and direct it against the USSR, they have decided on an obvious exaggeration, portraying a serious but obviously local accident as a global nuclear disaster," wrote Arbatov.[19]

When Gorbachev addressed his country and the world in his first and last Chernobyl speech on May 14, he picked up and developed many of Arbatov's points. Like Arbatov, he thanked Drs. Gale and Terasaki. He also mentioned the "objectivity" shown by Hans Blix but condemned the "unbridled anti-Soviet campaign" unleashed in the West, especially in the United States and West Germany. He attacked the statement issued in Tokyo by the leaders of the G-7 countries and laid out plans for increasing the role of the agency led by Blix, and he promised that a full report on the accident would be presented to the conference that was to be organized by Blix's agency. He also called on Ronald Reagan to meet with him, possibly in Hiroshima, to sign a treaty banning nuclear tests. Gorbachev was eager to turn the tables on the Americans by throwing Hiroshima into the debate over the handling of Chernobyl.[20]

Whatever the outcome of Gorbachev's counteroffensive in the West, at home his speech scored him very few political points. His interpreter, Pavel Palazchenko, who watched Gorbachev's address, admitted that he was in a difficult position, trying not to downplay the disaster while saying nothing that might arouse panic. The result was anything but the one desired by Gorbachev and his speech-writers. Moscow was "close to panic," recalled Palazchenko. "The city was rife with rumors, and few people believed the official version of events. . . . Government-run media was minimizing the disaster, both out of habit and fear of causing even greater panic. But the mood in Moscow was gloom and often anger. It was a mood of distrust of the authorities. In retrospect, I think, it [Gorbachev's address] caused a rift between the people and the government that never closed."[21]

If Gorbachev was distrusted in Moscow, in Kyiv people were angry, and many believed that it was the end of his career, which had begun with such promise. Among other things, distressed Kyivans resorted to black humor. Velikhov, who had figured prominently in the Blix propaganda coup, was greeted in Kyiv by Ukrainian colleagues with a new joke. A Chernobyl man and a Kyivan meet in heaven. "What brought you here?" asks the Kyivan. "Radiation," answers the Chernobyl man. "And what about you?" "Information," responds the Kyivan. Gorbachev was denying information not only to the world but also to his own people, as they knew better than anyone else, partly because of Western broadcasts.[22]

But Gorbachev would not give up. On the day after his television address, he met with Dr. Robert Peter Gale, whom he had praised in his speech, and the prominent American businessman Armand Hammer, a champion of improved relations with the USSR, who delivered American medication for the victims of Chernobyl. Hammer had begun his business dealings with the Soviets in the days of Vladimir Lenin and had actually met the founder of the Soviet state—a detail never lost on the Soviet media. Now the Soviets reported that Hammer had asked Gorbachev about the possibility of a summit meeting with Reagan, which had been discussed during their first meeting in Geneva in December 1985. Gorbachev said that he was interested in a summit on two conditions: it would have to yield tangible results, and the political atmosphere would have to be right. Soviet reporting left no doubt that "political atmosphere" meant the end of what Gorbachev had called "vicious anti-Soviet propaganda" in his address. The West was supposed to stop questioning his government's treatment of the Chernobyl accident.[23]

ON MAY 15, the day of his meeting with Gorbachev, Robert Peter Gale gave a press conference at which he provided figures of actual and potential victims of the disaster that dwarfed those cited by Soviet authorities, including Gorbachev himself. Gale followed Gorbachev in saying that 9 people had died up to that point, while 299 individuals were currently hospitalized with different degrees of radiation sickness. But he had other figures as well. Thirty-five people were in critical condition, according to Gale, and he and his team had operated on 19 of them. The American doctor predicted that the number of those affected by radiation poisoning might reach 50,000 to 60,000. He appealed for more drugs and equipment on top of those already brought to Moscow by Hammer. He was joined at the press conference by a Soviet colleague and announced their agreement to conduct joint research and produce a joint publication on the results of their work.[24]

The press conference was a propaganda success for the Soviets. They had shown the world, including those most affected by the disaster, that they were open and had nothing to hide. The actual outcome of Dr. Gale's efforts to help the Chernobyl victims was harder to assess. Before the end of the month, Gale gave another press conference in Moscow, stating that the death toll now stood at twenty-three. More disturbing for Gale and the American-Soviet cooperation effort was a statement by the Soviet deputy minister of health, Academician Yevgenii Chazov, who said that eleven of the bone-marrow-transplant patients had died. Later, the leading Soviet radiation sickness expert, Dr. Angelina Guskova, stated that bone-marrow transplants were doing more harm than good. Gale had assisted with those surgeries, and he was now fighting for his reputation. "A marrow transplant can only prevent you from dying of bone-marrow failure—it cannot prevent you dying from burns or radiation damage to the liver," he argued.

Gale estimated the success of his operations at 90 percent. This and similar statements of his were met with skepticism both in the Soviet Union and in his own country. But all that would come later. Whatever the practical outcome of his operations, at the critical moment in the East-West propaganda war over Chernobyl, Dr. Gale showed the Soviets that the Americans were there to render assistance, while helping the Soviets change the tone of the Chernobyl discussions in the international arena. At the time, Gale was a messenger of hope in a world divided by Cold War rivalries.[25]

SOVIET LEADERS were surprised by the international reaction to their cover-up of the accident and its consequences in the first hours, days, and weeks after it took place. In the United States, President Reagan created a special presidential task force on Chernobyl, and the White House press secretary, Larry Speakes, reported almost daily on its findings. Members of the administration attacked the Soviet government for its failure to release information in a timely manner.

Legitimate environmental and health concerns aside, the West was ready and even eager to engage in a new propaganda battle with the Soviet Union. "Economic issues generate nothing but yawns at home," stated a Reagan administration official who helped draft the radio address in which the president first raised the Chernobyl issue—almost two weeks ahead of Gorbachev.[26]

The Soviets fought back, trying to regain control of the Chernobyl narrative—the politically all-important story of who knew what and when, and what had been done about it. Cold War rhetoric served the Soviet regime well in its early attempts to mobilize the population and distract it from internal problems and economic hardships. Close to one-third of all Soviet media coverage of the Chernobyl accident during the first month after the disaster was dedicated to attacks on the West. Soviet propagandists were happy to point out inaccuracies and exaggerations in the first Western reports on the accident—inaccuracies caused by the Soviet information blockade. Gorbachev used the occasion to push for a nuclear-test ban—part of his foreign-policy offensive aimed at easing international tensions and freeing the struggling Soviet economy from the burden of continuing the arms race. But in the clash of Chernobyl narratives, Soviet authorities realized that they were losing the contest, and decided to loosen the grip of censorship over the Soviet press.[27]

Pressure from the West and the Soviet public's demands for accurate information had a major impact on Gorbachev's policy. Soviet journalists were suddenly granted access to people in the nuclear industry with whom they could not have hoped to speak earlier. The regime of secrecy was crumbling, and the era of glasnost, or openness, which would become a hallmark of Gorbachev's reforms only a few months later, was beginning. At the Harriman Institute for Advanced Study of the Soviet Union at Columbia University, Dr. Jonathan Sanders, who would later spend years as a CBS correspondent in Moscow, launched a new project, the Working Group on Soviet Television, using new technology to tape Soviet television broadcasts. He wrote in a conference paper that "coverage of the Chernobyl disaster marked a turning point in the history of Soviet communications. For the first time, television . . . began to meet people's demands for 'bad news,' to abandon the silence about domestic disasters."[28]

The turning point would prove critical for the development of the Soviet media, Soviet-American relations, and the incipient collapse of the USSR. The Soviet Union was living out its final years. There would be a great deal of bad news to come, and, after Chernobyl, no way for the Soviet regime to hide it from its own people and the world.

SARCOPHAGUS

YEFIM SLAVSKY, the minister of medium machine building, eighty-eight years old, but still standing tall, arrived in Chernobyl on May 21, almost a month after the explosion. He was considered the uncrowned king of the Soviet nuclear program. The reasons for his delayed arrival were numerous, but they weren't his fault. The Chernobyl accident had taken place at a plant not operated by his ministry, but everyone knew that the exploded reactor was his brainchild and that of the academic institutes he had helped to build and fund—a virtual empire that employed tens of thousands of civilian and military personnel. The Chernobyl-type RBMK reactors had first been built, run, and promoted by Slavsky's ministry. After the accident, many in the party and government tried to distance themselves from the once all-powerful minister, but ultimately they had no choice but to turn to him. Slavsky's experience in dealing with nuclear emergencies, coupled with the enormous human and technological resources his ministry commanded, were needed now as never before.

On May 15, the Politburo had put Slavsky and his ministry in charge of "burying" the exploded reactor, which had to be sealed permanently to stop the further spread of radiation. It was up to Slavsky to decide how to do it. He took charge immediately. Five days later, he created a special construction directorate within his ministry to deal with the problem and appointed a general to head it. The engineers and architects came up with a number of possible solutions. One of them called for burying the reactor under a mound of sand, concrete, and balls of metal. There were also proposals to erect an arch or an umbrella-like structure above the reactor. Eventually it was decided to build a concrete structure incorporating foundations, walls, and other elements of the reactor building that remained intact after the explosion. Time was of the essence—the Politburo wanted the reactor buried within four months, and building a protective structure that utilized the remaining parts of Unit 4 was the fastest way to deal with the problem. Officially, the new building above the reactor was referred to as the "shelter." Unofficially, it became known as the "sarcophagus." Slavsky became the main architect, priest, and undertaker of the concrete coffin.[1]

Coming up with quick, cost-efficient, and almost always temporary fixes for complex problems—achieved with the help of invariably limited technical resources and usually unlimited human ones—had been the essence of Slavsky's whole career and of the Soviet nuclear industry in general since its inception. There could have been no better choice than Slavsky to build the sarcophagus that would put not only a damaged reactor but also a whole epoch in the development of the Soviet nuclear energy program to rest. His first experience in dealing with the consequences of a nuclear accident had come in 1957 at the Maiak (Beacon) military nuclear plant in the closed city of Ozersk. At the time, Slavsky was just beginning his ministerial career, having been appointed barely two months earlier. It was on his watch that the Soviet method of fighting radioactive contamination—covering the affected areas with a thick layer of concrete—was born. Almost thirty years later, in Chernobyl, that was still the default solution.

In early June 1986, the Politburo approved plans for the construction of the sarcophagus, which was designed by a group of architects and engineers in Leningrad led by Vladimir Kurnosov. Slavsky now mobilized all the academic, industrial, and military cadres under his command. It was a military-style operation, with Slavsky as commander in chief. He was always eager to make an appearance on the front lines. Having dealt with numerous nuclear accidents without losing his unique capacity for work, the aged minister dismissed the negative effect of "small doses" of radiation. On May 21, his first day at the Chernobyl plant, he had flown over the damaged reactor on a helicopter and then approached the remains of the fourth unit on foot. He walked with two of his aides to the third reactor building, telling them, "We'll have a drink afterward, and it will all pass. But we have to take a good look and figure out what's going on here." Slavsky's subordinates recalled that the levels of radiation were "crazy" and that on walking up to the damaged reactor, Slavsky told his aides to stay behind: "I'm an old man and have nothing to fear, but you are still young."[2]

Slavsky's personnel divided the construction site into twelve sectors, each run by one of the many construction firms within his huge empire. Whole towns emerged around the damaged unit; new roads and railway lines, as well as entire concrete-producing plants, were built in the vicinity. First, as in Ozersk back in 1957, they covered the highly contaminated areas around the reactor with concrete, thereby turning them into relatively safe construction sites. Even so, trucks bringing concrete to the reactor had to be unloaded behind concrete walls, where the radioactivity level reached 50 roentgen per hour. Orders were issued to the major machine-building plants in Ukraine and the rest of the Soviet Union for new equipment and elements of the sarcophagus structure designed by Slavsky's engineers. The Ukrainian authorities already had their hands full, helping to fulfill the government commission's orders for people, materials, and equipment. Italian equipment was brought in to build the foundations of the sarcophagus, while powerful pumps from West Germany were used to supply concrete in order to build the walls sealing off the reactor.[3]

The first cadres whom Slavsky sent into battle, after doing the reconnaissance himself, were military men. General Yurii Savinov, a member of Slavsky's advance group, spoke of his task as one of preparing for a landing by a military unit with orders to defeat a new and invisible enemy—radiation. The military performed two functions: decontamination and construction. By early June, a total of 20,000 officers and men, mostly reservists, had been organized into construction battalions. The fact that they were being assigned to Chernobyl was concealed from many of them. Those who knew where they were going were often promised salaries five times higher than usual. Although the promises proved empty, the recruits worked with discipline and devotion. The only protests registered by the KGB had to do with overexposure to radiation. On June 2, 200 reservists refused to eat meals after the battalion commander and commanders of two companies, having sustained the maximum dose of 25 roentgen, left their units, but 170 soldiers who had already been exposed to that dose stayed in the area.[4]

Overexposure to radiation remained an issue until the very end of the building of the sarcophagus. Those who approached the reactor first had to deal with radiation levels ranging from 5 to 370 roentgens per hour. But Slavsky pushed on, and his generals and managers delivered results. By July 5, they had cleaned 800,000 square meters of territory around the nuclear plant and 24,000 square meters of building surfaces with special solutions. Twenty-six construction battalions comprising 80,000 people with 9,000 pieces of machinery and equipment took part in the first stage of the construction of a concrete wall 6 meters thick around the ruins of the reactor to allow relatively safe access to the area. By the end of July, they had built the foundations of the future sarcophagus. Fifteen thousand square meters of concrete had been poured, and it was estimated that 300,000 tonnes would be required to complete construction.

But not everything went as planned. Slavsky's pet project of covering the reactor with an 8-tonne aluminum cupola, to be lowered onto the freshly constructed walls of the sarcophagus by helicopter, went awry. As the helicopter transporting the cupola approached the reactor, the cupola dropped from the cable that secured it. "The flight took place at a height of 400 meters and a speed of 50 km per hour," read a KGB report describing the accident. "The 'cupola' fell to the ground and was shattered as a result." Fortunately, it did not hit the reactor or anyone on the ground. Rumor had it that Slavsky crossed himself and said, "Glory to God." The idea of trying again was abandoned on the spot. The ceiling of the sarcophagus would be built of concrete blocks, like the rest of the structure.[5]

Slavsky's designers, engineers, and military commanders, as well as reservists mobilized by military commissars from all over the

Soviet Union, worked in shifts, the first one lasting from mid-May to mid-July. A new shift arrived in mid-July and remained in place until mid-September. The third and last shift completed the construction of the sarcophagus in mid-November, only two months past the unrealistic deadline ordered by the Politburo in mid-May. By that time, approximately 200,000 workers had labored at Slavsky's construction site, building a 400,000-tonne concrete sarcophagus that sheltered their country and the world from raging radiation levels emanating from the damaged reactor.[6]

Slavsky would come to the sarcophagus construction site every two weeks to check on its progress. The Chernobyl power plant was only one of his numerous battlefields. Another, no less important, was the Kremlin, to which Slavsky was invited on July 3 for a Politburo meeting convened to look into the causes of the Chernobyl accident, draw conclusions from it, and punish the guilty. Who was responsible for the technological disaster of biblical proportions— the personnel in charge, who had ruined a supposedly perfect reactor through criminal disregard of rules and procedures, or the designers of the reactor in Slavsky's nuclear empire, which included the Kurchatov Institute? Depending on the answer to that question, Slavsky's position as head of the ministry, his reputation, and, most important, his legacy would either be reaffirmed or trashed. Also at stake was the future of RBMK reactors and the Soviet nuclear energy industry as a whole.

Slavsky was convinced that his subordinates could not possibly be at fault. When he had first heard of the accident, he had dismissed it as the doing, and, consequently, the problem, of a different ministry—that of energy, which operated the Chernobyl plant. Scholars at the Research and Development Institute of Power Engineering—directed by Nikolai Dollezhal, the designer of the Chernobyl-type reactors—which was also part of Slavsky's vast nuclear empire, blamed Ukrainian specialists for the accident. "The *khokhly* exploded the reactor," claimed one of the leading scientists upon hearing the shocking news of the explosion. He used a derogatory term to refer to the Ukrainians, in this case the managers and operators in Ukraine, where the Chernobyl plant was located. The fact that Slavsky, Dollezhal, and the director of the Kurchatov

Institute, Anatolii Aleksandrov, were all either Ukrainians themselves or came from Ukraine was of no significance. The thrust of the accusation was institutional rather than ethnic: Slavsky and his colleagues were seeking to deflect blame from his ministry and the Moscow institutes to the cadres in the periphery.[7]

The working group put together on April 29 by Boris Shcherbina, the first head of the government commission to investigate the causes of the disaster, was led by Slavsky's deputy, Aleksandr Meshkov, and consisted largely of representatives of Moscow research institutes—Dollezhal's institute, which had designed the reactor, and Aleksandrov's Kurchatov Institute, which had provided scientific support for the project. The working group began with six possible scenarios, but by May 2 it had pretty much decided on one: the reactor had exploded in the course of the turbine test as a result of violations of technical procedures by the operators of the plant.

That became the nuclear establishment's official line. Those holding a different opinion largely kept it to themselves. "The reactor exploded because the control rods were dropped during the emergency shutdown," one member of the group, Aleksandr Kalugin, quietly told another member, Valentyn Fedulenko, on April 29, the day on which the group began its work. He meant that the explosion had resulted from the sudden spike in power output caused by the lowering of the control rods—a scenario predicted in a paper circulated among the nuclear scientists sometime before the explosion. That explanation, which indicated that the designers were responsible, or at least partly responsible, for the accident, never had a chance of acceptance among the scientists representing institutions that had designed the reactor. In mid-May, they reported to the president of the Academy of Sciences, Aleksandrov, that the operators' procedural violations were solely responsible for the accident. Aleksandrov, the scientific director of the reactor, approved that explanation.[8]

The government commission headed by Boris Shcherbina took a similar approach. Shcherbina did not entirely dismiss problems involving the design of the reactor, but in his report to the Politburo, which discussed the matter at its meeting of July 3, he assigned primary responsibility to the operators of the reactor. "The accident took place as a result of the grossest violations of technical regulations on the part of the personnel in charge and in connection with serious flaws in the construction of the reactor," read Shcherbina's report. "But those reasons are not equally significant. The commission finds that errors made by the personnel in charge were the basic reason for the accident." That conclusion became the official line endorsed by the Politburo and fed to the domestic and foreign media, as well as to the global scientific community.[9]

The former director of the Chernobyl nuclear plant, Viktor Briukhanov, who had been dismissed from his position in late May, was the first to feel the brunt of the new party line. In early July, when he was invited to Moscow to field questions from the Politburo, Briukhanov was still depressed and, as he later recalled, his attitude was one of indifference. Nevertheless, he remembered the setting quite vividly: there was a gigantic table in the Walnut Room of the Kremlin. With the eye of an engineer who had spent much of his career in construction, Briukhanov estimated that it was about 50 meters long and 20 meters wide. At the head of the table sat Mikhail Gorbachev, with Politburo members to his left and right. The meeting lasted from 11:00 a.m. to 7:00 p.m. without a lunch break, although at some point waiters brought in sandwiches and drinks. Briukhanov was the third speaker on the agenda. He spoke for approximately fifteen minutes, describing what he thought had happened at the plant on April 26. Gorbachev asked him only one question: Did he know what had happened at the Three Mile Island power plant in the United States in 1979? Briukhanov responded that he did. No one asked Briukhanov anything else. They thought they knew what had happened and what his role in those events had been. He was there as a scapegoat.

After his presentation, Briukhanov sat down and listened to the rest of the discussion, which went on for hours. At the end of the proceedings, Gorbachev read a prepared draft resolution that proposed Briukhanov's expulsion from the Communist Party. The Politburo voted unanimous approval. Briukhanov, who had previously attended meetings of the Ukrainian Central Committee conducted by the tough boss of the republic, Volodymyr Shcherbytsky, found Gorbachev's manner fairly weak. He would later call the general secretary a spineless individual (*triapka*), an epithet often applied to Briukhanov himself by colleagues and subordinates. But ultimately Briukhanov was relieved that no one at the Politburo meeting tried to humiliate him, as had often happened during his directorship in high party offices, where party secretaries had threatened to "hang him by the balls" as they demanded the fulfillment of plan quotas.

No one now said anything of the kind to Briukhanov, but the media made him the primary culprit responsible for the disaster. The main Soviet television news program, *Vremia* (Time), announced his expulsion from the party to the whole country. Everyone knew what that meant: the start of criminal prosecution, which would land him in prison. Such were the unwritten rules of Soviet justice—expulsion, inevitably followed by imprisonment. In faraway Tashkent, Briukhanov's hometown, his brother would not allow their elderly mother to watch television, but neighbors told her what was going on. She suffered a fatal heart attack.[10]

As far as the outside world was concerned, Yefim Slavsky and the president of the Soviet Academy of Sciences, Aleksandrov, emerged unscathed from the ordeal, but behind the closed doors of the Politburo, the situation looked different. "What remains in my memory are strong impressions of general confusion: no one knew what to do," recalled Aleksandr Yakovlev, a key adviser to Gorbachev and an architect of perestroika. "The people responsible for that sphere—Minister Slavsky and President Aleksandrov of the USSR Academy of Sciences—said something incomprehensible. At one point there was an amusing exchange between them: 'Do you remember, Yefim [Slavsky], how many roentgens we got in Novaia Zemlia? And there's no problem: we're alive.' 'Of course I remember. But then we had a liter of vodka each.'" They discussed their exploits at the Novaia Zemlia archipelago in the Arctic Ocean, which had been a Soviet nuclear test site since 1954.[11]

While the two octogenarians recalled the good old days, Gorbachev demanded an answer to a simple question: Was the reactor reliable or not? The answer was critical to the future of the Soviet nuclear industry and his own perestroika plans. If all RBMK reactors were to be decommissioned—there were twelve of them in various parts of the country, not counting the damaged one in

Chernobyl—then the economic reforms Gorbachev was dreaming of would have to be suspended and alternative sources of energy sought. The Soviet treasury was already empty. Where could money be found to cover the losses generated by the Chernobyl catastrophe, to say nothing of decommissioning the rest of the RBMK reactors, which were responsible for generating 40 percent of all the electrical energy produced by Soviet nuclear plants? No one, including Gorbachev, knew how much it would cost to deal with all the different aspects of the disaster. Decades later, Belarusian economists estimated the overall price tag for their republic alone at US$235 billion, or thirty-two annual budgets of Belarus as of 1985.[12]

Nevertheless, Gorbachev pushed for an answer. He wanted to hear the opinions of nuclear scientists from Slavsky's ministry and institutes, but they were silent or evasive. Eventually, Gorbachev gave the answer himself: "The personnel are responsible for the fact that the accident took place, but the scope of the accident is due to the physics of the reactor." He asked Slavsky's subordinates whether RBMK reactors could still be built and operated. Slavsky's deputy, Aleksandr Meshkov, responded in the affirmative: "They can, if regulations are strictly adhered to." Gorbachev was not satisfied. "You surprise me," he told Meshkov. "All that's been collected about Chernobyl to date leads to a single conclusion—the reactor must be condemned. It's dangerous. But you are defending the honor of your uniform." Meshkov shot back: "No, I'm defending atomic energy." Gorbachev was quick with his own reply: "But which interests take precedence? That's the question we have to answer. That's what millions of people here and abroad demand of us."[13]

After reading the report on the causes of the accident, Gorbachev continued his attack on Slavsky's ministry. "But Meshkov is blaming everything on the personnel in charge. Where does such a disaster leave you? If we agree with you, what then? Continue as before? Everyone is out of step but Meshkov? In that case, better to get rid of Meshkov." Everyone knew that the attack was really directed against Slavsky, who tried to defend his deputy and himself. "The explosion was manmade," he told the Politburo. "The reactor is fine, with a long service life. But what did they do? After all, the experiment was carried out by a local engineer who had no right to do so."

Gennadii Shasharin, a deputy minister of energy responsible for the Chernobyl power plant, was fully on board, glad to shift as much blame as possible onto the designers. He told the Politburo: "The physics of the reactor determined the scope of the accident. People were unaware that the reactor might speed up in such a situation." He added: "The personnel are responsible for the accident. But its scope, I agree, is in the physics of the reactor." Shasharin was in favor of shutting down all existing RBMK reactors, as he could not guarantee their safety. That was a line Gorbachev was not prepared to take. "Shasharin's statement—to shut down atomic energy stations—is not serious," he said. Still, realizing that RBMK reactors were unsafe, he sought a compromise whereby they could be made safer instead of being shut down and decommissioned. "A cover?" he asked those present, referring to the concrete container around the reactor—a mandatory safety feature at all US reactors that was absent in Chernobyl. But building such containers was a very costly enterprise that the Soviet treasury could not afford, as Gorbachev was well aware. Probably for that reason, he immediately shot down his own proposal: "They say that if there had been a cover in Chernobyl, the emission would have been even worse."

The Central Committee secretary in charge of nuclear energy, Vladimir Dolgikh, also grappled with the question of what to do with existing RBMK reactors. "Reconstruction of the reactor from scratch makes it uneconomic," he said in response to Anatolii Aleksandrov's proposal that the reactors could be improved. "We are threatened with a huge loss of energy. After all, there are ten reactors [of the Chernobyl type] in Comecon [Council for Mutual Economic Assistance countries in Eastern Europe] and ten in our country. They are all out of date and dangerous." The party ideological watchdog, Yegor Ligachev, was in favor of reducing dependence on nuclear energy: "The structure of atomic energy must be changed completely. At present, irresponsibility is inherent in the very structure. Find alternative sources. Step on the gas!"

That Politburo meeting ended with the conclusion that the entire nuclear industry was badly in need of serious reform. While the operators were blamed in public, questions were raised privately about the safety of RBMK reactors in general. "Determine the type of reactor. Abolish the Chernobyl type," said Prime Minister Nikolai Ryzhkov to the Politburo. In drawing conclusions from the discussion, Gorbachev told his colleagues: "The [draft of the Politburo] resolution must be revised with regard both to energy output projections and the balance between atomic energy stations, gas, oil, hydroelectric stations, and coal. . . . The government must revise the program for the development of the energy sector up to the year 2000. We must consider whether it would not be worse to continue using atomic energy stations than to close them." That was anything but good news for Slavsky and his empire. But there was more: many held him indirectly responsible for what had happened at Chernobyl.

"We have come up against the super-secret character of the Ministry of Medium Machine Building," asserted Dolgikh, referring to the lack of external controls on Slavsky's nuclear empire. "The authority of Slavsky and Aleksandrov has become too great," said Ryzhkov, echoing Dolgikh. He was determined to dismember Slavsky's realm, proposing that "a ministry of atomic energy be created. Part of the Ministry of Medium Machine Building should be assigned to it. An interdepartmental council should be established, not under Slavsky but under the Academy of Sciences or the State Committee for Science and Technology or, better, under the Council of Ministers." When Gorbachev read the list of people responsible for the disaster and reprimanded in one way or another by the party—at the top was Briukhanov—Slavsky's name was missing. Gorbachev suggested that "Academician Aleksandrov be made aware of his responsibility in this whole business," and that Slavsky's deputy, Aleksandr Meshkov, be fired. Slavsky himself was being spared, for the time being: they still needed him to deliver the sarcophagus.[14]

THE POLITBURO battle pitched Gorbachev, his aides, and members of the Politburo against the nuclear scientists responsible for the design and construction of the reactor, who closed ranks around Slavsky. There was just one defector from their camp—Valerii Legasov, the chief scientific adviser to Boris Shcherbina's commission. At the

Politburo meeting, Gorbachev often addressed his questions about the reactor to Legasov instead of to the "father" of the Chernobyl-type reactors, Legasov's boss, Anatolii Aleksandrov. "Has the commission established why the unreliable reactor was approved for production? That type of reactor was rejected in the USA. Right, Comrade Legasov?" asked Gorbachev, looking for support among the nuclear scientists attending the meeting. Legasov responded that the United States had never produced or operated Chernobyl-type reactors. "The reactor fails to meet safety requirements according to the most important parameters," he told the Politburo. "In Finland in 1985, physicists . . . gave our atomic energy station high marks. But before doing so, we removed the automatic and technological components, substituting Swedish-American ones."[15]

Legasov later recalled Nikolai Ryzhkov's statement at the meeting to the effect that the Chernobyl accident was anything but random, as the Soviet nuclear industry had heading toward it it for a long time. Legasov was prepared to seek the causes in his own nuclear industry—the object of primary loyalty for his colleagues at the Kurchatov Institute. As far as they were concerned, Legasov, as deputy director of the institute, was there to defend its interests and those of the industry. But as a loyal communist and believer in the Soviet system, he put the interests of the system above those of Slavsky's nuclear empire. Many suspected him of careerism. He sided with the Politburo against his own people, divulging their internal secrets. They would never forgive him for that "betrayal."

In early July, Legasov, who had sustained a high dose of radiation during the first weeks after the explosion, was back in Moscow, working on his own document analyzing the causes of the explosion. Back in May, the Soviet government had promised Hans Blix, the head of the International Atomic Energy Agency (IAEA), a report on the accident at an international conference organized by the agency. The conference was scheduled for late August in Vienna, and the Soviet government appointed Legasov to head a commission responsible for drafting the report. He embraced the task with his usual enthusiasm, gathering a group of experts ranging from nuclear physicists to health and ecology experts, turning his own apartment into the commission's drafting room, and working on the report around the clock.

The forthcoming conference was getting a lot of attention in the West. The European political and scientific elites were extremely frustrated with the behavior of the Soviet government, which was slow in releasing any information about the accident, thereby jeopardizing the safety of the Central and Western European population. They were also extremely skeptical that the Soviets would say anything meaningful at the conference. Legasov, who was well aware of those attitudes, called in one of his advisers, Aleksandr Borovoi, and, swearing him to secrecy, showed him the draft program and resolution of the forthcoming conference, where Legasov was scheduled to speak for a mere thirty minutes.

"They presumed that in its report on the Chernobyl accident the Soviet Union would not say anything concrete," recalled Borovoi. "Since those reactors belong to the military type, everything would be kept secret, and the report would take only half an hour. After that, speeches were scheduled, the content of each stated in a phrase or two. It ended with a draft resolution of the IAEA for the Soviet Union to shut down all its RBMK-1000 atomic reactors, pay huge reparations to countries affected by the radioactivity, and ensure the presence of foreign observers at every atomic reactor in the Soviet Union." Legasov was eager to derail those plans. "We'll have to overcome that," he told Borovoi.[16]

Legasov and a group of his hand-picked advisers sat working on a comprehensive report that would contain a detailed chronology of the accident and its consequences. It was impossible to do so without talking about the construction of the reactor—a top-secret subject in the Soviet Union. As was to be expected, Slavsky and his aides refused to permit the release of such information to the international academic community, putting Legasov in a difficult position. At the Politburo meeting of July 3, the minister of energy, Anatolii Maiorets, had noted the absurdity of the situation caused by the outdated demand for secrecy. "It's apparent from foreign sources that the Chernobyl accident has already been modeled there," he told the Politburo. "So what should we do—present lies to the IAEA?"[17]

Legasov, who could not have agreed more, went directly to Premier Ryzhkov, who authorized him to proceed with the report. It would include not only information on the design of RBMK reactors, but also estimates concerning the amount of radioactivity released and its impact on agriculture and human health. Legasov was prepared to talk about everything. The Soviet report was 388 pages long, and Legasov got permission from the government to bring along experts on nuclear reactors who had been prohibited from leaving Soviet territory: this would be their first trip abroad. They were to answer specific in-depth questions dealing with their area of expertise. When Blix's aides asked the Soviet embassy in Vienna about the projected length of the Soviet presentation at the conference, they expected a report lasting about half an hour. They were told that the Soviet representative would be speaking for four hours. As it turned out, his presentation took even longer.[18]

The Vienna conference began on August 25. Legasov began his report with a discussion of the design of the reactor and a description of the Chernobyl power station. He continued with a description of the accident, an analysis of its causes, and a description of the impact, concluding with recommendations on how nuclear accidents could be predicted in the future. The report blew open much of the carefully designed shield of secrecy covering Soviet nuclear programs. The audience, consisting of close to six hundred nuclear scientists representing twenty-one international organizations and sixty-two countries, as well as two hundred journalists, was stunned. Legasov's presentation was greeted with a standing ovation.

"No one who attended the first day's session will soon forget it," according to a report on the conference in the *Bulletin of the Atomic Scientists*. The author of the report continued: "On August 25 the conference mood was bleak and tense; by August 29 [the last day of the conference] it had become cheerful, convivial, verging on the euphoric." Legasov became an instant celebrity, hailed by the Western media as one of the world's top ten scientists. His apparent openness about the causes and consequences of the Chernobyl disaster achieved the impossible, changing the image of the Soviet Union from that of an irresponsible perpetrator to that of a victim of unpredictable circumstances, open to sharing its experience and cooperating with the rest of the world to prevent similar accidents in the future.[19]

Despite such unprecedented openness, both about the Chernobyl nuclear plant and the Soviet nuclear industry in general, Legasov stuck to the party line in his Vienna report, blaming the reactor personnel for the accident. According to his report, "The basic reason for the accident was the extremely unlikely combination of violations of procedure and mode of operation that the personnel of the energy station allowed themselves."[20]

This was the "blame the personnel" line taken by Slavsky and his deputy, Meshkov, at the Politburo meeting the previous month. But many in Moscow's nuclear establishment—and, increasingly, in higher party circles—believed that Legasov had divulged too much information about the industry and immediately communicated their displeasure. Aleksandr Borovoi, one of the coauthors of the report, met Legasov when he entered his institute upon returning from Vienna. "Victory!" shouted Legasov to Borovoi, running up the stairs to his third-floor office. In high spirits, he left to meet with the Soviet leadership. Borovoi saw him again upon his return to the institute a few hours later. Legasov's demeanor had changed, his euphoria completely gone. "They understand nothing and even failed to grasp what we managed to accomplish," he told Borovoi in despair. "I'm going on leave."[21]

It is not clear whom Legasov met upon his return from Vienna, but there is little doubt that the country's highest officials, including Gorbachev himself, believed that he had pushed the glasnost (openness) envelope too far. At the July 3 Politburo meeting, Gorbachev had told his colleagues: "There are no interests that might compel us to hide the truth. It is our duty to all mankind to render complete conclusions." By early October, Gorbachev was breathing more easily. Not without satisfaction, he informed the Politburo that "since the meeting of member nations of the IAEA, Chernobyl has ceased to be an active element of anti-Soviet propaganda." Legasov had scored a major propaganda victory for the regime, but it was not appreciated by the leadership. Responsibility to humankind clearly did not entail the need to inform the world community of all that Legasov knew about the accident.[22]

Many expected that on September 1, 1986, his fiftieth birthday, Legasov would be given the highest Soviet wartime award, Hero of the Soviet Union, for his work in Chernobyl. He was denied it, as well as the highest peacetime award, Hero of Socialist Labor. Instead, he was presented with a Soviet-made watch—an obvious insult, given the expectations that he and many others had at the time. Clearly, Legasov had little support at the top of the Soviet power pyramid. He had every reason to feel betrayed. He had supported the political leadership against his own academic institution and industry, only to be repudiated by both for saying in public what he believed absolutely necessary to tell the world about the causes of the Chernobyl accident.

RUMOR HAD it that Slavsky, still running his nuclear empire, was opposed to bestowing the highest Soviet award on Legasov. If that was so, then it was Slavsky's last victory. By the fall of 1986, clouds were gathering on his horizon. Slavsky's closest ally, Anatolii Aleksandrov, stepped down as president of the Academy of Sciences in October (he had asked for that at the Politburo meeting in July, taking his part of the responsibility for what had happened in Chernobyl). Slavsky, who never accepted such blame, was forced out of his position at the top of the all-powerful ministry in the following month.[23]

The government commission deemed the sarcophagus ready on November 30. A few days earlier, Slavsky was visiting the construction site at the Chernobyl power plant when he received a call from Premier Ryzhkov, who asked him to come to Moscow the following day. When Slavsky responded that he was too busy overseeing the completion of the sarcophagus, Ryzhkov gave him one more day. "They are cooking something up," Slavsky told a subordinate who had overheard his telephone conversation with Ryzhkov.[24]

The meeting in Moscow lasted three hours. Ryzhkov assured Slavsky that he was satisfied with his work, but, given Slavsky's age, thought it would be good for him to retire. Slavsky, who dreamt of making history by staying in ministerial office until the age of one hundred, resisted to the very end. On leaving Ryzhkov's office, he asked the secretary for a piece of paper and wrote on it, in his usual blue pencil, "Please discharge me, as I am somewhat hard of hearing in the left ear." It was a sign of defiance, if not of hope that the premier would not sign a letter citing such a ridiculous cause for retirement. Slavsky did not hide his low opinion of the new leadership and its political course. He thought that his ministry needed no restructuring. As Slavsky saw it, he and his people had been outperforming all others without perestroika because he knew how to work better than anybody else. Slavsky's model of the Soviet economy was a militarized one. He saw no benefit in Gorbachev's reforms and despised his foreign initiatives aimed at easing East-West tensions. Not until a few weeks later was he convinced by his aides to write a proper letter of resignation.[25]

The era of the militarized economy was over. It had not only produced the Chernobyl disaster but had also been mobilized to clean up its consequences. In retirement, Slavsky would remember the good old days and recite his favorite poet's verses from memory. Quoting with gusto the father of the Ukrainian nation, Taras Shevchenko, the Romantic bard who glorified the pastoral beauty of Slavsky's homeland, he would intone: "A cherry orchard by the house. / Above the cherries, beetles hum. / The plowmen plow the fertile ground / And girls sing songs as they pass by. / It's evening— mother calls them home."[26] Back in the early 1960s, Slavsky had strained his influence to the utmost to name a new city built around a uranium mine in Kazakhstan after his favorite poet. His patriotism was as much Soviet as it was Ukrainian. He saw no difference between the two.

The Chernobyl disaster had destroyed the pastoral world celebrated by Shevchenko and remembered from childhood by Slavsky. The cherry orchards of northern Ukraine and parts of Belarus and Russia were now beaming radioactivity into the atmosphere, destroying life around them. There is no indication, however, that Slavsky ever felt himself or his industry to be responsible for the accident. He was prepared to take risks and deal with their consequences. Once before the Chernobyl accident, someone had asked him what would happen in case of a meltdown of a reactor core: "It will be very, very bad, but we'll manage even that." He did, indeed, take care of what had happened at Chernobyl, but at tremendous cost.[27]

The Soviet nuclear lobby hoped that the sarcophagus would bury not only the damaged reactor but also doubts about the whole nuclear program. Although party and government leaders were skeptical, they publicly accepted the narrative promoted by the nuclear lobby. The accident was blamed entirely on plant personnel. With the sarcophagus complete, Slavsky removed from his position at the top of the Soviet nuclear empire and its restructuring under way, and Aleksandrov retired as president of the Academy of Sciences, the government was ready to go after Viktor Briukhanov and his subordinates, who as far as the public was concerned was supposed to take all the blame for what had happened at the Chernobyl power plant on the night of April 26.

CRIME AND PUNISHMENT

ANATOLII ALEKSANDROV, the eighty-three-year-old veteran of the Soviet nuclear industry and chief academic adviser to the creators of the Chernobyl-type RBMK reactors, resigned as president of the Soviet Academy of Sciences on October 16, 1986. He was also ready to leave his post as director of the Kurchatov Institute of Atomic Energy. The all-important first page of the history of the Soviet nuclear program had been turned. The next page depended on who would succeed Aleksandrov as director in charge of 10,000 scholars and employees.

Aleksandrov looked to his first deputy at the institute, Valerii Legasov, as a possible successor, but others thought differently. The battle for the directorship began soon after Aleksandrov's resignation, when a large group of the institute's top scientists mobilized to prevent Legasov from moving into the director's office. They struck in the spring of 1987 during routine elections to the institute's main governing body, the academic council. Voting against allowing Legasov to join the council were 129 senior members of the institute. It was a devastating blow to the first deputy director, who had become used to running the institute while Aleksandrov was busy at the Academy of Sciences. Only 100 of his colleagues supported him.[1]

Legasov was a romanticist. He wrote poetry; in fact, in early youth he had aspired to become a professional writer, but had been dissuaded by Konstantin Simonov, a leading Soviet literary figure.

At the time, students argued about who was more important to the country, physicists or lyric poets. In 1959, the prominent poet Boris Slutsky had written, in one of his poems, "Somehow physicists are in vogue; somehow lyric poets are kept down," and proceeded to conclude that physicists were more important to society than those engaged in the humanities. Party ideologues who monitored the extensive discussion that followed the publication of Slutsky's poem offered a compromise: both were important. Legasov, a chemist by training, strove to pursue both callings.

Like Nikita Khrushchev, Stalin's successor at the top of the Soviet power pyramid and an architect of the ideological thaw that allowed discussions like the one about physicists and lyric poets to take place in a society traumatized by Stalin's purges, Legasov was a believer. Both believed in the Soviet system. Legasov had manifested his belief by joining the Communist Party while still a student at Moscow University, an act that many of his apolitical colleagues considered naïve or careerist. The Soviet scientific community would produce some leading political dissidents, including the physicists Andrei Sakharov and Yurii Orlov. During the Russian Revolution, Legasov's boss, Aleksandrov, had fought for two years in Ukraine in the ranks of the White Army; he had joined the party only at the age of fifty-nine, when that was absolutely necessary if he was to become director of the institute established by Kurchatov. Many scholars had no sympathy for the regime and kept their distance from the party whereas Legasov embraced its rule and ideals.[2]

Legasov also believed in the power of Soviet science and the safety of the nuclear reactors produced with the help of his institute. Only two years before the Chernobyl disaster, he went on record to support them, writing, "It is safe to say that nuclear power is considerably less harmful to human health than coal power, which is equally potent. . . . Specialists are of course well aware that it is impossible to produce an actual nuclear explosion at a nuclear power station, and only an improbable concatenation of circumstances could bring about the semblance of such an explosion, which would be no more destructive than an artillery shell." It appeared that Legasov was blindly following the party and industry line. His belief in the safety of reactors came with his post as deputy to Aleksandrov. Yevgenii Velikhov, Legasov's colleague and competitor both at the institute and on the site of the Chernobyl accident, would later recall that Legasov had no involvement in the construction of the reactor or inside knowledge of its physics. One physicist called him "a boy from the chemical periphery." Legasov promoted RBMK reactors in his official capacity as first deputy of the institute's director.[3]

As at the institute, on the site of the Chernobyl accident Legasov was in the forefront, manifesting not only his belief in the system but also his leadership qualities and readiness for self-sacrifice. "Legasov was the only competent person there," said a Ukrainian colleague, remembering Legasov's days in Chernobyl. "He poked around everywhere. In the early days he spent time on the 'bookcase' [the exhaust pipe that remained intact after the explosion]. Like everyone else, he was afraid of radiation. But he had to have the moral right to send others there, so he first went himself." Legasov soon realized that the whole Soviet Union was facing a disaster of global proportions. At stake was the future of millions of people, if not of the entire world. Without much hesitation, he put his own health and life on the line to save others. Like many who would come to Chernobyl with him and in his wake, Legasov did not understand the full extent of the risks he faced, but he would grasp them more quickly and completely than anyone else.[4]

Legasov would compare the situation created by the nuclear accident with that of World War II, as did many others at the Chernobyl power station. But his comparisons with the Great Patriotic War (as the Soviet-German conflict was known in the USSR), which had been highly mythologized by Soviet propaganda, went beyond the self-sacrifice shown by Red Army soldiers and Chernobyl liquidators. He would also talk about the unpreparedness of the Soviet system to deal with both disasters—the nuclear accident and the military catastrophe of the Nazi invasion of the Soviet Union in the summer of 1941. "There was such lack of preparation at the plant!" recalled Legasov. "Such disorder! Such fright! As in '41. Exactly. Nineteen forty-one, but in an even worse version. With the same 'Brest,' the same courage, the same desperation, but also with the same unreadiness." The latter reference was to the heroic defense of the Brest fortress in western Belarus by Red Army soldiers during the first weeks of the German invasion.[5]

As the chief scientific adviser in Chernobyl in the critical days immediately after the accident, Legasov was responsible for many key decisions made by the government commission. One of them was his proposal to bury the reactor under thousands of tons of sand, clay, and lead. The task was achieved at the cost of the health and, ultimately, the lives of helicopter pilots, but some of Legasov's colleagues considered the whole exercise nothing but a waste of lives and resources. The mound of sand over the reactor's burning "mouth" did nothing to decrease radiation levels, and many believed that it increased the danger of the reactor overheating and melting down a second time. Legasov would defend his decisions to the end, but Chernobyl not only devastated his body—he sustained the maximum dose permitted by Soviet regulations, 25 roentgens, many times over—but also burdened his mind with disturbing thoughts about his role in endangering the health and lives of others.[6]

Legasov's work on the report that he delivered in Vienna in August 1986 lifted his spirits and made him believe that he could improve the safety of Soviet nuclear power plants, but his reception in Moscow upon his return was devastating. Officialdom was unhappy: he had gone too far in divulging the secrets of the Soviet nuclear program. Not only were party leaders dissatisfied, but, even more important, so were his bosses in the nuclear industry and his colleagues at the Kurchatov Institute. The latter were angry with Legasov, feeling that he had betrayed them. Legasov, for his part, believed that he had done the right thing for his country and the world. Privately he began to regret that he had not gone far enough. He would later tell his friends: "I told the truth in Vienna, but not the whole truth." In his report, Legasov had blamed the accident almost exclusively on the faults and errors of those in charge of the plant. He had not spoken about the defects of the reactor itself, which helped to turn what would otherwise have been a serious accident into a nuclear catastrophe. That was one more burden on his conscience.[7]

Officially, Legasov had sustained 100 rem of radiation, but neither he nor his doctors knew the actual extent of his exposure.

At Chernobyl, when he went to the most dangerous areas of the plant, he often left his dosimeter behind. He began to feel the first symptoms of disease that summer, as he prepared his report for the Vienna conference. In November 1986, he was invited to join Politburo members atop the Lenin Mausoleum during the Red Square parade marking the anniversary of the Bolshevik Revolution—one of the highest honors ever to be bestowed on a Soviet scientist—but his health was already so compromised that he could not attend. Legasov's wife, Margarita, began to record his symptoms in a medical diary: nausea, headaches, exhaustion. Tests showed an increase of white cells in his blood, a sign of the suppression of bone-marrow activity and acute radiation sickness.

In May 1987, doctors found myelocytes in Legasov's bloodstream—young cells that were supposed to be in the bone marrow but had now gotten into his blood, indicating the risk of developing cancer. With his health deteriorating and his spirits at an all-time low after his colleagues refused to elect him to the Kurchatov Institute's academic council, Legasov checked into a hospital for the treatment of his radiation sickness. In despair, he decided to end his life by overdosing on sleeping pills. Only the vigilance of the medical personnel saved his life: doctors pumped his stomach before the sleeping pills could kill him. Legasov would try to consign Chernobyl to the past and start his life anew. It was easier said than done.[8]

As LEGASOV underwent treatment in a Moscow hospital, the managers of the Chernobyl nuclear plant, whom the Politburo deemed solely responsible for the nuclear disaster—a judgment that Legasov had announced to the world in Vienna—went on trial.

It was decided to hold the trial in Chernobyl itself, in the heart of the 30-kilometer exclusion zone. The formal reason for that decision was a Soviet legal provision requiring trials to be held in the place where the alleged crime had been committed. Law or no law, it was a very strange choice. Levels of radiation in Chernobyl were still extremely high. Although asphalt had been removed from roads and sidewalks and then buried, radioactivity was everywhere, especially on the shoulders of the newly built roads. In the town center, journalists could see permanently installed dosimeters. Anyone entering the building in which the trial was taking place had first to wash his footwear in containers of water placed near the entrance. The security regime established in the exclusion zone helped the authorities to exercise complete control over what was happening in the building where the trial took place and its vicinity.

The hall of the local cultural center was turned into a makeshift courtroom by fitting the windows with metal bars, separating the stage from the seating area with a curtain, and removing rows of seats. The room had a seating capacity of two hundred and was always full—operators and other personnel of the nuclear plant attended the proceedings when they were not otherwise occupied. One of them, Nikolai Karpan, made notes that he would publish years later. The trial lasted eighteen days, from July 7 to 29, 1987, with journalists permitted to attend only on two of them. On the opening day they listened to the prosecutor's statement; on the closing day they heard the verdict. The journalists quipped that it was an "open" trial in a "closed" zone.[9]

Six managers and safety officers of the Chernobyl nuclear plant were put on trial for violation of safety rules and negligence of duty. The defendants included the former director of the plant, Viktor Briukhanov; the former chief engineer, Nikolai Fomin; and Fomin's deputy, Anatolii Diatlov, who on the morning of April 26 had overseen the shutdown of Unit 4 that led to the explosion. They were the prime suspects, arrested long before the trial and now seated at the same desk in the center of the room. The other defendants— the chief of the reactor division of the plant, Oleksii Kovalenko; the chief of the plant's shift on the night when the accident occurred, Boris Rogozhkin; and an official of the nuclear safety administration, Yurii Laushkin—were seated separately.[10]

Viktor Briukhanov, the principal defendant, was not at all surprised to find himself on trial. His first thought on seeing Unit 4 destroyed by the explosion had been about going to prison. Long experience as a Soviet industrial manager had taught him that the first to be blamed for any large-scale accident was the director—and if the accident was serious enough, the director would go to jail. The Politburo decision of July 1986 to expel him from the Communist

Party "for major errors and shortcomings in work leading to an accident with serious consequences" had opened the door to his arrest. The KGB officer who ordered Briukhanov's arrest in August had told him that he was better off in prison. The scope of the disaster had shocked the public, which was looking for scapegoats, and the former director of the plant was an easy target. During one of Briukhanov's interrogations, a KGB officer unknown to him entered the room and said: "I would shoot you myself." Briukhanov responded: "Well, go ahead, stand me up and shoot!" By that point, he was ready for almost anything.[11]

Briukhanov had already spent almost a year in the KGB prison in Kyiv. Only after his arrest had he been examined by doctors, who told him that he had sustained approximately 250 rem, ten times the dose allowed the liquidators. Briukhanov suffered from acute radiation syndrome and was no stranger to regular headaches and attacks of excruciating pain behind the ears. He spent most of the pretrial months in solitary confinement—the worst kind of imprisonment imaginable. Only once was he allowed to see his wife, Valentina. The family, which included a teenage son and an older daughter who had given birth to a baby girl four months after the Chernobyl accident, was suffering a stunning reversal of fortune. Previously one of the most respected families in Prypiat, the Briukhanovs were now shunned by many of their former friends and neighbors. After the emergency evacuation of Prypiat, they had almost no possessions. Not until August, after Briukhanov's arrest, was Valentina allowed to visit their apartment in Prypiat and take a few of their belongings. "The dosimetrist went in first," she recalled. "He gave permission to take a few items and books. We cleaned every volume with a cloth soaked in a weak vinegar solution. It was thought to counteract radiation." To make things worse, the court authorities froze Briukhanov's bank account, to which he had transferred his monthly salary and vacation pay prior to his arrest.[12]

Valentina concealed the fact of her husband's arrest from their daughter, who was breastfeeding her baby. To support herself and her teenage son, Valentina returned to work at the Chernobyl nuclear plant, helping to run the remaining reactors, which went back onstream in the fall of 1986. She was allowed to work without weekend breaks and holidays because she wanted to keep busy to avoid thinking about the disaster and what now awaited her and her husband. The strategy backfired: Valentina's blood pressure climbed dangerously high, and one day her colleagues called an ambulance to take her to the hospital.

Valentina had to seek various ways of coping with the stress and trauma of Chernobyl. Eventually she found a new purpose in life by recommitting herself to her family. The turning point came when a female doctor took her by the shoulders and told her to pull herself together—she still had a family to take care of. Valentina took those words to heart. There was also another episode that helped her survive the crisis. "I am very grateful to a simple woman from Prypiat," she recalled. "Once, when I was walking from the bus stop and sobbing, she came up to me, embraced me, and said: 'Valiusha, why are you crying? After all, Viktor is alive, and that's the main thing! Just look how many graves there are after Chernobyl.'" Valentina would now fight not only for herself but also for her husband. She convinced him to get a lawyer—at first, depressed and fatalistic, he had refused to do even that.[13]

The trial, originally planned for the spring of 1987, had been postponed because of the unstable mental condition of one of the accused, the former chief engineer, Nikolai Fomin. He had been arrested on August 13, 1986, the same day as Briukhanov. This was shortly after his release from Hospital No. 6 in Moscow, where he had been treated for radiation poisoning. The trauma of the arrest, coupled with the effects of acute radiation syndrome, drove Fomin to depression. In March 1987, while in prison, he broke his eyeglasses and tried to cut his veins with the shards. The doctors saved him, and by July 1987 they had made him sufficiently stable to face trial.[14]

In the hall of the Chernobyl cultural center, Briukhanov and Fomin were seated on either side of the man considered by many to be the true cause of the catastrophe—Fomin's deputy, Anatolii Diatlov, who had been in charge of the ill-fated turbine test on the night of the explosion. Like Fomin, Diatlov had been treated in Hospital No. 6, but he had not been released until early November 1986. The doctors estimated that he had absorbed 390 rem. He left the hospital with open wounds on his legs—the result of radioactive burns sustained on the night of April 26. From the start, in the eyes of the criminal investigators, Diatlov was the main culprit of the disaster. They also opened criminal investigations of the chief of the Unit 4 shift, Aleksandr Akimov, and the reactor operator Leonid Toptunov, but had to close them because of the deaths of both in May 1986. Diatlov was ill but still alive. He was arrested on December 4, a month to the day after his release from the hospital.

In June 1987, when everything was ready for the trial, Briukhanov, Fomin, and Diatlov were moved from the KGB prison in Kyiv to the regional prison in Ivankiv, a town 50 kilometers south of Chernobyl that served as headquarters of the state commission dealing with the consequences of the disaster. From there they would be taken daily to Chernobyl for sessions of the trial. Briukhanov and his former subordinates were accused of violating three articles of the Soviet Ukrainian criminal code. The first dealt with safety rules at enterprises subject to explosion hazards; the second with abuse of power, which allegedly manifested itself in withholding information about the true scope of the accident; and the third with negligence, as a result of which the managers supposedly failed to train plant personnel properly.[15]

Viktor Briukhanov refused to plead guilty to the first two charges. He claimed that no instruction had ever defined a nuclear plant as an enterprise subject to explosion hazards: neither lawmakers nor industry leaders, who issued operating instructions, had contemplated the notion that a reactor might explode. On the second count, he claimed that he had informed the authorities of what was going on at the power plant to the best of his ability, but his suggestion to evacuate the population of Prypiat had been ignored. The most damning evidence against him was the letter he had signed on the morning of April 26, giving the lowest level of radiation at the power plant from the readings then available to him. To the prosecutor's question, "Why was there no information about 200 roentgens per hour in the letter to party and Soviet organs?" Briukhanov responded: "I looked at the letter inattentively; of course that should have been added." Although Briukhanov tried to defend himself as best he could, he knew that whatever he said or did at the trial, his fate had already been decided in Moscow. "It was apparent before-hand that I would be punished," he said later, recalling his thoughts at the time of the trial.

Briukhanov admitted that he was guilty of negligence, as the accident had happened on his watch. "I am guilty as manager of having missed something, of having been careless or inefficient in some way. I understand that this is a serious accident, but everyone bears some blame for it," he told the court. Briukhanov's assumption of partial responsibility for the accident made a positive impression on the judges, although everyone knew that he had not been di-rectly involved in the events of April 26. "You know, this is my first encounter with such a calm, self-possessed defendant," one of the judges said privately to Valentina Briukhanova, "although you can feel that he's upset. A real man!"[16]

Nikolai Fomin adopted a completely different strategy. Saved and brought back to life by prison doctors after his botched suicide attempt, he was now fully revived and not above shifting responsi-bility for the accident to his subordinates. His line of defense was quite simple: the program of the turbine test he had approved was sound, and the explosion at Unit 4 would never have taken place if Diatlov and Akimov had followed it. "I'm convinced that it was not the program that caused the accident," he told the court. When the prosecutor asked, "Who, in your opinion, is mainly responsible for the accident?" Fomin responded: "Diatlov and Akimov, who al-lowed deviations from the program."[17]

Anatolii Diatlov, the manager directly responsible for the vio-lation of conditions for the turbine test set by the program, refused to take a page from Fomin's book and blame his subordinates, who by that time were deceased and might well have been made the ulti-mate scapegoats. Diatlov chose a different and much nobler course, but one that was also dangerous for the authorities. He admitted guilt for a number of violations of the operating instructions, such as leaving fewer than fifteen control rods in the active zone of the reac-tor; failing to restore the power level of the reactor to 700 megawatts of thermal energy (MWt) after the accidental loss of power, as sug-gested by the test program; and delaying the use of the emergency shutdown (AZ-5) button to stop the reaction.

But Diatlov was adamant that none of those violations would have caused the explosion if the reactor had been in sound operating condition. "If we had pressed the button earlier, the accident would have taken place earlier," argued Diatlov. "That is to say, the accident was conditioned by the state of the reactor. I gave the order to stop the power level of the reactor at 200 megawatts [of thermal energy], since I considered that the reactor met the safety standards accepted in the USSR." In effect, Diatlov was pointing a finger at the designers of the RBMK reactors, which turned out to be anything but explosion-proof because of their positive void effect, meaning their capacity to accelerate a reaction when control rods were inserted into the active zone. Diatlov was going public with an accusation against the designers of the reactor that many in the industry and the political elite knew to be justified.

Diatlov eventually concluded that neither the principal presiding judge, a member of the Soviet Supreme Court, Raymond Brize, nor the main prosecutor, Yurii Shadrin, a senior aide to the general prosecutor of the Soviet Union, was interested in uncovering the truth about the explosion. Indeed, they shielded the designers of the RBMK reactors from responsibility by removing all materials pertaining to reactor design from the case against the plant managers and assigning them to a separate criminal case requiring further investigation. The commission of nuclear experts summoned by the court to look into the causes of the explosion was dominated by representatives of the institutions responsible for the design of the RBMK reactors, and evidence given by witnesses from among the operators and engineers of the Chernobyl nuclear plant was often ignored by the judges.[18]

Briukhanov believed that the new management of the station deliberately called a meeting of senior personnel at the plant to coincide with the reading of the trial's verdict in order to avoid a scandalous protest against it. Nonetheless, more than five hundred employees of the plant signed an appeal to pardon Briukhanov. As Diatlov later wrote: "By July 1987 it had become clear to many that bringing charges against personnel was illegitimate. The witnesses knew what measures were being taken to modernize the remaining reactors, considered them, and drew their own conclusions." He was referring to the modifications made to RBMK reactors after the Politburo meeting of July 1986. Well aware that responsibility for the disaster could not be ascribed solely to the operators and designers, the senior authorities decided to make them scapegoats nevertheless. Commenting subsequently on the verdict, Briukhanov observed: "After all, it was necessary to show the Central Committee of the party and the whole world: look, we've found the culprit. And how could science be lagging behind in the Soviet Union? It was the best in the world."

The court ruled that Briukhanov and his subordinates were guilty because they "had not ensured that the power station personnel would observe technological discipline; moreover, they themselves systematically violated official instructions and ignored the directions of supervisory organs." Briukhanov was also blamed for delays in the evacuation of personnel. "Having manifested confusion and cowardice, Briukhanov did not take measures to limit the scope of the accident, did not implement a plan to protect the personnel and the population from radioactive emissions and, in the information he presented, deliberately minimized the data on radiation levels, which hindered the timely evacuation of people from the danger zone," read the verdict.[19]

When the judge read out the verdict, Briukhanov was shocked by the severity of the punishment meted out to him—ten years of incarceration, the same as the sentence given to Diatlov and Fomin. Ultimately, it did not matter what the three top managers had done at the time of the accident or what attitude they took in court—all were sentenced to the same term. Three other defendants received sentences ranging from two to five years of imprisonment. "The judge of the Supreme Court rendered the verdict that he was ordered to give," said Briukhanov in retrospect. "I think that if they had found an article whereby I could have been shot, they would have shot me. But they didn't find it." The prison authorities suspected that because the sentence came as a shock, Briukhanov might commit suicide. "On the night after the verdict, a guard placed a chair by my bed and spent the whole night there, lest I do something to myself," recalled Briukhanov. "But he only interfered with my sleep." Briukhanov had no thought of killing himself: he was cut from different cloth. As he later told a journalist, "leaving this life is simple enough, but what would you prove by doing so, and to whom? What would it accomplish?"[20]

NOT EVERYONE was so resilient. Valerii Legasov succeeded in his second suicide attempt on April 27, 1988, one day after the second anniversary of the Chernobyl nuclear disaster. He hanged himself in his own apartment when the members of his family left for work. There was no mistaking the seriousness of his intent: the knot on the rope was so tight that even the police officers who investigated his death found it difficult to untangle. Legasov left no suicide note but put in order the verses he had dedicated to his wife over his lifetime. On the previous day he had brought home personal items from his office, including his favorite photo of two Chernobyl storks—a symbol of life returning to the disaster area.

Since his first suicide attempt in the summer of 1987, Legasov had been attempting to put his life back on track, dedicating himself to his new obsession—the security of Soviet nuclear reactors. In October 1987 he wrote an article for *Pravda* arguing the priority of science over the demands of industry and production quotas. "When it became necessary not to improve on the old but to create the new, the voice of principle was accorded to science," maintained Legasov, presenting an idealized picture of the early development of the Soviet nuclear program. He continued: "And then there is the opposite example, so tragically demonstrated in Chernobyl. When science was constrained . . . , and decisions that were less than optimal began to be made."[21]

Legasov tried to be as diplomatic as possible in his criticism of the Soviet nuclear industry, but his diplomacy did not work as planned. In the fall of 1987, he was subjected to yet another humiliation. Anatolii Aleksandrov, who had kept on supporting Legasov, announced to the staff of the Kurchatov Institute that the Politburo had decided to bestow the highest Soviet peacetime award, the Star of Hero of Socialist Labor, on Legasov for his work at Chernobyl. But the Politburo decision was reversed at the last moment, apparently by Gorbachev himself. It was a severe blow to Legasov's prestige and morale. His health was failing, and his hopes for the realization of his scientific plans were fading with it. On April 26, 1988, the second anniversary of the Chernobyl accident, Legasov's colleagues in the Academy of Sciences rejected a new idea of his, that of creating an interagency committee on chemical research. On that day he removed his personal effects, including the Chernobyl photos, from his office. A day later, he was dead.

In the period leading up to his death, Legasov had begun to question perestroika—Gorbachev's policies aimed at the transformation of Soviet society. He told one of his colleagues that he believed the wrong people were in charge. It is hard to imagine that he had anyone other than Gorbachev himself in mind. In the interviews and memoirs about Chernobyl that Legasov recorded on tape during the months prior to his suicide, he expressed concern about the security record of the nuclear industry. He was critical of many features of the RBMK reactors built in the USSR, especially the lack of a concrete shelter above the reactor, which was required by international standards in hopes of preventing the spread of radiation in the case of an accident. In his view, the regulation of reactor activity by means of control rods with graphite components was inadequate. Legasov was critical of the all-powerful minister of medium machine building, Yefim Slavsky, but had good things to say about Premier Nikolai Ryzhkov, who had supported him as he worked on his report for the International Atomic Energy Agency conference in Vienna in the summer of 1986. It was Ryzhkov who attended Legasov's funeral on behalf of the Soviet leadership. Gorbachev was nowhere to be seen.[22]

Valerii Legasov, who had told the world much of the truth about the consequences of the Chernobyl disaster in his Vienna report but would not tell the whole truth about its causes, was now dead, crushed by radiation-induced depression that heightened his sense of betrayal and guilt. Viktor Briukhanov, Nikolai Fomin, Anatolii Diatlov, and three of their colleagues from the management of the Chernobyl nuclear plant were serving time in prison. As far as Soviet public opinion was concerned, Legasov's death was an unfortunate accident, while the guilty were being justly punished. With the reactor buried and the uncomfortable truth about the explosion apparently buried with it, Mikhail Gorbachev was finally able to launch his political and economic reforms. The future looked promising, if not entirely bright. Unanticipated by many in Moscow and in other capitals all over the world, the Chernobyl disaster would come back in a most unexpected way to affect the lives of those already influenced by it.

VI

NEW DAY

WRITERS' BLOCK

In January 1988, the leaders of the Writers' Union of Ukraine sent a memorandum to the Communist Party bosses in Kyiv proposing an international conference on the health impact of the Chernobyl catastrophe. The Ukrainian writers volunteered to organize such a conference together with the Academy of Sciences of Ukraine and their colleagues in the Writers' Union in Moscow. The party officials indicated their preference for postponing the event until the following year, citing a busy schedule for the fall of 1988. The officials also alluded to what they claimed had been the Writers' Union's lack of cooperation with the party.

Although the Communist Party of the Soviet Union and the KGB allowed the Writers' Union to exist, they also closely monitored its activities. The Ukrainian writers did their best to persuade the authorities that they had no sinister intentions, noting that the idea of holding such a conference had emerged from discussions at the all-Union writers' conference held in Leningrad in the fall of 1987, which had been dedicated to the seventieth anniversary of the Bolshevik Revolution of 1917. The Kyiv authorities decided to play for time—issues related to the Chernobyl disaster were rapidly becoming a political hot potato.[1]

SINCE THE first weeks after the disaster, the KGB had been busy monitoring attitudes toward it among Ukrainian political dissidents.

In early June 1986, the secret police reported to the party authorities on the ethnonational interpretation of the disaster developing among people suspected of nationalist convictions or sympathies. I. Z. Shevchuk, a former member of the nationalist underground that had fought the Soviets in western Ukraine after World War II, had allegedly told a KGB agent that he believed "the Russians are deliberately building such stations on Ukrainian territory, knowing that if an accident should happen, it would be basically Ukrainians who would suffer."

Although that was not a widespread view, there was no doubt among the dissidents that the Chernobyl accident had brought about a national catastrophe. Mykhailyna Kotsiubynska, a niece of a famous Ukrainian author of the turn of the twentieth century, Mykhailo Kotsiubynsky (she was named in his honor), was close to the Ukrainian Helsinki Group. She allegedly said the following to an acquaintance, who reported her words to the KGB: "We have been struck by a disaster from which we will not soon recover. The nation is in danger of dying out, of physical annihilation. The disaster that has struck us is a disgrace of global proportions. It is a disgrace first and foremost to the shortsighted leaders who gave the order to build atomic energy stations in densely settled areas, precisely in Ukraine, which is endowed with unbelievably rich lands."[2]

The KGB was doing its best to stop the spread of such views both at home and abroad. It was bad enough that they could influence Western public opinion. But they could also be transmitted back into the Soviet Union via broadcasts of the Voice of America, Radio Liberty, and other Western radio outlets. To keep up the appearance of openness, Soviet authorities allowed foreign correspondents into Ukraine, and even into the Chernobyl zone. But their visits were carefully choreographed, and contacts with dissidents and other "undesirable elements" were either forestalled or monitored.

In the fall of 1986, the secret service paid special attention to two Americans, Mike Edwards and Steve Raymer, who came to Ukraine to work on articles for a special issue of *National Geographic* on the Chernobyl disaster. "The measures taken have staved off the Americans' efforts to make contact with Ye[vhen] O[leksandrovych] Sverstiuk, O[lha] P. [Ivanivna] Stokotelna, I[rina] B[orisovna]

Ratushinskaia, and a number of other individuals known in the West for their nationalist and anti-Soviet activity, from whom they might receive news of a tendentious nature," reported KGB officers on the results of their work. The KGB also kept a close eye on the movements and contacts of Tania D'Avignon, a Ukrainian American photographer and interpreter affiliated with the Ukrainian Research Institute at Harvard University, who accompanied the two journalists. The KGB suspected her of ties with Ukrainian nationalist centers abroad as well as with the CIA. "On behalf of Intourist, T. D'Avignon was officially cautioned against violating established norms for the presence of foreigners in the USSR," noted the KGB report. "According to operational data, this had a positive restraining influence on the Americans' activity in collecting negative information."[3]

Many Ukrainian dissidents, including Yevhen Sverstiuk, a member of the Ukrainian Cultural Club and one of the "nationalists" whom the KGB prevented from meeting with the *National Geographic* team, were writers, poets, and artists who had allies in the Ukrainian Writers' Union. The regime treated members of the union as valuable propaganda assets who could help indoctrinate and control the masses. The top writers, who were extremely well paid through the government-regulated system of royalty payments for publications, could get away with politically incorrect opinions as long as they kept them out of their published works. Writers were among the earliest and strongest supporters of Gorbachev's perestroika, pushing the limits of glasnost with every passing day and month. In a country that outlawed political opposition and jailed critics of the regime, writers, long regarded as the "conscience of the nation," since tsarist times, became the substitutes for a loyal opposition, often voicing the concerns and frustrations of their colleagues in the ranks of the dissident movement.

SOVIET WRITERS in general, and Ukrainian writers in particular, had been raising concerns about ecological issues for decades. The subject had first entered Soviet literary discourse in the late 1940s and had gained prominence in the 1960s. In Russia, Aleksandr Solzhenitsyn had been among those addressing such concerns. In his writings and in the works of other nationally minded writers, the subject was closely associated with concern for the preservation of historical and religious traditions. As their works attested, the writers who were early ecologists were also Russian nationalists, and the nationalist critique of the Soviet regime included elements of ecological activism. Chernobyl made that link more pronounced than ever, galvanizing nationalists in a number of Soviet republics.

In Belarus, the republic that had suffered the most from the radioactive fallout of the Chernobyl disaster, the ethnonational interpretation was formulated in the first weeks after the accident by one of the republic's leading writers, Ales Adamovich. As a young boy, Adamovich had joined the partisan struggle against the Nazis in the forests of Belarus; afterward, he made a name for himself as a novelist with books based on his experience of the war. In June 1986, in an appeal to Mikhail Gorbachev highlighting the impact of the Chernobyl catastrophe on his republic, he wrote: "We are not about to make an 'all-European' stir, but we realize that Belarus is experiencing something that can be compared only to the tragedy that it underwent during the years of the past war. The very (physical) existence of our people of 10 million is in question. Radiation struck our republic first and foremost." Research on Chernobyl became Adamovich's ruling passion. He visited territories contaminated by the explosion and interviewed people who were willing to talk about the accident, including Academician Valerii Legasov, who granted him a number of interviews before his suicide in April 1988.[4]

In Ukraine, the most prominent author writing about what many saw as the government-sponsored destruction of nature was Oles Honchar, one of Ukraine's most celebrated writers and a recipient of the Stalin Prize for literature in 1948. He had focused on that theme long before the Chernobyl accident. "The heat of war hovers over and the soot falls on the gardens that chip away. Poison everywhere," Honchar wrote in *The Cathedral*, a 1968 novel describing the impact of industrial development on his native land. The book was heavily criticized by Soviet authorities. In the aftermath of the Chernobyl disaster, Honchar was appalled by the treatment the Ukrainian rulers had meted out to their own people, believing that they had sacrificed the health of Kyivans in order to manifest their loyalty to the regime, especially at the May 1 parade. In June 1986, Honchar delivered an emotional speech at the ninth congress of the Writers' Union of Ukraine, claiming that Chernobyl had changed how Ukrainian writers "related to the world."[5]

In the same month, at a meeting of Ukrainian writers with the party boss of the republic, Volodymyr Shcherbytsky, Honchar pushed for a complete shutdown of the Chernobyl nuclear plant. "I asked whether a petition might not be presented to decommission that station as technically incompetent and built for some reason in the Polisian marshes, right beside a city of several million," wrote Honchar in his diary. Shcherbytsky did not seem to comprehend the problem. "In answer, he somehow waved his arms in agitation and, almost choking in his haste, started giving me explanations about rockets and the future of atomic energy," wrote Honchar. Rhetoric about the common good of humanity made no impression on Honchar. His immediate concern was the welfare of his nation, Ukraine. "They say," he wrote, "that neighbors have to be supplied with energy. But why should the Ukrainian land fall victim? Why should Ukrainian children suffer those satanic doses?"[6]

HONCHAR'S STAND on Chernobyl and the future of nuclear power plants in Ukraine signaled a major change of attitude on the part of Ukrainian political and cultural elites toward nuclear power and nationhood. Back in the mid-1960s, Ukraine's communist leaders had rushed to jump on the nuclear bandwagon as an emblem of modernity—their republic had finally joined the exclusive nuclear club. Writers were prepared to overlook the fact that modernity was coming to Ukraine in the garb of the Russian language and culture, undermining the cultural foundations of their imagined modern nation. As construction of the Chernobyl plant began, a Russian-speaking enclave took shape in the heart of the Ukrainian Polisia region. Like most Ukrainian cities in the twentieth century, Prypiat was taking in Ukrainian-speaking peasants from the countryside and turning them into Russian-speaking urbanites who would gravitate toward Russian culture.

Ukrainian intellectuals faced a Hobson's choice: unless their nation embraced modernity, it would have no future, but embracing it meant abandoning their national distinctiveness. Ukrainian writers decided to claim the Chernobyl power plant for their republic without inquiring too closely into the language and culture of the people running it, as they had done earlier with the Soviet industrial giants of the interwar period, from the metallurgical enterprises of the Donbas to the machine-building factories in Kharkiv and Dnipropetrovsk—all in eastern Ukraine. In their writings they pretended that those in charge were Ukrainians who spoke Ukrainian.

The first to claim the Chernobyl nuclear plant for the Ukrainian nation and culture was the sixty-five-year-old patriarch of Ukrainian playwrights and screenwriters, Oleksandr Levada. He was the foster father of the freethinking sociologist Yurii Levada, who would go on to establish the Levada Center for the study of Russian public opinion in Moscow. In the spring of 1974, two years before the launch of the first reactor at the Chernobyl nuclear plant, a leading Kyiv theater staged Levada's play *Hello, Prypiat*, which deals with the construction of the plant. Levada sidestepped the issue of the de facto Russian cultural appropriation of the Ukrainian region. Almost all the characters in the play, even those coming to Prypiat from Moscow or other parts of Russia, are ethnic Ukrainians. The play's conflicts between modernity and tradition, industrialization and the environment, take place within Ukrainian society, which is by definition linguistically and culturally Ukrainian. This cultural idyll ignored the problem of Russification implicit in the modernizing projects that the authorities in the all-Union capital were planning and carrying out.[7]

Levada defined the main theme of his play as the relationship between progress and the environment. The play promotes nuclear energy as the cleanest source of electricity, asserting its compatibility with environmental protection. Any notion that nuclear power might pose a threat to people or the environment is dismissed. Those expressing such concerns in the play are negative characters—either wartime collaborators with the Nazis who are still hostile to the Soviet state or backward-looking peasant women. Ironically, one such character all but predicts the creation of the exclusion zone and the resettlement of its inhabitants: "For people are saying, one to another, you know, that when that station starts working, in twenty-four hours they'll take us all 50 versts [33 miles] away because some kind of atoms will start flying and butting heads like rams, and there'll be no place for people here."

All such concerns are dismissed by the positive characters as alarmist and unfounded. Academician Mazurenko, one of the Ukrainian scientists, points to nuclear energy as an alternative to the destruction of nature caused by traditional industries, lauding the Chernobyl station as a model for the future. Nuclear power is hailed as a solution to the concerns about the ecological impact of industrial development raised in Honchar's *Cathedral*. Only after the explosion would people discover early "warnings" of the catastrophe in the words of Levada's negative characters.[8]

Unlike his dissident stepson, Yurii, Oleksandr Levada was a stalwart communist who believed in and advanced Soviet propaganda clichés. But it was not only party loyalists like Levada who embraced the "nuclearization" of Ukraine. Among the early proponents of the construction of the Chernobyl plant were a number of young Ukrainian writers who were close to dissident circles. The most prominent of these was a rising star of Ukrainian poetry, Ivan Drach. In his mid-thirties when construction of the station began, Drach belonged to the generation of the 1960s—young and ambitious writers and intellectuals who came to the fore at the time of Khrushchev's ideological thaw and promoted the Ukrainian language and culture as part of a larger liberal agenda for the transformation of the Soviet Union.

Drach and his fellow poets and writers fell on hard times after the dismissal of Nikita Khrushchev in 1964, which spelled the end of the cultural "thaw." But in 1976, with the thaw long gone, repression against intellectuals on the rise, and many of Drach's friends in prison, he published a collection of poems that finally met with the authorities' approval. It glorified Vladimir Lenin and celebrated the official policy of the "friendship of peoples" that served as a cover for the Russification of the non-Russian peoples of the Soviet Union. Among the themes prominently featured in Drach's collection was that of technological progress, embodied in the construction of the Chernobyl power plant.

In a poem titled "The Legend of Polisia," the river Prypiat, personified as a Ukrainian girl, marries a newcomer to the area called Atom. Her personal feelings aside, Prypiat believes that the marriage will serve her people well. "I'll give myself to him in marriage, give myself right away," declares the river-turned-girl. "The time has come for me to serve the people just that way / Let my Atom help the Dnieper and the Donbas." The industrial Donets Basin in eastern Ukraine stood as a symbol of economic development that needed electrical energy to move ahead. For his short book of poetry, Drach was awarded the Shevchenko Prize in literature, the highest Ukrainian award of its kind. A few years later Drach would receive the State Prize, one of the highest all-Union awards, for his subsequent work.[9]

There is no indication that Drach did not believe in the positive power of nuclear energy when he wrote the verses that helped improve his standing with the authorities. But there is ample evidence that he regretted his enthusiasm for nuclear power and the nationalization of the atom following the accident. The Chernobyl disaster put an abrupt and tragic end to his erstwhile fascination. In May 1986, his son, Maksym Drach, a student of medicine at the time of the accident, was sent along with classmates to the exclusion zone in order to establish medical control over the area and check the traffic moving in and out of the zone for radioactivity. Without proper protection, he sustained a high dose of radiation, fell ill, and found himself in a Kyiv hospital undergoing emergency treatment for radiation sickness. He would suffer from the effects of exposure for years to come.

Drach now returned to the theme of Chernobyl and nuclear power with a different set of feelings and images. For him, what had happened to Ukraine was a nuclear apocalypse. The image of the local beauty that had personified the Prypiat River in his earlier work gave way to the image of the Virgin Mary—in the Ukrainian tradition, the Mother of God—in a poetic novel titled *The Madonna of Chernobyl*. "That fiery cross, and on it and in it / My son is burning in a circle of fire," says the Mother of God in one passage. "For atomic nails have been driven into his hands / For his lips are burning with hellish pain." With regard to his own and his generation's

earlier belief in the power of nuclear energy to bring about a benefi-
cent transformation of the world, Drach wrote: "The salt of recognition is the fruit of repentance."[10]

And repent he did. Drach's new view of Chernobyl appeared
in print in early 1988. In the next two years he would become one
of the leaders of Rukh (Movement), a grassroots organization that
would propel Ukraine into a democratic revolution. That revolution would help lead to the disintegration of the Soviet Union and
the creation of an independent Ukrainian state. "Chernobyl roused
our souls, showing us in real terms that we were on the edge of a
precipice, an abyss, and that all our cultural efforts were a vanity
of vanities, a waste of effort, a rose under a bulldozer," wrote Drach
later, recalling the role that the Chernobyl disaster had played in the
"awakening" of Ukrainian society.[11]

THE TURNING point in the writers' efforts to break into the public
sphere with their concerns about the harmful effects of the Chernobyl disaster on public health and the environment came in June
1988. That month, Mikhail Gorbachev convened a major party forum, the Nineteenth Communist Party Conference, which adopted
a program of political reform, opening the door to the first relatively free Soviet elections since the Russian Revolution of 1917.
The elections would take place in the following year, but the scope
of glasnost, or openness, was expanded right away, focusing public
attention on the consequences of the Chernobyl accident and the
question of the responsibility of senior party authorities for what had
happened during and after the explosion.

Borys Oliinyk, a delegate to the party conference and one of
the leaders of the Writers' Union, came to Moscow with a petition
signed by 6,000 people from his native Ukraine. They wanted party
officials in Moscow to stop the construction of nuclear plants in
Ukraine, especially the one near the town of Chyhyryn, the capital
of the seventeenth-century Cossack state that served as a symbol of
Ukrainian national history and identity. From the podium of the
party conference, Oliinyk declared: "The arrogance and scorn of
some all-Union agencies . . . with regard to the fate of Ukraine verge
not only on some kind of merciless cruelty but also on an insult to

national dignity." He demanded the punishment of those responsible for the construction of nuclear power plants in Ukraine. "I recall how, demanding the construction of the Chernobyl atomic energy station, some . . . would often say that it was so safe that you could install a reactor under the bed of newlyweds," stated Oliinyk. "We shall not demean ourselves by recommending that the mockers place their beds beside reactor No. 4. But we have the right to demand that the designers who committed the gravest miscalculations in choosing sites for atomic energy stations in Ukraine be held personally responsible."[12]

Only one year earlier, in the summer of 1987, the KGB had been going after people who distributed leaflets and wrote graffiti calling for an end to the construction of nuclear power stations in Ukraine. Now the text of Oliinyk's address was published in the Soviet media. The taboo on the discussion of the Chernobyl disaster was lifted by the mere fact that Oliinyk had spoken from the podium of the all-Union party forum. The responsibility of the officials in Moscow, and not just of the Chernobyl plant managers then serving their prison sentences, was now up for public discussion. Writers such as Oliinyk, Honchar, and Drach were in the forefront of change. For a time they all but abandoned their occupation of writing novels and poems and focused on the task of blocking the construction of new nuclear reactors on Ukrainian soil.[13]

In November 1988, a colleague of Oliinyk's in the Writers' Union, Yurii Shcherbak, helped organize the first truly mass rally in the city of Kyiv that was not under party control. A medical doctor by training and a distinguished medical scholar, Shcherbak had spent three months in the Chernobyl Exclusion Zone soon after the explosion, interviewing scientists, plant operators, and liquidators. He began to publish chapters of his documentary novel about Chernobyl in the liberal Moscow journal *Iunost'* (Youth) in the summer of 1987 and completed it a year later. The novel detailed the tragic events of 1986, stressing the heroism and self-sacrifice of those involved in fighting the manmade disaster. In December 1987, Shcherbak, along with some fellow writers and scientists concerned about plans for the construction of more nuclear power plants, organized the first ecological organization in Ukraine, called Green

World (*Zelenyi Svit*). It became the key organizer of the mass rally on the environment in Kyiv in November 1988.[14]

SHCHERBAK'S WAS the second rally to be organized in Kyiv in connection to the Chernobyl disaster. The first one took place on April 26, 1988, the second anniversary of the disaster, the Ukrainian Cultural Club, the republic's first nongovernmental organization, staged a commemorative rally. This was the second attempt of the club's chairman, Serhii Naboka, a thirty-two-year-old graduate of the Kyiv University school of journalism, who had been sentenced for "anti-Soviet propaganda" and imprisoned for three years, to hold such a rally. The previous one had been planned for the first anniversary. Naboka and his friends, known to the KGB at the time as members of a "network," had drafted a letter demanding that the Soviet leadership in Moscow shut down the Chernobyl power plant, stop the construction of new nuclear plants in Ukraine, hold public consultations, and perhaps a referendum, on the development of the nuclear industry, and designate April 26 as a day of mourning and remembrance. Naboka and his followers had hoped to use the 1987 rally in downtown Kyiv to collect signatures for the letter. The KGB had learned of these plans, however, and successfully used their agents to dissuade Naboka and his followers from proceeding. The KGB considered it to be part of an effort by Western intelligence services to exploit the nuclear accident to get movements and organizations ostensibly concerned with ecology to engage in subversive activities and wrest themselves free of party control.[15]

Naboka had revived the idea of a rally in the spring of 1988. Now his group was organized as the Culture Club and was determined to stage a rally. Among its organizers was Oles Shevchenko, a forty-eight-year-old dissident and member of the Ukrainian Helsinki Group. That organization, established in 1976, had set itself the task of monitoring the Soviet government's compliance with the human rights obligations it had assumed by signing the 1975 Helsinki Final Act of the Conference on Security and Cooperation in Europe. For its pains, the Helsinki Group was banned and persecuted by the authorities, who not only continued to violate their citizens' constitutional rights but also jailed those who monitored and protested abuses. Shevchenko was among the human rights activists who were arrested and imprisoned. Not until 1987 was he released from exile in Kazakhstan as part of Gorbachev's policies of glasnost and perestroika. Ukrainian activists placed their people's national rights at the top of their agenda. Now Shevchenko and his supporters had found a new cause—Chernobyl.

The posters prepared for the demonstration read: "Atomic energy stations out of Ukraine," "We do not want dead zones," and "Referendum on nuclear power stations!" The demonstration was to take place on the city's main square, which was then named for the October Revolution, and later generally known as the Maidan. The authorities tried to persuade the organizers, including Naboka and Shevchenko, to cancel the rally. When they failed, they asked party leaders to mobilize police officers and members of citizens' patrols, created by the police, to disperse the demonstration by force. In the days leading up to the Chernobyl anniversary, Kyiv's authorities began repair work on the pavement in part of the square, surrounded the rest with a construction fence, and brought students to the area to practice marching for the upcoming May 1 parade.

When members of the Ukrainian Cultural Club appeared on the square, the police and KGB agents attacked the participants, as well as some passersby and onlookers, and forcibly packed them into buses—about fifty people in all. They were taken to a nearby police station and strip-searched. Shevchenko was arrested for carrying a sign and shoved into a bus as he tried to cite an article of the Ukrainian constitution to the police: "Citizens of the Ukrainian SSR have the right to freedom of speech and demonstration." He would spend fifteen days in prison for alleged "hooliganism."[16]

An estimated 10,000 people gathered in the city center for the demonstration on November 13, 1988. The authorities, who had suppressed the first Chernobyl rally only a few months earlier, now backed down and gave their formal consent for the gathering. For Yurii Shcherbak, his Green World, and everyone else concerned about ecology and the impact of Chernobyl, this was their first chance to appear in public before a crowd of thousands. Shcherbak was among the first to speak. He attacked the atmosphere of secrecy surrounding the long-term impact of radiation on public health and suggested the creation of a commission to study the effects of the disaster, as well as the establishment of public control over the Ministry of Health and its activities. He also called for the erection of a monument to the victims of Chernobyl; a special day, April 26, dedicated to their memory; and a public referendum on the further construction and continuing use of nuclear power stations in the republic.[17]

The party authorities had given permission for a mass rally on the condition that its program would be confined to ecological questions. By then, ecology was generally considered a legitimate subject for public discussion: after all, not only the population at large but also the political and cultural elites were affected by the Chernobyl fallout. But it was difficult, if not impossible, to keep any public event within the limits of a purely "ecological" format, which was already political by definition. The Ukrainian writers decided to exploit that opening. Ivan Drach, then head of the Kyiv Writers' Union, and his colleague Dmytro Pavlychko, a secretary of the Writers' Union of Ukraine, had been working since November 1, 1988, on the creation of a working group to launch a "People's Movement of Ukraine." It was conceived as an umbrella organization for Ukrainian cultural groups that would perform a role similar to that of the popular fronts then active in the Baltic republics of the USSR, which were pushing for the economic and political sovereignty of their homelands. According to KGB reports, Drach and Pavlychko were planning to use the rally to declare the creation of a "Popular Movement of Ukraine." They failed to do so at this time, because they could not control the list of speakers or the microphone, which were in the hands of government officials. Ultimately, they did form such a group, the one that would be called Rukh.[18]

The rally still became a turning point, however. Approximately two and a half hours into the event, the microphone was suddenly seized by Ivan Makar, a physicist who had just been released from prison for his role in organizing political rallies in Lviv, the cultural center of western Ukraine. That city, which had been annexed to the Soviet Union on the basis of the Molotov-Ribbentrop Pact in 1939, had been a hotbed of Ukrainian moblilization at the time of perestroika. With Makar attending the Kyiv rally, KGB officers managed to turn off the sound system, but Makar delivered his speech anyway. Warning the attendees who could hear him that the central authorities were about to pass laws making it impossible for the republics to gain economic sovereignty, Makar called on Ukrainians to organize a popular front like those in the Baltics, and fight together with the Baltic groups for sovereignty against the center. Although Makar's speech was not transmitted through the sound system, it had long-term consequences. The Kyivan crowd demanded that the KGB switch on the sound system during Makar's speech, chanting the word "Mi-cro-phone," and this became the title of a documentary film about the ecological consequences of Chernobyl that was shot with the help of Yurii Shcherbak. It also became a battle cry for glasnost in Ukraine about the Chernobyl disaster and much else.[19]

Ukrainian writers formed a group to work on the creation of Rukh ten days after the rally, on November 23. Drach was elected chairman, and Honchar and Shcherbak also joined the group. A few weeks later, at a writers' forum, Drach declared that the need for Rukh had been "dictated above all by the idea of Chernobyl." He continued: "The only alternative to those atomic stations is the People's Movement of Ukraine." Years later, Drach recalled that "Chernobyl was the stimulus of all the democratic processes in Ukraine, Polish Solidarity the example, and the Writers' Union the cradle."[20]

The Ukrainian writers mobilized their supporters around the theme of Chernobyl and petitioned the highest authorities in Moscow to institute changes. In late 1988, Honchar, Oliinyk, and other prominent writers addressed themselves to the all-Union party's Central Committee, pushing ahead with the agenda proposed at the November rally. They demanded the creation of a special commission to investigate the consequences of the Chernobyl disaster, bringing the activities of the ministries of energy and health under public control, and conducting referenda on the construction of new power plants. The Ukrainian authorities sounded a retreat. Suppression alone was no longer an option.

"It must be admitted that for a long time, due attention was not paid to implementing measures for the protection of the environment," wrote the party boss of Ukraine, Volodymyr Shcherbytsky, to Moscow. He reminded his colleagues and bosses in the Soviet capital that the Ukrainian government had petitioned the center to stop construction of the Chyhyryn nuclear power plant—it had been one of the main demands of the November rally in Kyiv. Shcherbytsky also argued against the construction of any other nuclear power plants in Ukraine. "According to the data of the Academy of Sciences of the Ukrainian SSR," he wrote, "as much as 90 percent of the territory of the Ukrainian SSR is characterized by complex geological and hydrological conditions basically unfavorable to the building of atomic power stations."[21]

The tide had turned. Through his policy of glasnost, Gorbachev had encouraged local cultural elites to rebel against local party authorities in the name of perestroika. Rebel intellectuals such as Serhii Naboka, veteran dissidents such as Oles Shevchenko, and writers formally loyal to the regime, such as Ivan Drach, all found in Chernobyl a new cause to add to their previous agenda of political freedom, human rights, and the development of the Ukrainian language and culture. Appeal to this new cause turned out to be more effective in mobilizing broad support for their agenda than anything they had promoted earlier. It was the issue of Chernobyl that allowed the dissidents and the rebel intellectuals to break the common front of the communist authorities, pitting regional elites against their bosses in Moscow.

Before they knew it, Gorbachev and the reformist leaders in Moscow were facing a common front of rebel intellectuals and dissatisfied and disoriented republican officials. Nowhere was that change more obvious than in the field of nuclear energy. Republican political leaders and local cultural elites demanded that Moscow stop construction of new nuclear plants and shut down those already in operation. And that was just the beginning. In 1989, popular movements all over the Soviet Union would write on their banners slogans linking nuclear safety with national liberation.

NUCLEAR REVOLT

MIKHAIL GORBACHEV first visited the Chernobyl nuclear plant on February 23, 1989, almost three years after the catastrophe. He never explained why it had taken him so long. On his first and last visit to Chernobyl, the Soviet leader was accompanied by his wife, Raisa. The photos published in Soviet newspapers featured both of them dressed in white gowns, speaking to similarly dressed plant managers and a score of party and government officials in one of the reactor units of the troubled power plant.[1]

Three of the four units of the plant were producing electricity in February 1989, but the problems caused by the explosion were far from over. In December 1988, the KGB reported to party officials in Kyiv on a number of problems with Unit 4, the sarcophagus that covered it, and the decontamination efforts. Scientists and engineers were still unsure how much radioactive fuel remained in the damaged reactor and what state it was in. They could not investigate more closely for lack of equipment capable of withstanding radioactivity levels above 200 roentgens per hour. Civil and military crews worked around the clock removing contaminated soil and burying radioactive trucks and equipment. But here, too, appropriate equipment was unavailable, and much of the work was done with primitive machinery, endangering the health of the workers and soldiers present and slowing down the whole process. The bulldozers that were removing contaminated soil often mixed it with "clean" soil, further polluting the area and spreading radiation instead of containing it.

Then there were continuing problems with the sarcophagus itself. The shelter over the damaged reactor was partially built on the sections of the old unit's walls that had survived the explosion. This had been considered a brilliant architectural solution because it would save the lives and the health of construction workers, but now it was seen to display all the hallmarks and drawbacks of the Soviet "quick fix" approach. The foundations of the reactor unit, which were never designed to bear large loads, were slowly sinking into the ground under the weight of the new concrete structure above it. The pouring of concrete onto the approaches to the sarcophagus, in order to suppress radiation from the soil, as well as the construction of the underground concrete wall that was supposed to prevent radio-nuclides from contaminating the groundwaters of the Dnieper River Basin, changed flow of groundwater beneath Unit 4 and thus made the foundations of the sarcophagus less stable.[2]

Mikhail Gorbachev had little to offer the struggling power plant by way of new funds, as the Soviet economy was in free fall, accelerated by declining oil prices on world markets—the main source of hard-currency earnings for the state budget. He placed his hopes for improving Soviet economic performance in market reforms. In May 1988, nine months before coming to Chernobyl, he managed to obtain parliamentary approval of a law on "cooperatives"—collective enterprises in the industrial sector and the service industry—that broke the state monopoly on economic activity in urban areas and allowed non-state ownership of small businesses. But those were marginal changes. Partial reforms failed to revive the economy, which had always suffered from shortages of foodstuffs and consumer goods but now went into a tailspin, leaving the shelves of Soviet stores half empty. The behemoth of the Soviet economy was showing fewer and fewer signs of life.

Inspired by a vision that dated back to the Prague Spring of 1968, when the Czech communists tried to create a communism with "a human face," Gorbachev believed that economic reform was impossible without some form of democratization. What Gorbachev saw around him seemed to confirm his view that the two aspects of reform were interdependent. His perestroika initiative undermined the state monopoly on ownership of property and thus the economic foundations of Soviet socialism; consequently, it met with stiff resistance on the part of the old party and managerial cadres. Gorbachev responded by pushing political reform, challenging the party elite's power monopoly, and introducing elements of electoral democracy into the political system, which was still basically that of the Stalin era. With these measures, he hoped to outmaneuver his conservative opponents while mobilizing his liberal supporters. He gave the latter a degree of political freedom that was supposed to compensate for empty shelves and economic hardships.[3]

On February 24, 1989, the day after his visit to Chernobyl, Gorbachev met in Kyiv with Ukrainian writers who were eager to launch "the Movement for Perestroika," which was in preparation since the November 1988 Kyiv rally. It would later become known as Rukh. They wanted to help him with his economic and political reforms and sought his help in return. Volodymyr Shcherbytsky, the party boss of Ukraine and a protégé of the long-deceased Soviet leader Leonid Brezhnev, was present at the meeting. There was no love lost between Gorbachev and Shcherbytsky, who was one of the few conservatives remaining in the Politburo. The quarrel over the parade of May 1, 1986, in Kyiv, when Gorbachev had forced Shcherbytsky to conduct the mass rally despite rising levels of radioactivity, was only one of the many causes of the strained relations between them. Shcherbytsky had faith neither in Gorbachev's reforms nor in his management of them. Convinced that the new general secretary was driving the country to ruin, he once remarked to his aides: "What idiot invented the word 'perestroika'?"[4]

Shcherbytsky had used all the resources at his disposal, including party organizations and the KGB, to prevent the Rukh movement from forming. The Ukrainian writers placed their last hopes in Gorbachev's visit. Oles Honchar, Ukraine's most celebrated writer, pulled no punches when, in his presentation to Gorbachev, he called the Shcherbytsky-orchestrated campaign against Rukh "persecution." Gorbachev did not interrupt Honchar, allowing him to vent his frustration at the Ukrainian party boss. He pretended to know nothing about the predicament in which Rukh found itself.

That, at least, was the impression that Ivan Drach, an anti-nuclear convert and a future Rukh chairman, took from the meeting, which he considered a success. "[Gorbachev] wanted to hear our opinion. That is the approach of a real leader who does not rely only on outside assessments," Drach told one of his acquaintances who turned out to be a KGB informant. "We are now hopeful that the hounding of Rukh will stop."[5]

The KGB reported Drach's words back to Shcherbytsky. There is little doubt that after the meeting with Gorbachev, Shcherbytsky ordered the media to stop its campaign against Rukh. Although the resistance of the Ukrainian party apparatus to Rukh continued, its "persecution" of it subsided considerably after Gorbachev's visit, allowing the Rukh founders to begin preparations for their first congress. That congress met in September 1989. A few weeks before it began, Gorbachev forced Shcherbytsky to resign. A new era was about to begin in Ukraine, significantly affecting the fate of the Chernobyl nuclear plant and the exclusion zone around it.

THE RUKH leadership would take no prisoners in the war on the nuclear lobby and its supporters in Ukraine. On February 16, 1989, a few days before the writers' meeting with Gorbachev, the Rukh program had been published in *Literaturna Ukraïna* (Literary Ukraine), the newspaper of the Writers' Union of Ukraine. The program included an extensive section on ecological issues, with special attention to the Chernobyl disaster and its consequences. The ecology section followed the section on social justice and preceded sections on the nationality question, culture, and language, which were all-important to the writers. The Rukh program called for the shutdown of the Chernobyl nuclear plant and of all the other RBMK reactors in Ukraine; a halt to the construction of new nuclear power plants in Ukraine, no matter what reactor type they were designed to use; medical examinations for the entire population of Kyiv and other regions adjacent to the Chernobyl power plant; and rehabilitation measures for those adversely affected by the disaster.[6]

The Rukh program was published in time for the first semi-free elections to take place in the Soviet Union since the Revolution of 1917. In order to speed up reforms of the Soviet political system,

Gorbachev had decided to create a new institution, the Congress of People's Deputies—a "super-parliament" of 2,250 representatives. One-third of the deputies would be appointed by the party, while two-thirds would be elected by the voters. The elections were scheduled for late March and early April 1989, with the congress to begin work in May. The elections turned into a battle between the Communist Party appointees and the representatives of the new democratic movements, who treated ecological concerns as seriously as they did the political and economic problems that were besieging the rapidly declining Soviet system.

Throughout the Soviet Union, the leaders of the now awakened civil society, distressed by economic hardship but encouraged by Gorbachev's political reforms, turned to eco-activism. It soon took on the features of eco-nationalism, a political movement whose leaders linked concerns about environmental protection with ethnonational agendas, presenting their republics as the principal victims of the center's environmental policies. Nuclear power plants were depicted as embodiments of Moscow's eco-imperialism. In Lithuania, controversy centered on the Ignalina Nuclear Power Plant, which ran on Chernobyl-type reactors. In September 1988, Sajudis, the Lithuanian popular front, which served as one of the models for the Ukrainian Rukh, mobilized close to 20,000 people to form a human "Ring of Life" around the Ignalina plant. The plant was perceived as not only an ecological but also a cultural threat to the Lithuanian nation: like the Chernobyl plant, it was staffed largely by Russians and representatives of other nonindigenous peoples. In Armenia, a December 1988 earthquake triggered mass protests leading to the closure of the Metsamor Nuclear Power Plant, which was built in an area of high seismicity only 36 kilometers from the capital of the republic, Yerevan.[7]

In Ukraine, Rukh was the largest of several new organizations demanding the closing of the Chernobyl power station and taking care of victims of the disaster. The Ukrainian writer and pioneering environmental activist Yurii Shcherbak, who was among the early supporters of Rukh, continued to lead his Green World association. In April 1989, the association published its own program, centered on Chernobyl-related issues. Shcherbak ran in his native Kyiv for a seat in the Congress of People's Deputies. A scholar and writer who had never been a member of the Communist Party, he faced stiff competition from the party-backed candidates, who accused Shcherbak of being both a Ukrainian bourgeois nationalist and a Zionist. If that were not enough, they pointed out that his wife was Polish. But Shcherbak, who stuck to his ecological platform, was elected with 57 percent of the vote—a major accomplishment in a constituency with six candidates on the ballot.[8]

In late May 1989, when Yurii Shcherbak arrived in Moscow to attend the first session of the Congress of People's Deputies, he was by no means the only member of the newly created "super-parliament" to put ecological concerns at the top of his agenda. Of the 2,250 deputies, about 40 represented environmental organizations of various kinds, while at least 300 included environmental issues in their programs.

But there was one deputy who stood out in the small but active group of "Greens" in the first semi-freely elected Soviet parliament. Her name was Alla Yaroshynska, spelled in Russian as Yaroshinskaya. A young journalist elected to the congress from the Zhytomyr region of Ukraine, which lay in close proximity to the Chernobyl Exclusion Zone, she became the new face of the movement that demanded the whole truth about the Chernobyl disaster and its consequences. She ran on a program focused on the fate of one of the regions most affected by radiation fallout, the Narodychi district, located less than 80 kilometers west of the Chernobyl nuclear plant. "It is imperative to make public the consequences of radioactive contamination in the Narodychi district, hitherto carefully concealed from the people," read Yaroshinskaya's electoral program. Once in Moscow, she made sure the entire Soviet Union heard about the tragedy of Narodychi and took notice of the long-term effects of the Chernobyl disaster.

Yaroshinskaya was known as a rebel. A native of Zhytomyr, a city of some 250,000 people located 140 kilometers west of Kyiv, she had earned a journalism degree from Kyiv University in the early 1970s, when the KGB was destroying whatever liberal spirit still remained from Khrushchev's de-Stalinization campaign of the 1950s and early 1960s. She saw fellow students expelled from the university for placing flowers at the foot of the monument to the Ukrainian national poet, Taras Shevchenko, on the wrong day: it was fine to celebrate his memory in March, but not in May, as doing so in May was considered an expression of Ukrainian nationalism, which was to be suppressed. Then her husband was expelled from the university for writing a paper in which he praised the revolutionary-era anarchist leader Nestor Makhno, whom he called a hero of Ukraine: the Soviets considered him a dangerous counterrevolutionary warlord. Yaroshinskaya collected signatures under a letter protesting her husband's expulsion, but to no avail.

A talented writer, Yaroshinskaya landed a job with the Zhytomyr regional newspaper but refused to join the Communist Party and remained the only non-party reporter on the party-run paper. She was an idealistic believer in social justice and the overall soundness of the Soviet system, and she knew that she could not publish articles exposing "deformations" and abuses of the system. She thought it would be acceptable, however, to write personal letters to Moscow authorities alerting them to abuses of power by local party bosses. Some of her letters were addressed to Leonid Brezhnev, the head of the party and of the Soviet state. One day she was summoned to the regional party headquarters and ordered to stop writing letters if she wanted to keep her job. She was also called in to the KGB to discuss her political views, and on one occasion she was even abducted for a few hours: a KGB colonel drove her to the outskirts of the city as part of an attempt at intimidation. Yaroshinskaya was not cowed by those tactics.

The election of Gorbachev as general secretary of the party revived Yaroshinskaya's hopes of working for a better and more just society. In 1986, she and a fellow journalist, Yakiv Zaiko, established a club called "For Perestroika" in Zhytomyr. The communist media accused them of trying to create a political party—a crime tantamount to high treason in the pre-perestroika USSR. There was no hope of publishing anything critical of corrupt local party officials in Zhytomyr's newspapers, so she tried more-liberal media outlets in the capital. It worked: glasnost had yet to reach Zhytomyr, but it was taking its first steps in Moscow. In June 1987, the second most important Soviet newspaper, *Izvestiia* (News), published an article by

Yaroshinskaya attacking the Zhytomyr party bosses for their intolerance of criticism and persecution of opponents. The party bosses fought back, forwarding a letter to the Moscow Central Committee signed by nineteen local reporters who attacked Yaroshinskaya. Her own newspaper called a party meeting at which the authorities grilled her for six long hours. She was demoted to a part-time position but refused to abandon her efforts.[9]

In the fall of 1987, Yaroshinskaya, then thirty-four years old and a mother of two, asked her male boss for a day off. He refused. She insisted and got her way, but only after she told him that she needed that day to have an abortion. Instead of going to a clinic, Yaroshinskaya drove to the small town of Narodychi, where in the office of a local official she studied a secret map of radioactive contamination in the region. She had become interested in the problem on hearing that the authorities had begun building new housing and facilities there for people resettled from the exclusion zone near Chernobyl. It seemed to Yaroshinskaya that the new location was still too close to the disaster site to be safe. When she asked her editor for permission to visit the new settlements, he had refused the request, telling her that it was none of their business. Theirs was a regional newspaper, while decisions on resettlement were made at the republican and all-Union levels. It was up to Moscow- and Kyiv-based newspapers to look into the matter. But Yaroshinskaya would not take no for an answer.

The first village in the Narodychi district that Yaroshinskaya visited in complete secrecy from her editor was called Rudnia Ososhnia. It was under constant radioactive monitoring. The local school was closed when radiation levels in the building exceeded 1.5 rem per hour. But the authorities kept their construction program going, expecting people to keep living there. The workers whom Yaroshinskaya met at the local construction site were tiring easily and developing headaches. They were paid supplements of thirty rubles per month, allegedly for better nutrition. The workers called it "coffin money." Neither they nor the locals could tell her why a new kindergarten was being built in the radiation-affected village, where there were hardly any children. A new bathhouse was also going up, apparently to make it easier for the locals to cleanse themselves of radioactive dust.

Yaroshinskaya was appalled. For the next month, she and her husband spent their weekends going from one contaminated village in the region to another. The situation was the same in all of them: there was new construction, and radiation levels were being kept secret, endangering the health of both the locals and the evacuees who were being resettled there from the exclusion zone. There was also a growing health crisis, particularly among children. Yaroshinskaya learned from local medical personnel that 80 percent of the children in the region had enlarged thyroid glands—a sign of abnormally high levels of radiation exposure. Before the Chernobyl disaster, only 10 percent of the children had had such symptoms.

If Yaroshinskaya was to help the inhabitants of the region, she had to publicize their plight. Her own newspaper was not prepared to publish the results of her investigation, but she could always try Moscow, where her criticism of the local party elite had already appeared. But Chernobyl turned out to be a question too sensitive even for the central media. *Izvestiia*, where she had published previously, turned her down, claiming that the topic was classified. The party's leading newspaper, *Pravda* (Truth), did likewise after half a year of procrastination, saying that a similar article had already been approved for publication. Yaroshinskaya's appeals to Vladimir Gubarev, a *Pravda* reporter and author of numerous articles on Chernobyl, proved fruitless. The mouthpiece of perestroika, the journal *Ogonek* (Little Light), headed by the Ukrainian writer and poet Vitalii Korotych, whom Yaroshinskaya managed to meet in person, also passed on the article after numerous promises to publish it. So did another liberal newspaper, *Literaturnaia gazeta* (Literary Newspaper).[10]

All these rejections were mainly due not to the wishes of the editors but to the strict censorship imposed on anything related to Chernobyl, short of glorification of the efforts of the liquidators. Vitalii Karpenko, the editor of the popular newspaper *Vechirnii Kyiv* (Evening Kyiv), later recalled having been reprimanded simply for publishing a photo of Kyiv with very few people on the streets in one of his issues for May 1986. His case was examined by the secretary of the Ukrainian Central Committee. Anything the newspaper published on the matter had to be approved either by the Moscow authorities or by the secretary of the Kyiv city committee of the Communist Party.

Later, Yaroshinskaya got ahold of secret resolutions adopted by party and government agencies classifying anything dealing with Chernobyl. The KGB had taken the lead as early as May 1986, putting the stamp of secrecy on everything from the causes of the accident to its impact. In June, the Ministry of Health issued a directive classifying all information about the medical treatment of the population in contaminated areas and radiation levels sustained by the liquidators. In July, the Ministry of Defense ordered its human resources officers not to register service in the Chernobyl zone in the personal files of officers and soldiers dispatched there. There was to be no mention of radiation exposure in the files unless it reached 50 rem—ten times the usual norm and twice the "Chernobyl norm."[11]

Yaroshinskaya had hit the wall, or so it seemed at the time. There were limits to glasnost. Blaming local authorities for corruption and shortcomings was part of Gorbachev's attack on the conservative party apparatus, but telling the truth about the Chernobyl plant, for which the central government was responsible, was an entirely different matter. That would have required an admission of guilt on the part of the central authorities, including Gorbachev himself, for hiding the truth from the people and then spending tens of billions of rubles to rehabilitate the affected areas and inhabitants. Gorbachev simply did not have the money. The economy was in decline, with long-awaited reforms disrupting the existing system and putting an even greater strain on dwindling government coffers. Yaroshinskaya was reduced to distributing copies of her article to her friends. In the era of glasnost, she had no choice but to return to *samizdat*—the practice, perfected by Soviet dissidents of the previous era, of copying prohibited texts on private typewriters and circulating copies among friends and acquaintances.[12]

But then, suddenly, everything changed. At the Nineteenth Communist Party Conference, called by Gorbachev in the summer of 1988 to approve his plans for semi-free elections, the Ukrainian writer Borys Oliinyk broke the taboo on speaking publicly about the responsibility of the central authorities for the Chernobyl disaster. In September 1988, the liberal Moscow journal *Novyi mir* (New World) published an article by the Belarusian writer Ales Adamovich titled "Word of Honor, It Won't Blow Up Again—A Non-Specialist's Opinion." Adamovich not only wrote about the responsibility of senior officials for the accident but also argued that huge areas around the Chernobyl power plant were much more contaminated and dangerous to live in than had yet been admitted. The truth had been concealed, claimed Adamovich, so as not to endanger the construction of new power plants. In fact, he went on, the situation was catastrophic: for example, the Belarusian city of Bragin, 90 kilometers from Chernobyl, was considered too dangerous for medical doctors to be stationed there on a permanent basis. Accordingly, they worked in shifts, while the residents, including women and children, stayed permanently, wrote Adamovich.[13]

Adamovich's article pushed the envelope of glasnost further than Oliinyk's address, making the fate of contaminated areas a topic of legitimate discussion. In September 1988, the month in which Adamovich's article was published, Yurii Shcherbak brought a small group of filmmakers to Narodychi, the subject of Yaroshinskaya's utmost concern. The documentary film produced with his help showed calves born without eyes and with other deformations caused by high radiation levels at local farms—in one case, levels 150 times those of the city of Kyiv. Over the course of the year after the accident, sixty-four deformed animals were born at that particular farm, compared to three during the previous five-year period. But the twenty-minute documentary would not be shown on television or the big screen, leading one of the filmmakers to publish an article describing his findings. The authorities were still doing their best to keep the truth under wraps, but their monopoly on information was crumbling.[14]

In the late summer of 1988, Yaroshinskaya was allowed for the first time to speak in front of large audiences about her findings in the Narodychi district. The response was enormous, and on the part of almost everyone except party officials, extremely positive. The citizens of Zhytomyr wanted to hear more and would prevail upon their bosses at factories and research institutes to make halls available for her presentations. By the time the elections to the Congress of People's Deputies came along in the spring of 1989, no hall was big enough to accommodate all those wishing to hear Yaroshinskaya speak. She was now addressing crowds in city squares and sports stadiums. Party officials tried to intimidate her with threatening phone calls and hate mail. They tried to pressure her husband, who served in a firefighters' unit of the Ministry of the Interior, to divorce her. They harassed her son and opened criminal investigations against dozens of her supporters, accusing them of planning an attack on Communist Party headquarters. Still she gathered crowds, with the numbers of people coming to hear her reaching 20,000 to 30,000. Eventually, the authorities gave way. The local newspaper, for which she had worked all those years, and which had turned against her under political pressure, found itself obliged to print her electoral program. She was elected with more than 90 percent support.

In Moscow Yaroshinskaya quickly established a bond with like-minded deputies. Among them was her fellow ecological activist Yurii Shcherbak; the author of an explosive documentary, *Zapredel* (Beyond All Norms), about the Narodychi district, Mykhailo Belikov; and the ecological activist Serhii Koniev from Dniprodzerzhynsk, an industrial town known chiefly as the birthplace of Leonid Brezhnev but later reviled as one of the most polluted cities in the Soviet Union.

Once Mikhail Gorbachev opened the discussion at the Congress of People's Deputies, Yaroshinskaya and Shcherbak put themselves on the speakers' list, hoping to get a chance to talk about the ecological consequences of the Chernobyl disaster, but neither Gorbachev nor any of the officials chairing sessions invited them to speak. Yaroshinskaya then took the initiative, approaching Gorbachev at the dais to ask him to allow her to speak on the Chernobyl issue, and Gorbachev obliged. She used her three minutes at the podium to talk about the Narodychi district, the contaminated villages, and the lies told by Ukrainian health officials, who had claimed that levels of radioactivity in the region had done no harm to those who continued to live there. She characterized the situation as scandalous and handed Gorbachev Belikov's documentary about Narodychi.

The veil of official silence on the consequences of the disaster was slowly being lifted. After her presentation, Yaroshinskaya received dozens of telegrams and letters congratulating her on her courage. In the following days, other deputies raised their voices against the government's concealment of information on the issue. Among them was a medical doctor from the Mahilioŭ region of Belarus. The party boss of Belarus, Yevgenii Sokolov, went to the podium to declare that 18 percent of Belarusian territory had been contaminated by the Chernobyl fallout. Yaroshinskaya was appalled that the most senior official in the Ukrainian delegation, Prime Minister Vitalii Masol, remained silent. But when she returned to her native Zhytomyr and held her first meeting with her electors at the packed city stadium, and later met with the people of Narodychi, she could tell them that things were changing.

Before the end of summer, a government commission headed by the Soviet deputy prime minister was already in Narodychi, accompanied by the deputy head of the republican government and the party boss of the region. The KGB informed Ukraine's party boss, Volodymyr Shcherbytsky, about the session of the Academic Council on Radiobiology of the All-Union Academy of Sciences held in Narodychi. Close to eight hundred people had gathered at the Narodychi cultural center on June 13 to meet with the scientists. Some members of the audience called for strikes to force the authorities to proceed from finally acknowledging the problem to doing something about it. But aside from reporting on such demands, there was little that the KGB could do: the political situation was changing.[15]

By the fall of 1989, not only Ukraine but also neighboring Belarus was in ecological revolt against the communist authorities. In Belarus, the first mass demonstration took place in the capital, Minsk, on September 30. Government officials tried to stop the buses that were bringing people from contaminated areas in the Homel and Mahilioŭ regions to take part in the rally, but they failed. Close to 30,000 people listened to the leaders of the Belarusian Popular Front, an analog of the Ukrainian Rukh, which had been created a few months earlier in virtual exile in the Lithuanian capital of Vilnius. No Belarusian party or state official addressed the people. The authorities were in denial, if not of the grim consequences of

the Chernobyl disaster, then of the fact that they had lost their monopoly on information and political action.[16]

THE ELECTIONS to the Congress of People's Deputies and the work
of the congress helped lift the veil of secrecy over the ecological
effects of the Chernobyl disaster, but they did little to bring the
antinuclear activists closer to their ultimate goal of shutting down
Chernobyl and other nuclear power plants using RBMK reactors.
The activists got their chance, however, with the elections to the
republican parliaments that took place in early March 1990—the
second stage of Gorbachev's political reform. This time there were
no party-appointed deputies—everyone who would occupy a seat
in the new republican parliaments had to be elected—and the results came as a major surprise to Gorbachev and the other officials
in Moscow. The deputies to the all-Union congress in the summer
of 1989 had only been able to name and shame; the deputies to the
republican parliaments in the spring of 1990 turned from words to
actions.

The voters in a number of republics elected proponents of national independence who associated it closely with the idea of de-
nuclearization. But the degree of antinuclear zeal among the new
deputies differed from one republic to another, depending on the
distance it wanted to establish between itself and Moscow. Lithuania, the site of the Ignalina Nuclear Power Plant, where mass rallies
in support of shutting it down had been held in the fall of 1988, was
the first republic to declare its independence from the Soviet Union.
The declaration was made in March 1990, soon after the newly
elected republican parliament met for its first session. Alarmed, Gorbachev responded with an economic blockade of the rebellious republic. The newly elected leadership of the Lithuanian parliament
now had to consider the importance of an energy source independent of Moscow. The economic blockade, coupled with a number of
temporary shutdowns of two of the Ignalina plant's RBMK reactors,
forced the leaders of the antinuclear movement to change course and
start thinking not about shutting reactors but building new ones to
ensure their country's independence. Nuclear activism was turned
on its head.[17]

In Ukraine, where the political elites were still far from seriously considering the possibility of separation from Moscow, but where the impact of the Chernobyl disaster had become an urgent issue, anti-nuclear activism not only survived the elections of 1990 but grew in strength. In February 1990, in the middle of the electoral campaign, and in an attempt to remove the Chernobyl issue from Rukh's political arsenal, the communist leadership of Ukraine orchestrated a decision of the outgoing communist-controlled parliament to shut down the Chernobyl power plant by 1995. The authorities also refused to register Green World as a separate party and prevented its members from running on a party platform.

But these desperate moves to eliminate Chernobyl as a central political issue had little impact on the outcome of the elections. Among the most popular election slogans was a rhyme that blamed the Communist Party for the Chernobyl disaster and looked forward to seeing it expire at the nuclear plant: "Khai zhyve KPRS na Chornobyl's'kii AES!" (Long live the CPSU at the Chernobyl atomic energy station!). Three-quarters of the leaflets produced in Ukraine during the electoral campaign included references to Chernobyl and ecology, which appeared to be more important to the electorate than questions of the economy and social justice. More than a hundred deputies elected to parliament, roughly one-quarter of the total, joined the pro-Rukh parliamentary caucus known as the People's Council—a stunning blow to the seemingly all-powerful party machine.[18]

In April 1990, the fourth anniversary of the Chernobyl disaster was marked by rallies in towns and settlements near all the nuclear power plants in Ukraine. In the town of Netishyn, next to the Khmelnytskyi Nuclear Power Plant in western Ukraine, about 5,000 people rallied; in the city of Rivne in the same region, some 3,000 demonstrated—they demanded the closure of the Khmelnytskyi and Rivne plants, which were located roughly 100 miles apart. The protesters, largely local, spoke Ukrainian and carried signs denouncing the party and attacking the mostly Russian and Russian-speaking personnel of the two power plants. They tried to break through the gates of the plant in Netishyn but were stopped by guards. The Netishyn rally supported the workers of one of the local concrete-producing facilities who had stopped supplying concrete for the construction of a new unit at the Khmelnytskyi plant. Those workers called on other workers to join the strike. The KGB was alarmed and informed party officials in Kyiv about the danger of mass unemployment in Netishyn if the authorities caved in to pressure and shut down the Khmelnytskyi plant.[19]

The Ukrainian authorities sought a compromise. In the summer of 1990, the new parliament, continuing the trend started by its predecessor, which had decided to shut down Chernobyl, introduced a five-year moratorium on the construction of new reactors on Ukrainian soil. The parliament also created a special commission to deal with problems related to the Chernobyl disaster and to investigate the role of the Moscow and Kyiv authorities in withholding information about the dangers presented by the explosion. The main goals of the ecological movement had been achieved. Ahead was the difficult task of finding and managing scarce resources to rehabilitate the people and territories affected by the disaster. But the political leaders of the mass movement that had been launched into political orbit by the Chernobyl-inspired antinuclear mobilization were now turning their attention elsewhere.

In October 1990, Rukh held its second congress in Kyiv. The early ecological activist Ivan Drach, now a parliamentary deputy, was reelected as its leader, but Rukh changed its program, dropping the words "for Perestroika" from its name and declaring the independence of Ukraine as its principal goal. The ecological segment of the program underwent minimal change, but it was no longer ecology alone that was causing people to mobilize for mass action. With the political freedoms that had been won under the banner of eco-nationalism, the new political leaders could openly address the issue of gaining independence for their republic. The shock wave of Chernobyl was about to destroy the foundations of the Soviet Union.[20]

INDEPENDENT ATOM

O N THE warm summer morning of August 24, 1991, thousands of people gathered around the Ukrainian parliament building in downtown Kyiv. It was a Saturday, and many Kyivans who would otherwise have been at work were demonstrating. There were also out-of-towners from almost every part of Ukraine. They carried posters that read: "Down with the CPSU!," "Ukraine is leaving the USSR!," and "No to the Fascist Union!"[1]

A few days earlier, on August 19, hardliners in Moscow, led by the KGB chief, Vladimir Kriuchkov, had ousted Mikhail Gorbachev and created an emergency committee that sought to reverse the Soviet president's democratic reforms. The plotters isolated Gorbachev at his summer mansion in the Crimea but failed to arrest his political rival, the charismatic president of Russia, Boris Yeltsin, who mobilized Muscovites in defense of democratic freedoms. The military refused to suppress the mass revolt, and by the evening of August 22, the coup had all but evaporated. The triumphant Yeltsin brought Gorbachev back to Moscow but refused to yield to him the powers he had acquired during the coup. He forced Gorbachev, who was still in shock, to dismiss his security ministers and accept those whom Yeltsin recommended. The Russian president also banned the activities of Gorbachev's last bastion of power, the Communist Party.[2]

Yeltsin's countercoup made him the most powerful figure in Moscow. The Ukrainian party elite became worried, and the leaders of the opposition forces in the republic were concerned about the possibility of a repeated coup by the hardliners. Neither group relished the notion of remaining under Moscow's control. As in the case of the Chernobyl disaster, the rulers and the opposition found a common enemy in Moscow, but they could not agree on whether they should move toward autonomy or independence. If they became autonomous, they would have some control over their local affairs but still remain part of the Soviet system; independence would mean a complete break from the Soviet Union. Rukh's supporters demanded independence, while the party elite vacillated. The Kyivans, Rukh activists, and other supporters of democratic reform who gathered outside the Ukrainian parliament on the morning of August 24 demanded that the communist majority in parliament declare complete independence from Moscow. They were growing increasingly restless and calling for punishment of the Ukrainian communist elite, which had temporized during the coup instead of supporting Yeltsin.

As tension grew outside the parliament building, Leonid Kravchuk, the speaker of parliament, who was under direct attack for his passivity during the coup, called on the forty-nine-year-old writer and Rukh activist Volodymyr Yavorivsky, the head of the parliamentary commission on the consequences of the Chernobyl disaster, to speak. Earlier in the day the leaders of the People's Council, the democratic opposition in the communist-controlled parliament, had given Kravchuk a number of draft bills on which they wanted the parliament to vote. Among them was a declaration of Ukrainian independence drafted by Levko Lukianenko, a dissident who had spent twenty-five years in the Gulag and internal exile. The communist deputies claimed that they had not seen the draft and refused to discuss it. Yavorivsky decided to use the opportunity presented to him by Kravchuk to read the text.

He began with a call for unity: "Honored deputies, honored guests, our dear Ukrainian people! Let us say that this is not a moment of revenge but a moment of truth. Let us say that those who have gathered here are not victors and vanquished but that, in fact, we have all been vanquished. This is our chance to leave those disputes behind." He then read the draft declaration with its key statement: "The Supreme Soviet of the Ukrainian Soviet Socialist Republic solemnly declares the independence of Ukraine and the creation of an independent Ukrainian state—Ukraine." This came as a severe shock to many members of the communist majority. Nevertheless, they did not rebel or protest but requested a break for consultations. When they returned to the chamber, they were prepared to support Yavorivsky's proposal. Kravchuk put the question of independence to a vote. The result was stunning: 346 deputies voted in favor, with only 2 opposed and 5 absentees. Subject to a referendum, scheduled for December 1, 1991, Ukraine, the second-largest Soviet republic after Russia and the home of the Chernobyl nuclear plant, would become a sovereign country.[3]

Levko Lukianenko, the author of the declaration, later claimed that the honor of being the first to read it out in parliament went to Yavorivsky because Kravchuk, who wanted an affirmative vote, believed that Yavorivsky, who had been a member of the Communist Party until 1990, had a better chance of convincing the communist majority. "Lukianenko is a nationalist, twice sentenced, and a recidivist," said Lukianenko, recalling the reasoning conveyed to him by those close to Kravchuk. "Communists hunted nationalists and imprisoned them; they look upon the nationalist Lukianenko as an enemy, while Yavorivsky is close to the majority. If Lukianenko reads the act, then many communists may not vote for it, but they would see Yavorivsky as one of their own, and a constitutional majority would be attained more quickly." Kravchuk later denied that he had chosen Yavorivsky over Lukianenko for political reasons. One way or the other, it was Yavorivsky who got to the microphone first.[4]

THE WRITER-TURNED-ACTIVIST Volodymyr Yavorivsky first emerged on the Ukrainian political scene in the fall of 1989 as the principal organizer of the first congress of Rukh. In the spring of 1990 he ran for election to the Ukrainian parliament, condemning the damage done to Ukraine by the Chernobyl disaster.

Yavorivsky was an unlikely champion of the antinuclear cause. An early proponent of nuclear power in Ukraine, he had published a novel, *Chain Reaction* (1978), that celebrated the construction of the Chernobyl power plant as a triumph of communist modernity and a sign of Ukrainian entry into the modern age. He dismissed concerns about the safety of nuclear reactors as an overreaction to the American bombing of Hiroshima and Nagasaki, maintaining that only in the capitalist world were there legitimate grounds for such concerns. Like a number of Ukrainian writers, including the leader of Rukh, Ivan Drach, Yavorivsky began his evolution from nuclear enthusiast to antinuclear crusader soon after the catastrophe. In the novel *Maria with Wormwood at the End of the Century* (1987), which Yavorivsky wrote after spending more than two months of symbolic repentance in the Chernobyl Exclusion Zone, he showed a growing preoccupation with the idea of personal responsibility for one's actions. More concretely, as Yavorivsky saw it, the individual was responsible to the family, to the homeland, and, ultimately, to the nation. Oles Honchar, Ukraine's best-known writer at the time, wrote to Yavorivsky upon the publication of his book: "Through your words, Ukraine itself addressed the world from the Chernobyl era with its pain and its hopes."[5]

Once in parliament, Yavorivsky focused on the Chernobyl issue and became chairman of its Chernobyl commission, which was created at his initiative. The commission investigated the role of the Ukrainian party and government elite in hiding the truth about the scope of the disaster and the harm it had done to the population. The KGB, which had been involved in the cover-up, now found itself reporting to Chairman Yavorivsky on the problems that continued to beset the damaged reactor and the territories contaminated by the explosion. Among Yavorivsky's early targets was the former director of the Chernobyl plant, Viktor Briukhanov. A fiery speaker and writer whom his opponents accused of populism, Yavorivsky often made controversial statements in his written and oral interventions. He claimed, for example, that on the night of the explosion, Briukhanov had been relaxing with his lover in a forest lodge and then neglected his duties as director, withholding information about the scope of the accident from his superiors and the public at large.[6]

Briukhanov, released from imprisonment in September 1991 after serving half of his ten-year sentence, was outraged. He recalled later: "Yavorivsky created an image for himself with the hasty publication of a book. He wrote all kinds of inventions. Since then, I cannot abide the name Yavorivsky." Briukhanov served most of his term in a prison facility in eastern Ukraine. He was a local celebrity in prison: when he arrived, inmates left their quarters to take a look at the individual charged with responsibility for the world's worst technological catastrophe. The prison authorities offered him a managerial position, but he refused to take it and worked as a mechanic. They would later transfer him to a halfway house in central Ukraine. It was only after his release that Briukhanov first met his granddaughter, who was already five years old: she was born soon after the Chernobyl disaster.[7]

Scarred by his imprisonment—he later claimed that "about 95 percent of those whom I saw there could hardly be called human beings"—Briukhanov was still in relatively good health. He went to work for the Chernobyl nuclear plant and then the national agency dealing with the nuclear industry. Many of his codefendants were less lucky. The former chief engineer, Nikolai Fomin, never recovered psychologically from the shock of the disaster. In 1988 he was transferred from a regular prison to a psychiatric hospital and then released. His deputy, Anatolii Diatlov, who had run the turbine test at Unit 4, managed quite well psychologically, never admitting his guilt, but suffered from acute radiation syndrome and was released from prison in 1990 on health grounds.[8]

By the late summer of 1991, when Ukraine declared its independence from the crumbling Soviet Union and Briukhanov was released on parole, few believed that Briukhanov, Fomin, and Diatlov were the main culprits responsible for the disaster. In November 1991, a commission of Soviet nuclear scientists, led by the new director of the Kurchatov Institute of Atomic Energy, Yevgenii Velikhov, concluded that the managers and operators of the Chernobyl plant were not entirely to blame for the accident. They accepted the conclusion drawn by a commission of the USSR State Committee for the Supervision of Safety in Industry and Nuclear Power, the Soviet nuclear industry watchdog, which declared as follows: "The flaws in the construction of the RBMK-1000 reactor, which was used in Unit 4 of the Chernobyl atomic energy station, predetermined the grave (catastrophic) consequences of the Chernobyl accident. The reason for the Chernobyl catastrophe is that the designers of the RBMK-1000 reactor chose a conception that turned out not to have taken sufficient account of problems of security."[9]

By the fall of 1991, Yavorivsky and his commission were going after bigger fish than Briukhanov in their search for those who were responsible, targeting the leadership of the Communist Party of Ukraine and the government. Volodymyr Shcherbytsky was dead, having succumbed to cancer in February 1990. But other principals were still around, including the former head of government and chair of the Chernobyl commission of the Ukrainian Politburo, Oleksandr Liashko. He stepped down as prime minister in July 1987, a year after the accident, in the month when Briukhanov and others were put on trial. Liashko was proud of his role in dealing with the consequences of the accident: he had insisted on the evacuation of Prypiat before anyone else, contrary to the position taken by the head of the government commission, Boris Shcherbina, and his own party boss, Volodymyr Shcherbytsky.

Yavorivsky took a different view of Liashko's role in the Chernobyl story. "What actions were taken by the government and the Central Committee of the Communist Party of Ukraine when they learned of the accident?" he asked Liashko when the former premier responded to the commission's call and came to testify about his role and that of the government in dealing with the disaster. "It was at night, when I was at home," responded Liashko, referring to the call he had received that night from the Soviet premier, Nikolai Ryzhkov. "What actions could have been taken?" Dissatisfied with the answer, Yavorivsky mocked the former premier: "Fine. So you slept soundly and then went off to work?" Liashko shared with the commission what he remembered about his efforts to mobilize city transport for the evacuation of the city of Prypiat and the activities of his government in the first days and weeks after the accident. The questioning lasted two hours. Liashko and the commission parted on a friendly note: two members of the commission accompanied the seventy-five-year-old Liashko to his apartment building in downtown Kyiv. The former premier believed that for him, at least, the investigation was over. He was in for a surprise.[10]

ON DECEMBER 11, 1991, ten days after Ukrainians voted en masse in a national referendum to endorse their parliament's declaration of independence, and one day after parliament ratified the agreement signed by Boris Yeltsin of Russia, Leonid Kravchuk of Ukraine, and Stanislaŭ Shushkevich of Belarus to dissolve the Soviet Union and create the Commonwealth of Independent States, the Ukrainian parliamentarians gathered again to hear the report of Yavorivsky's commission on its investigation into the cover-up of the consequences of the Chernobyl disaster.

The agreement signed by Yeltsin, Kravchuk, and Shushkevich declared that "the parties acknowledge the planetary nature of the Chernobyl disaster and commit themselves to join and coordinate their efforts to minimize and overcome its consequences." But for the time being, each country was dealing with the consequences of the disaster in its own fashion. That applied to the degree of media publicity given to Chernobyl-related economic, social, public health, and ecological problems, programs for dealing with the effects of the disaster, and the level of political and legal responsibility now being assigned to the communist government leaders who were in power at the time of the accident.[11]

In Ukraine, the parliament went for harsh treatment of the former Soviet officials who had dealt with the Chernobyl catastrophe—much harsher than in any other post-Soviet republic affected by the accident. Ukrainian eco-nationalism was still strong, its mobilizing power seemingly undiminished. Eco-nationalism had earlier fueled the drive for Ukrainian independence, but it was now becoming a weapon in the struggle over the country's future between the rising national democrats and the disoriented former communist apparatchiks who still controlled the government. Yavorivsky pulled no punches in his attack on the former communist government leaders for their real or alleged role in the Chernobyl cover-up. His ultimate targets were the Soviet imperial system of government, but for now he went after the Ukrainian post-communist elite, which was still clinging to power, as was no longer the case in Lithuania, the birthplace of eco-nationalism in the Soviet Union.

Yavorivsky began his report by characterizing the Chernobyl accident as a Ukrainian national tragedy first and foremost. "The biblical Star of Wormwood fell upon our earth and poisoned not only our grain, water, and air, but also your blood and ours," he asserted, developing some of the themes first addressed in the works and pronouncements of Ukrainian writers—his own and others—published after Chernobyl. He went on to claim that in grisly fashion the catastrophe had made Ukrainians a chosen people. "It is now clear that we are a people chosen by God," continued Yavorivsky. "Perhaps our neighbors will not deny us at least that."

Who was responsible for the national tragedy? Yavorivsky blamed the Communist Party of the Soviet Union and its branch in Ukraine. He called that branch "Little Russian," using the name for Ukrainians in the Russian Empire of the tsars and thus employing a powerful trope of anticolonial discourse. As far as Yavorivsky was concerned, communism, its empire, and militarism had come together to destroy the Ukrainian nation, which had been ruled as a colony. How was that accomplished? By placing a reactor with major design flaws 130 kilometers from Kyiv, at the intersection of three Ukrainian rivers—the Dnieper, Prypiat, and Desna. Moving on to the issue of responsibility for the accident, Yavorivsky rejected the interpretation endorsed by Gorbachev, according to which the plant operators alone had been responsible. Contrary to his own earlier claims, Yavorivsky exonerated the managers and engineers who had been tried by the communist regime in the exclusion zone in 1987, treating them as victims rather than perpetrators. Instead, he lashed out against "officials who hushed up the scope of the disaster, hid it from the people, and failed to take measures to protect the people from radiation, thereby committing a crime of huge proportions."

Yavorivsky read out excerpts from the commission's interviews with former Ukrainian officials, including Premier Oleksandr Liashko; the former head of parliament, Valentyna Shevchenko; and the former minister of health, Anatolii Romanenko. Romanenko's public silence on the health consequences of the disaster in late April and early May 1986 had made him one of the most hated officials in Ukraine. According to Yavorivsky, the commission had established that the republican authorities had had access to information about radiation levels in Chernobyl and other areas from the first days after the accident, but that they had failed to inform the population about the dangers. He argued that even if the authorities did not understand the information they had been given, and were unable to assess it correctly, that did not assuage their guilt—it only made them guiltier. "And they are guilty not only of utter incompetence as heads of a nuclear state; they are also guilty of not desiring, not wishing to know the truth, to say nothing of not wanting to tell the people about it," declared Yavorivsky. "For leaders of such rank, this is a crime that verges on genocide of their own people."

Yavorivsky charged not only former but also current leaders of the Ukrainian state and its officials, many of them present in parliament, with having committed a crime against the Ukrainian nation. Speaking after Yavorivsky's report, they fought back, defending their record and shifting responsibility to Moscow. The former attorney general of Ukraine, Mykhailo Potebenko, who had been hospitalized with symptoms of radiation poisoning after spending the first post-accident weeks in the exclusion zone, claimed that he had done everything in his power to prosecute those responsible for the accident, but supreme authority at the time lay with the attorney general of the USSR in Moscow. General Boris Sharikov, the deputy commander of the Kyiv military district, who had been active in dealing with the consequences of the disaster, accused Yavorivsky of being too emotional in his assessments. A crime had indeed been committed, but it had to do with the actions of certain individuals before the explosion, not after it.

Ivan Pliushch, the new speaker of the Ukrainian parliament (he had taken over from Leonid Kravchuk, who had been elected president of Ukraine a few days earlier), took the floor to calm the parliamentarians. "We were all raised in the belief that we would soon be heating our apartments with atomic boilers," said Pliushch, explaining the former leaders' ignorance of the security risks posed by nuclear energy. "I am for the commission continuing its work and finding out even more about the Chernobyl disaster, its causes and consequences," said Pliushch, "not just to deprive someone of his freedom but to remind leaders of every rank about their extraordinary responsibility to their people." He concluded: "It has turned out that those who actually caused the disaster became judges of others. Therefore, in order to dot the i's and determine who is to blame and establish everyone's degree of responsibility, we have to ask: who will judge?"

The chamber exploded in applause. Pliushch had in mind the guilty verdict delivered to Viktor Briukhanov, Nikolai Fomin, and Anatolii Diatlov by Moscow-appointed judges in the summer of 1987. But it could also be taken as a rebuke to Yavorivsky. Who was he to judge Pliushch, who had gone to the disaster site immediately on April 26, organized the evacuation of Prypiat, and received twice the permitted level of radiation—50 rem? Had not Yavorivsky himself welcomed the arrival of nuclear power in Ukraine in his literary works? Yavorivsky was clearly on the defensive when he returned to the podium. "Honored colleagues!" he said. "We have prepared the resolution [to be adopted by parliament], and it seems to me that it is dispassionate. Let the procuracy investigate. We have given only a political evaluation."[12]

The person who found Yavorivsky's words most offensive was not in parliament that day. The former premier, Oleksandr Liashko, learned of Yavorivsky's speech from the newspapers a few days later. On the day before the speech, Liashko had buried his daughter, who had died of cancer, and he was in mourning when news of the speech reached him. Liashko was particularly offended by Yavorivsky's claim that he had gone peacefully to sleep after Ryzhkov called him in the middle of the night on April 26, and that later in the day Liashko had called the Ministry of Foreign Affairs to find out whether they knew anything about the accident, instead of calling his minister responsible for nuclear energy. Yavorivsky had indeed misquoted Liashko's testimony to the commission, in the course of which Liashko had referred to the Ministry of the Interior, not the Ministry of Foreign Affairs.

But that was not the end of Liashko's troubles. He was soon summoned to the prosecutor general's office, which had opened a criminal investigation on the basis of the materials prepared by Yavorivsky's commission. Liashko wrote a long letter to the speaker of parliament, Ivan Pliushch, claiming that he had had nothing to do with ordering the May Day demonstration in Kyiv and had been in the dark with regard to the levels of radiation in the city—his wife, children, and grandchildren had also attended the parade. He blamed the deputy minister of health for the lack of information and stressed his own leading role in organizing the evacuation of Prypiat, as well as the subsequent evacuation of pregnant women and children from Kyiv. Liashko asked the speaker to put the question of responsibility for the Chernobyl disaster back on the parliamentary agenda and conduct hearings with his participation and that of other Ukrainian officials who had been in office at the time of the Chernobyl accident.

Ivan Pliushch did not respond to Liashko's letter but met with him when the former prime minister threatened to make the letter available to every parliamentary deputy. "I know that Yavorivsky talked a lot of blather that deserves no attention, and the matter will come to nothing," Pliushch assured the former prime minister. He arranged for Liashko to meet with the deputy general prosecutor of Ukraine, who also told him that the criminal case was pro forma and would not lead anywhere. The prosecutor went on to ask Liashko for advice: Would it make sense to open a criminal case against the members of the all-Union commission headed by Boris Shcherbina? Liashko was skeptical—the commission had made its decisions on the basis of information available at the time. Besides, the Soviet Union was no longer in existence, the USSR prosecutor's office had rendered its conclusions years earlier, and the court had sentenced those responsible for the accident. Liashko expected that the Russian prosecutor's office would not prove helpful in reopening the case.[13]

The Ukrainian prosecutor's office followed Liashko's advice and never opened a case against the all-Union officials. It did, however, proceed to charge Liashko and other former Ukrainian leaders, including the deceased Volodymyr Shcherbytsky and the former speaker of parliament, Valentyna Shevchenko, with "misuse of power and official position, entailing grave consequences." Liashko refused to admit guilt and soon learned that the case had been closed because of the statute of limitations. The criminal case initiated against Liashko and other former leaders on February 11, 1992, was closed on April 24, 1993, almost seven years to the day after the accident. The five-year limit on the prosecution of government officials for dereliction of duty had expired on April 26, 1991, even before the prosecutor's office opened the case, which amounted to a mere public relations gesture to pacify the parliamentary opposition and the concerned Ukrainian public.[14]

THE ATTEMPTS of the Rukh activists and eco-nationalists represented by Yavorivsky to not only condemn but also prosecute the Soviet-era Ukrainian leaders for complicity in the Chernobyl cover-up had failed. But they did succeed in making the government of independent Ukraine tell the truth about the impact of the disaster on the republic's territory and population.

Ukraine had embraced the public debate over the causes and consequences of the Chernobyl disaster as a tool of state- and nation-building—a means of mobilizing opposition to the former imperial center, promoting social solidarity, and establishing the legitimacy of the new state in the eyes of its citizens and the world at large. As a result of public pressure and the activity of Yavorivsky's commission, Ukraine adopted the most liberal post-Chernobyl social welfare laws of any former Soviet republic. It recognized almost 90,000 of its citizens as Chernobyl invalids of the most severely affected category, and thus most in need of social assistance. In Russia, that category included only 50,000 people, and in Belarus, the republic that had suffered the most from the Chernobyl fallout, only 9,000 individuals were similarly categorized. Ukraine also recognized more than 500,000 individuals as liquidators, another social welfare category, as opposed to 200,000 in Russia and somewhat more than 100,000 in Belarus. Moreover, the Ukrainian legislators rejected the Soviet post-Chernobyl standard of 35 rem as the acceptable level of lifetime radiation exposure. Like legislators in Belarus and Russia, they adopted a threshold of 7 rem—the average that a citizen of the United States might absorb over a lifetime.

The social and economic ramifications of those decisions were enormous. To deal with the new expenditures, the legislators had introduced a Chernobyl tax of 12 percent on corporate income. Ukraine began life as a sovereign state in the 1990s with high expectations but a more than modest GDP of $1,300 per capita, which was crushed by the post-Soviet economic crisis and high inflation. The economy was contracting at a rate of between 10 and 23 percent per annum, and by 1994 GDP had fallen to slightly more than half of what it had been during the first year of independence. In the mid-1990s, 5 percent of the Ukrainian budget was allocated to fund projects dealing with the consequences of the Chernobyl accident, and 65 percent of social welfare resources were devoted to assisting the 3.3 million citizens categorized as victims or "sufferers" of the Chernobyl accident.[15]

For many, gaining the status of a "sufferer" whose health had been adversely affected by the Chernobyl disaster became a means of dealing with the enormous hardships presented by economic decline, rising unemployment, and the reduction of government funding for the numerous social welfare programs of Soviet times. "If a person needs medicines, a person needs money. The diagnosis we write is money," commented a medical doctor attending to people who had been resettled from the contaminated areas. The Chernobyl social protection programs expanded the welfare system that had been inherited from the Soviet period, creating a major additional impediment to the economic recovery of newly independent Ukraine.[16]

While liberation from the empire helped to expose the truth about Chernobyl, the empire nevertheless left Ukraine with an enormous unpaid bill. It seemed at the time that the only way to start paying it off was to go back to the use of nuclear energy. That is indeed what happened in Ukraine soon after the declaration of independence. The same parliament of eco-nationalists that had passed a number of extremely liberal laws defining the status of Chernobyl disaster victims and created Yavorivsky's commission to investigate responsibility for the accident was now ready to turn its back on ecological concerns and put economic objectives at the top of its agenda. It saw no other way of saving the people from starvation and the newly independent country from collapse.

In February 1990, the Ukrainian parliament had voted to close the Chernobyl power plant by 1995. A fire in October 1991 at Unit 2, caused by a faulty switch in the turbine, destroyed part of the roof of the plant's turbine hall and made the task even more urgent. Parliament voted to close Unit 2 immediately, and Units 1 and 3—which were still operating—in 1993, two years ahead of schedule.

But in the fall of 1993, a few months before the scheduled early closure, the very same parliamentarians reversed their earlier decision. They annulled the resolution of August 1990, which had imposed a five-year moratorium on the construction of new nuclear power stations in Ukraine, and decreed that the Chernobyl Nuclear Power Station would be closed only after the existing reactors there reached the limit of their operating life. Reactors 1 and 3 would remain in operation, and plans were made to restart Unit 2—the site of the 1991 fire. Eco-nationalism was now giving way to nuclear nationalism. The story of the Chernobyl disaster had taken another abrupt turn.[17]

and asked for American and Western assistance. The West was responsive, but Western leaders wanted Ukraine to give up its nuclear arsenal. Kuchma would be going to Washington to discuss the conditions under which his country might do so.[2]

Shcherbak had his work cut out for him—the ecological activist found himself plunged into nuclear arms negotiations. When the Soviet Union fell apart in December 1991, Ukraine inherited 1,800 nuclear warheads that had been deployed with Soviet armed forces stationed there and agreed to dismantle and send them to Russia. Ukraine was supposed to complete the task by the end of 1994, but parliamentarians in Kyiv soon put forward a number of conditions, including monetary compensation for weapons-grade uranium in the nuclear warheads. After the US government promised financial aid, the Ukrainians agreed to get rid of their nuclear arsenal, the third largest in the world after those of the United States and Russia. The agreement on the transfer of weapons was signed in January 1994, but the Ukrainian parliament asked for guarantees of territorial integrity and security after the weapons left its territory. What it got in return were legally nonbinding assurances. In November 1994, a few days before his planned visit to Washington, Kuchma finally convinced parliament to accept the deal—security assurances and financial assistance in return for giving up the nuclear weapons.

President Bill Clinton was happy to welcome the Ukrainian president to Washington on November 22, 1994. Kuchma was greeted with a sixteen-gun salute and compared to President Franklin D. Roosevelt, who had led America at a time of economic hardship—a reference to the deep economic crisis in Ukraine and the rest of the post-Soviet space. Clinton praised Kuchma's courage in "removing the threat of nuclear weapons and laying the groundwork for an era of peace." The United States was providing Ukraine with an aid package of up to $200 million. A few weeks later, on December 5, 1994, Clinton and Kuchma put their signatures to the Budapest Memorandum, a document that provided security assurances from the United States, Russia, and Britain against threats or use of force against Ukraine, as well as against Kazakhstan and Belarus, two other post-Soviet states that were giving up their Soviet-era nuclear arsenals. China and France would provide their own assurances in a separate protocol. Ukraine and other post-Soviet states joined the nuclear weapons nonproliferation treaty as nonnuclear states.[3]

In the long run, the deal turned out to be disastrous for Ukraine. Twenty years later, in March 2014, nuclear-free Ukraine became an object of aggression by one of the signatories of the Budapest Memorandum when Russia, now led by President Vladimir Putin, annexed Ukraine's Crimea and unleashed a hybrid war in Ukraine's eastern Donbas region. The Ukrainian parliament appealed to the signatories of the Budapest Memorandum but did not get very far, as the memorandum did not require any military action on the part of the signatories. The United States and its European allies limited their response to introducing economic sanctions against Russia. It was too little, too late to reverse Russia's aggression and restore Ukraine's territorial integrity.[4]

It all looked very different at the time of the signing of the agreements in 1994. Ukraine received significant diplomatic endorsement, financial assistance, and assurance that in return for the nuclear warheads, Russia would continue to provide fuel for Ukraine's nuclear plants, including the one in Chernobyl, all of which ran on Russian-produced enriched uranium. The Chernobyl plant received considerable attention in the Washington negotiations of November 1994 between Kuchma and Clinton. "President Clinton drew attention to the significant resource commitments . . . and to the importance of receiving early assurances that the Chernobyl reactors would be shut down," read the statement. The United States wanted Ukraine to stick to its parliamentary decision of 1990 to shut down the Chernobyl plant by 1995. But Ukraine, faced with a deepening economic crisis, had since reversed its decision and wanted the plant to run indefinitely. According to the same joint statement, Kuchma refused to budge under pressure from the American president. Instead, he merely "assured President Clinton that Ukraine takes seriously the international community's concerns about the continued operation of the Chernobyl nuclear power plant." He pointed to the need for "minimizing the social impact on the plant's personnel and ensuring that sufficient economically priced electricity is available to meet Ukraine's domestic needs." Clinton got the message: cash-strapped Ukraine could not afford to lose two working nuclear reactors in Chernobyl unless it received financial compensation. The two presidents agreed to work on those issues together with the members of the G-7 group of the world's largest economies.[5]

THE PRIMARY responsibility for the safety of nuclear reactors lay with the countries that owned them, but the G-7 governments of the world's richest economies instructed the World Bank and the European Bank for Reconstruction and Development (EBRD)—where they established the Nuclear Safety Account in 1993 to accumulate funds designed to help the Eastern European countries still using Soviet-era reactors to ensure their safety—along with other institutions, to work on the issue. The directors of the nuclear power companies in the West were in a panic: another accident in the East could damage the reputation of nuclear power in the West beyond repair and potentially put them out of business. They lobbied their governments to launch a program for upgrading Eastern European reactors using Western technology and money. Armed with new technology and backed by government funds, Western nuclear firms took steps to ensure that another major nuclear accident would not happen in the East.[6]

Despite the handsome financial incentives and political pressure from the West, the Ukrainian government continued to drag its feet. The economy was in free fall, and high inflation had eaten up the savings of the population, making the situation serious enough for the G-7 summit in Naples to include a special section on the Ukrainian economy in its official communiqué of August 1994. The Ukrainian government argued that it could not simply shut down a power plant that produced up to 6 percent of the country's electrical energy. That would result in a layoff of personnel in the midst of an economic crisis whose severity not only paralleled but almost dwarfed the Great Depression of the 1930s. Kyiv would give up its nuclear arms but would not budge on Chernobyl. To the rest of the world, this seemed incomprehensible. The young nation was among those that had suffered the most from the accident; nevertheless, it was not only keeping the rest of its nuclear plants and reactors in operation but also refusing to close a plant in a highly contaminated zone that presented a significant risk to the operating personnel. And beside the operating reactors in Chernobyl was the deteriorating sarcophagus, which, as Kuchma told Clinton in November 1994, was in need of repair.

The Western governments would not give up. The European Union indicated how seriously it took the situation when it suspended $85 million in economic assistance to Ukraine unless progress was made on the closure of the Chernobyl power plant. Badly in need of money, Kuchma announced in April 1995, in time for his meeting with an EU and G-7 delegation led by the French minister of the environment, Michel Barnier, that he was committed to closing down the plant. But his officials, including Sergei Parashin, party secretary of the Chernobyl plant at the time of the 1986 accident and now its director, were skeptical. Speaking to television reporters, Parashin complained about political pressure from the West, claiming that his personnel knew for certain that the Chernobyl station was no less safe than any other in Ukraine.[7]

For Parashin and his subordinates, the proposed closure of the plant meant the prospect of personal economic disaster—the loss of wages and salaries that were high by Ukrainian standards, making it possible to survive in the new market economy with its extremely high prices. Workers remained at the plant even though they absorbed high doses of radiation—information that they withheld from their doctors. "They hang on to the Zone," said a local doctor to an American graduate student conducting research on the consequences of the Chernobyl disaster. As long as they were employed by the Chernobyl plant, engineers and workers could pay their bills: if it were closed, they would end up on the street.[8]

The leaders of the G-7 countries tried to find money to deal with the economic and social rehabilitation of the Chernobyl workers. "Recognizing the economic and social burden that the closure of Chernobyl will place on Ukraine, we [will continue] efforts to mobilize international support for appropriate energy production, energy efficiency and nuclear safety projects for Ukraine," read the communiqué issued by the Western leaders in Halifax, Canada, in June 1995. "Any assistance for replacement power for Chernobyl will be based on sound economic, environmental and financial criteria." Western experts opposed the construction of a new gas-fired power station in the contaminated area—a project advocated by the Ukrainian government—and let their Ukrainian counterparts know that they could not count on an unlimited line of credit. Money would be available only for projects approved by Western institutions.[9]

In December 1995, representatives of the G-7, the European Union, and Ukraine signed a memorandum that promised Western assistance for the decommissioning of the Chernobyl plant, coupled with money for completing the construction of two reactors at other Ukrainian plants and the reconstruction of a number of coal-fired electrical power stations to offset the loss of energy caused by the closure of Chernobyl. The Ukrainian government hoped to get $4.4 billion for those programs, but Western governments and financial institutions pledged $2.3 billion instead. Of that sum, close to half a billion was to come in the form of grants for the closure of plants and $1.8 billion as credits for the construction of new reactors at the Khmelnytskyi and Rivne nuclear power plants in western Ukraine. Chernobyl had to be shut down by the year 2000.[10]

The December 1995 memorandum did precious little to eliminate persisting tensions between representatives of Ukraine and its Western donors. Ukrainian government officials complained that the grant portion of the deal was too small, and that the Western powers had almost entirely neglected the question of building a new shelter above Unit 4. International institutions and Western countries (as well as Japan in the Far East) were reluctant to deliver funds for the construction of the two additional nuclear reactors. Experts from the European Bank for Reconstruction and Development, one of the main funders of Chernobyl-related projects, argued that it was more cost-effective to upgrade existing reactors than to build new ones. Besides, they argued, because of the economic crisis, the Ukrainian economy was consuming much less energy than before, industry needed less energy, and additional energy sources would make it less urgent to reform the country's energy market or introduce energy-saving measures. Antinuclear nongovernmental organizations (NGOs) in the West endorsed that idea.[11]

But President Kuchma and his officials insisted that the Chernobyl plant would not be closed until the additional two reactors were built and became operational. Many in the West thought he was bluffing. This perception grew stronger when, after long procrastination, Ukraine shut down Unit 1 in the fall of 1996. In June 1997, Unit 3 was closed for maintenance. Given that Unit 2 was not reactivated after the fire in the fall of 1991, the entire station had in fact been nonoperational since the summer of 1997. It looked as if the Ukrainians had shut it down without waiting for funds to build the other two reactors. But the Ukrainian nuclear industry did not give up on Chernobyl. In October 1997, it solemnly marked the plant's twentieth anniversary. When the former director, Viktor Briukhanov, came to the podium to address the gathering of plant workers, he was met with a standing ovation. "The whole auditorium rose; they applauded so loudly that my ears rang," recalled Briukhanov's wife, Valentina.[12]

To show that they were serious about the continuing operation of the plant, Ukrainian officials restarted the reactor of Unit 3 and reconnected it to the grid in June 1998, claiming that it could work safely until 2010. The Ukrainian government then turned to Russia for help in completing the construction of the two reactors in western Ukraine that the West was reluctant to fund. Ukraine's sudden eastward turn petrified Western governments, which now faced the possibility that the Chernobyl power plant would go on operating indefinitely, putting safety standards at other Ukrainian nuclear plants into question and threatening the commercial interests of Western companies that hoped to get the job of completing the two reactors Ukraine had asked for.

The West, however, was anything but united. The French and Finnish governments, lobbied by the leaders of their nuclear industries, indicated their readiness to complete the construction of the two reactors, but the German government was hampered by a parliamentary resolution, adopted under pressure from the Green Party, that prohibited funding of nuclear-related projects. Despite opposition in Germany and a number of other countries, the European Bank for Reconstruction and Development decided on December 7, 2000, to provide a $215 million loan for the construction of the two

Ukrainian reactors. The bank's decision opened the door for a loan of more than $500 million from the European Commission. Chernobyl could now finally be closed.[13]

On December 15, 2000, eight days after the EBRD decision, President Kuchma officially announced the decommissioning of the Chernobyl nuclear plant. In a speech delivered in Kyiv on the occasion of the closure, he assured the world that there would be no further nuclear threat from Ukraine. He then added: "We believe that Ukraine will not find it necessary to repent of the decision it has made." The decision was highly controversial in Ukraine itself. Ten days earlier, parliament had voted to extend the operation of Unit 3 into 2011 to deal with increased demand for electricity during the winter. The head of the powerful communist faction in parliament declared that the closure of the Chernobyl power station was "not a policy decision but a purely political one, directed to the detriment of the country's national interests." But the brunt of criticism came from the personnel of the power plant. On the eve of the closure, when Kuchma visited Unit 3 in the company of the prime ministers of Russia and Belarus, as well as the US energy secretary, Bill Richardson, the personnel, dressed in white as usual, put on black armbands as a sign of sorrow and protest against the decision.

Aleksandr Novikov, then head of the department of nuclear safety at the plant, later recalled:

> On that day, the emotional state of those in the control room of Unit 3 was very depressed. Men who had gone through fire, water, and copper pipes were sobbing. . . . And I'm not afraid to admit that there was a feeling of confusion: I myself had no idea what I would do next. . . . The other feeling was resentment. . . . The third element was emotional emptiness, because it was all done with some kind of indecent enthusiasm, boisterously, one might even say festively, but in my view, the black armbands worn by the staff in charge of operations put everything in its place as far as our attitude to that event was concerned.[14]

Opponents of the closure claimed that the power plant had been upgraded with regard to technology and safety and could continue working successfully until the year 2011, earning hundreds of millions of dollars in revenue for the electrical energy produced, while the completion of the replacement reactors at the Khmelnytskyi and Rivne nuclear plants was nowhere in sight. Indeed, they would not be finished and connected to the grid until 2004. But one way or another, with the closure on December 15, 2000, the saga of the Chernobyl power plant was over.

THE WORLD was entering the new millennium without Chernobyl, but much of the plant's legacy remained. Russia, Ukraine, and Belarus, the three newly independent countries most affected by the Chernobyl disaster, estimated their overall losses from it in the hundreds of billions of dollars.

In Ukraine alone, close to 38,000 square kilometers—approximately 5 percent of the country's entire territory, inhabited by up to 5 percent of its population of 54 million (1991)—were contaminated by the explosion. Even harder hit was Belarus, with more than 44,000 square kilometers of land severely contaminated, accounting for 23 percent of the republic's territory and 19 percent of its population. But of all the countries affected by Chernobyl, Russia had the largest contaminated area, close to 60,000 square kilometers. Given the country's size, that area constituted only 1.5 percent of its territory, with 1 percent of its population. All three countries had to bear the cost of resettlement and deal with the health problems caused by the disaster, not only among those residing in or resettled from the contaminated areas, but also among the hundreds of thousands of liquidators exposed to high doses of radiation in the first days, weeks, and months after the explosion.

In terms of immediate deaths attributable to the accident, the Chernobyl nuclear disaster turned out to be anything but a highly destructive force. Whereas the nuclear bombings of Hiroshima and Nagasaki claimed close to 200,000 immediate victims—more than 100,000 killed and the rest injured—the Chernobyl explosion caused 2 immediate deaths and 29 deaths from acute radiation sickness in the course of the next three months. Altogether, 237 people were airlifted from Chernobyl to Moscow and treated in the special clinic there. Out of these, 134 showed symptoms of acute radiation syndrome. It has been claimed that a total of 50 people died of acute radiation syndrome, and that 4,000 may die in the future of radiation-related causes. But the ultimate Chernobyl mortality toll, though difficult to estimate, may yet turn out to be significantly higher. Current estimates place it between the 4,000 deaths estimated by United Nations agencies in 2005 and the 90,000 suggested by Greenpeace International.[15]

In Ukraine, in the first five years after the disaster, cases of cancer among children increased by more than 90 percent. During the first twenty years after the accident, approximately 5,000 cases of thyroid cancer were registered in Russia, Ukraine, and Belarus among those who were younger than eighteen at the time of the explosion. The World Health Organization estimates that approximately 5,000 cancer deaths were related to the Chernobyl accident, but this figure is often challenged by independent experts. In Ukraine in 2005, 19,000 families were receiving government assistance owing to the loss of a breadwinner whose death was deemed to be related to the Chernobyl accident. Other consequences include genetic damage to people born after the disaster. Scientists are particularly concerned about cases of microsatellite instability (MSI), a condition that affects the ability of DNA to replicate and repair itself, which has been detected in children whose fathers were exposed to radiation after the accident. Similar changes were found earlier among children of Soviet soldiers who absorbed radiation during nuclear tests.[16]

The cost of the disaster was enormous, and all three East Slavic countries had to deal with it in one way or another. They adopted largely similar formulas, defining the most contaminated areas whose inhabitants were in need of resettlement or assistance and then establishing categories of citizens who were considered to have been most severely affected, making them eligible for financial compensation and privileged access to medical facilities. Altogether close to 7 million people would receive some form of compensation for the effects of the Chernobyl fallout. But the size of the groups eligible for subsidies and the amount of financial compensation differed in the three states, depending on the interplay of politics and economic circumstances.

Russia's oil and gas riches helped it deal with the post-Chernobyl crisis, while resource-poor Ukraine and Belarus had nothing comparable. Those two countries introduced a special Chernobyl tax in the early 1990s, amounting in Belarus to 18 percent of all wages paid in the nonagricultural sector. In general, however, the Belarusian government dealt with the enormous challenge by continuing the Soviet tradition of suppressing investigations of major disasters. Although Belarus was the post-Soviet country most affected by Chernobyl fallout, its antinuclear movement never attained the proportions of its Ukrainian counterpart. Nor did the Belarusian Popular Front exercise influence comparable to that of the Ukrainian Rukh. The Belarusian parliament and government lacked the political will and, more importantly, the resources to admit the full scope of the disaster and deal effectively with its consequences. In 1993, the Belarusian parliament adopted laws reducing the levels of soil contamination considered dangerous for human habitation. Even then, with significantly less territory and population covered by social welfare laws, the government only managed to allocate less than 60 percent of the funds approved by legislators for Chernobyl-related programs.[17]

When it comes to Western assistance, Ukraine got most of the attention and resources, largely because it inherited the Chernobyl nuclear plant and its devastated Unit 4. The first priority identified by Ukraine as requiring Western help after the closure of the Chernobyl plant was the construction of a new shelter over the sarcophagus that had been hastily built to cover the damaged fourth reactor in the first months after the explosion. The Ukrainian government announced an international competition for the construction of the new shelter in 1992. In June 1997, the G-7 countries pledged $300 million toward the realization of the project, whose total cost was then estimated at $760 million. A special Chernobyl Shelter Fund was created at the European Bank for Reconstruction and Development to collect the rest of the funds. That turned out to be a major challenge.

Originally, it was expected that the new shelter would be built by 2005. But it was not until 2007 that the French Novarka consortium, which included Vinci Construction Grands Projets and Bouygues Construction, won the contract to erect a 30,000-tonne sliding steel

arch, 110 meters in height and 165 meters in length, with a span
of 257 meters, over the old sarcophagus. Construction of the arch,
which had to endure for the next one hundred years, began in 2010;
the deadline for completion, originally scheduled for 2005, was later
postponed to 2012, and then to 2013, 2015, and, finally, 2017. Its
cost has been estimated at 1.5 billion euros, with the total cost of the
New Safe Confinement Project exceeding 3 billion euros.[18]

It took nine years after the fall of the USSR to close the Chernobyl Nuclear Power Station and more than a quarter century to
build a new shelter over the damaged reactor. The international community emerged victorious in the contest of security priorities. Relations between the two main actors in the post-Chernobyl drama,
the Western funding agencies and the Ukrainian government, were
not unlike those in a family with a teenager who promises not to
behave dangerously if given an ever larger allowance. Some scholars
referred to it as environmental blackmail.[19]

But the closure of the Chernobyl power plant and the construction of the new shelter is more than just a story of nuclear extortion
of funds by a poor country from rich ones. More than anything else,
it is a story of the clash between the demands of individual nations
for economic development and the security of the world, as well as
of the threat posed to the latter by the political and economic decline of the nuclear powers and the uncertain future of the post-
imperial states.

Moscow, the former capital of the empire responsible for the design and operation of the damaged reactor, disappeared behind the
borders of the Russian Federation, leaving it to Ukraine and the international community to clean up the mess. But the Russian invasion of Ukraine in 2014 brought the fighting within 322 kilometers
of the city of Enerhodar, the site of the Zaporizhia nuclear power
plant, Europe's largest, which operates six reactors. The war also interrupted the nuclear cycle whereby Ukraine received its nuclear fuel
from Russia and sent its spent fuel back there. In 2016, Ukraine began the construction of its own spent-fuel facility and declared plans
to reduce its almost total dependence on Russian fuel by covering
40 percent of its needs with purchases from the US-based Westing-
house Electric Company. While the war and the disruption of the

traditional nuclear cycle brought new challenges to the struggling Ukrainian economy, the nuclear industry of the land of Chernobyl took another important step away from its Soviet legacy.[20]

What remained unchanged and impervious to remedy by any amount of internal mobilization or outside assistance were the long-term consequences of the Chernobyl disaster. While the actual impact of radiation exposure on the health of the population is still debated, there can be little doubt that the society as a whole was left traumatized for decades to come. Every sixth Ukrainian adult reports being in poor health, a significantly higher percentage than in neighboring countries, and those affected by the Chernobyl radiation have lower levels of employment and fewer working hours than the rest of Ukraine's population. And then there is the environment. The new shelter over the damaged reactor No. 4 notwithstanding, the area around the nuclear plant will not be safe for human habitation for at least another 20,000 years.[21]

In April 2016, when the world marked the thirtieth anniversary of the disaster, there was a temptation to breathe a sigh of relief. The half-life of cesium-137, one of the most harmful nuclides released during the accident, is approximately thirty years. It is the longest "living" isotope of cesium that can affect the human body through external exposure and ingestion. Other deadly isotopes present in the disaster have long passed their half-life stages: iodine-131 after eight days, and cesium-134 after two years. Cesium-137 is the last of that deadly trio of isotopes. But the harmful impact of the accident is still far from over. With tests revealing that the cesium-137 around Chernobyl is not decaying as quickly as one would wish for, scholars believe that the isotope will continue to harm the environment for at least 180 years—the time required for half the cesium to be eliminated from the affected areas by weathering and migration. Other radionuclides will also remain in the region for a long time. The half-life of plutonium-239, traces of which were found as far away as Sweden, is 24,000 years.[22]

EPILOGUE

THESE DAYS, European tour operators offer trips to Chernobyl from Brussels, Amsterdam, or Berlin for under 500 euros. Visitors are promised safety, comfort, and excitement while visiting the place where on April 26, 1986, the explosion at reactor No. 4 ended one era and initiated another. The city of Prypiat and the exclusion zone as a whole is a time capsule.

In 2015, in the wake of the Revolution of Dignity, the Ukrainian parliament voted to remove statues of Vladimir Lenin and other communist leaders from the country's streets and squares. Overnight, the exclusion zone became a communist preserve. The monument to Lenin still stands in the center of Chernobyl, and when the Ukrainian president visited the city on the thirtieth anniversary of the disaster, the authorities covered up a sign on the approach to the city that depicts the Order of Lenin. The sign is now painted in the Ukrainian national colors, blue and yellow. It was red back in 1986, when the entire Chernobyl nuclear plant was named after Vladimir Lenin.[1]

The new high-tech shelter over the old sarcophagus of reactor No. 4 that visitors see on their trips to the exclusion zone stands today as a monument to the failed ideology and political system embodied in the Soviet Union. But it is also a warning to societies that put military or economic objectives above environmental and health concerns. In April and May 1986, the firefighters, scientists,

engineers, workers, soldiers, and police officers who found themselves caught in the nuclear Armageddon called Chernobyl did their
best to put out the nuclear inferno. Some of them sacrificed their
lives, many their health and well-being, in this effort. They tried as-
tounding measures. They dropped thousands of tonnes of sand from
helicopters onto the open reactor. They dug a tunnel with little more
than their bare hands to freeze the land beneath it. They built dams
along the riverbanks to prevent contaminated water from flowing
into the Prypiat River, and from there into the Dnieper, the Black
Sea, the Mediterranean, and the Atlantic.

These measures achieved the seemingly impossible: they put the
reactor to sleep. But even today we do not know which of the strategies the Soviets tried and the technical solutions they implemented
actually worked. Could some of them have made things worse? The
eruption of the nuclear volcano stopped for reasons that the scientists and engineers could not comprehend, just as they were initially
at a loss to explain why the reactor had exploded in the first place.
Although the reason for the explosion was eventually discovered,
we are still as far from taming nuclear reactions as we were in 1986.
Unpredictable events keep happening, causing new nuclear disasters, like the one at Fukushima, Japan, in March 2011, when not one
reactor, as at Chernobyl, but three sustained a partial meltdown of
their cores and released radiation directly into the Pacific Ocean as a
result of an earthquake followed by a tsunami.[2]

THE WORLD is growing bigger but not safer. The planet's population was close to 5 billion in 1986; today it numbers more than 7
billion, and it is projected to reach 10 billion by 2050. Every twelve
to fourteen years, the planet gains another billion residents. As the
physical world shrinks for its growing population, its resources and
energy reserves dwindle. As the population of Europe declines, and
that of North America shows only modest growth, Asia and Africa
are expected to show dramatic increases, with the African population more than doubling by mid-century, reaching a total of over 2
billion. Thus, most of the globe's population growth will occur in
countries already struggling to feed their hungry and replenish their
sources of energy.

Nuclear power seems to provide an easy way out of the growing demographic, economic, and ecological crisis. Or does it? Most new reactors under construction today are being built outside the Western world, which is known for the relative safety of its reactors and operating procedures. A whopping twenty-one new reactors are under construction in China, plus nine in Russia, six in India, four in the United Arab Emirates, and two in Pakistan. Five new reactors are currently being built in the United States, and one in Britain. The next great nuclear-power frontier is Africa. Volatile Egypt is currently building two reactors—its first in history. Are we sure that all these reactors are sound, that safety procedures will be followed to the letter, and that the autocratic regimes running most of those countries will not sacrifice the safety of their people and the world as a whole to get extra energy and cash to build up their military, ensure rapid economic development, and try to head off public discontent? That is exactly what happened in the Soviet Union back in 1986.

The immediate cause of the Chernobyl accident was a turbine test that went wrong. But its deeper roots lay in the interaction between major flaws in the Soviet political system and major flaws in the nuclear industry. One such flaw at Chernobyl was the military origin of the nuclear power industry. Chernobyl-type reactors had been adapted from the technology that had been created to produce nuclear bombs. Moreover, although it was volatile under certain physical conditions, the Chernobyl-type reactor had been pronounced safe. It was actively promoted by leaders of the Soviet military-industrial complex, who later refused to take responsibility for the disaster. Another flaw was the violation of procedures and safety rules by station personnel, who bought into the myth of the safety of nuclear energy and adopted a reckless "we can do it no matter what" type of attitude. It was the same attitude that characterized the country's desperate attempts to catch up with the West in the economic and military spheres. Immediately after the accident, as panic spread, the authoritarian Soviet regime imposed control over the flow of information, endangering millions of people at home and abroad and leading to innumerable cases of radiation poisoning that could otherwise have been avoided.

Today, the chances of another Chernobyl disaster taking place are increasing as nuclear-energy technology falls into the hands of rulers pursuing ambitious geopolitical goals and eager to accelerate economic development in order to overcome energy and demographic crises while paying lip service to ecological concerns. While world attention is focused on the nonproliferation of nuclear arms, an equally great danger looms from the mismanagement of "atoms for peace" in the developing world. The story of Chernobyl points to the need to strengthen international control over the construction and exploitation of nuclear power stations as well as to develop new nuclear technologies, such as those now being explored by Bill Gates and his TerraPower Company, to produce cheaper, safer, and ecologically cleaner reactors. Investment in such research and development, with the clear understanding that financial benefits may be years, if not decades, away, is imperative if the world is to survive the current energy crisis and the continuing demographic boom.[3]

Is another nuclear Armageddon like the one at Chernobyl possible today? As the disaster receded in time, the loudest voices were those of optimists who denied such a possibility. Safety procedures at nuclear power plants have indeed improved, the old Soviet-era RBMK reactors have been decommissioned, and new reactors offer a level of safety that nuclear engineers could only have dreamed of in 1986. But then, a quarter century after Chernobyl, came Fukushima. The nuclear disaster of 2011 at the Fukushima Daiichi Nuclear Power Plant indicated a different vulnerability of nuclear reactors. Future accidents like those at Chernobyl or Fukushima may have various causes—a lapse in personnel discipline, a fault in reactor design, or an earthquake. There is also the growing danger of terrorist attacks on nuclear plants—one such attempt was investigated by the Belgian authorities in March 2016. Cyberattacks by hackers are another danger, as evidenced by the one that shut down the Chernobyl radiation monitoring systems in June 2017: the Ukrainian authorities believe that it had originated in Russia.

The decommissioning of the Chernobyl nuclear plant and the construction of a new sarcophagus over the damaged reactor helped to close the most tragic page in the history of the nuclear power industry, but it is still imperative that we draw the right lessons from the Chernobyl disaster. The most crucial lesson is the importance of counteracting the dangers posed by nuclear nationalism and isolationism and of ensuring close international cooperation between countries developing nuclear projects. This lesson is especially important today, when the forces of populism, nationalism, and anti-globalism are finding more adherents in a world that relies increasingly on nuclear technology for the production of energy.

The world has already been overwhelmed by one Chernobyl and one exclusion zone. It cannot afford any more. It must learn its lessons from what happened in and around Chernobyl on April 26, 1986.

ACKNOWLEDGMENTS

My gratitude goes first to those who helped me most with this book. My wife, Olena, and my literary agent, Jill Kneerim, believed in this project from the very beginning. Hennadii Boriak of the Institute of Ukrainian History in Kyiv and Olha Bazhan of the Central State Archives of Public Organizations of Ukraine helped me gain access to Communist Party and government documents. Andrii Kohut and Maria Panova were equally helpful in providing access to KGB files pertaining to the catastrophe. Orest Hrynewych, managing director of the Ukrainian Institute of Modern Art in Chicago and a nuclear engineer of many years' experience, helped me grasp the basics of the industry to which he has dedicated a good part of his life. Olga Bertelsen, a postdoctoral fellow at the Harvard Ukrainian Research Institute in 2016–2017, read the entire manuscript and made excellent suggestions on how it could be improved. Myroslav Yurkevich, as always, did an excellent job of editing my prose. Financial support granted by the Harvard Department of History, the Ukrainian Research Institute, and the Davis Center for Russian and Eurasian Studies helped me complete the project, and Cory Paulsen at the Department of History and M. J. Scott at the Ukrainian Institute did their best to make those funds available.

I was very glad to work once again with Lara Heimert and her wonderful Basic Books team, including Brian Distelberg, Roger

Labrie, and Alia Massoud. At Perseus Books my thanks go to Collin Tracy and Kathy Streckfus. They did an outstanding job of editing the book and guiding it through the publishing process. Needless to say, I alone am responsible for any shortcomings, of which I hope there are not too many.

NOTE ON RADIATION IMPACT
AND MEASUREMENTS

RADIATION—THE EMISSION or transmission of energy—comes in a variety of forms. The explosion of the Chernobyl nuclear reactor caused the spread of ionizing radiation, which carries enough energy to detach electrons from atoms and molecules. It combines electromagnetic radiation, including gamma rays and X-rays, with particle radiation, which consists of alpha and beta particles and neutrons.

There are three different ways of measuring ionizing radiation. The first assesses the level of radiation emitted by the radioactive object, the second determines the level absorbed by the human body, and the third estimates the amount of biological damage caused by the absorption of radiation. Each of these categories has its own unit of measure, and in all cases old units are gradually being replaced by new ones in the International System of Units (SI). A unit of emitted radiation, formerly known as the curie, has been replaced by the becquerel, with 1 curie equaling 37 gigabecquerels (GBq). The older unit of radiation absorption, the rad, has been replaced with an SI unit called the gray (Gy), which equals 100 rad. Biological damage, formerly measured in rem, has been replaced by the SI unit known as the sievert (Sv).

"Rem" stands for "roentgen equivalent man" and is equal to 0.88 of a roentgen, a legacy unit used to measure exposure to the ionizing electromagnetic radiation produced by X-rays and gamma rays. At the time of the Chernobyl disaster, Soviet nuclear engineers used the old units to measure exposure to radiation and the biological damage caused by it. The first dosimeters they were able to obtain measured radiation exposure in microroentgens per second. The doctors who treated the first victims of the accident measured their patients' radiation dosage in units called ber, a Russian-language equivalent of rem that stands for "biological equivalent of roentgen." Converting old units of measure into new SI units is cumbersome, with rem being a welcome exception: 100 rem equal 1 sievert and, in measuring gamma and beta radiation, amount to 1 gray.

Since the rem is closely associated with the roentgen and easily convertible into sieverts, it is the best unit for expressing the impact of radioactivity on the human body as measured at the time of the accident. Today, 10 rem, or 0.1 Sv, is the standard five-year limit of biological damage sustained by nuclear industry workers in the West. The maximum dosage allowed cleanup workers at the Chernobyl power plant in the summer of 1986 was 25 rem, or 0.25 Sv. The biological damage sustained by evacuees from the Chernobyl Exclusion Zone is assessed today at 30 rem (0.3 Sv). Radiation sickness, which is accompanied by symptoms such as nausea and the "nuclear tan"—a darkening of the skin burned by radiation, which does not necessarily lead to death—starts at 100 rem (1 Sv). Half of those who sustain biological damage lethal to bone marrow, which occurs at a level of 400–500 rem, or 4–5 Sv, die within one month. The Chernobyl operators and firefighters who died within a month of exposure sustained 600 or more rem (6 Sv). That corresponds to 6 grays of gamma or beta radiation. Aleksandr Akimov, the leader of the Unit 4 shift at the time of the explosion, absorbed an estimated 15 grays of radiation. He died fifteen days after the accident.

Soviet officials and engineers used kilometers to measure distance and metric tonnes to measure weight. I followed my sources in both cases, while providing equivalents in miles and US tons in parentheses where these units are first mentioned. For those wishing to calculate US measurements, 1 kilometer equals 0.62 miles and 1 metric tonne equals 1.1 US tons. User-friendly calculators are available for most units of measure by typing the two types of units (for example, "kilometers to miles") into an online search box.

NOTES

PROLOGUE

1. "25 Years After Chernobyl, How Sweden Found Out," Radio Sweden—News in English, April 22, 2011, http://sverigesradio.se/sida/artikel.aspx?programid=2054&artikel=4468603; Serge Schmemann, "Soviet Announces Nuclear Accident at Electric Plant," *New York Times*, April 29, 1986, A1.

CHAPTER 1: CONGRESS

1. "XXVII s'ezd KPSS," YouTube, 1986, published May 1, 2015, https://www.youtube.com/watch?v=DFtuqNiY4PA.

2. Mikhail Posokhin, *Kremlevskii dvorets s"ezdov* (Moscow, 1965); Aleksandr Mozhaev, "Vtoraia svezhest'," *Arkhnadzor*, March 2, 2007, www.archnadzor.ru/2007/03/02/vtoraya-svezhest.

3. William Taubman, *Khrushchev: The Man and His Era* (New York, 2004), 507–528; Mark Harrison, "Soviet Economic Growth Since 1928: The Alternative Statistics of G. I. Khanin," *Europe-Asia Studies* 45, no. 1 (1993): 141–167.

4. Archie Brown, *The Gorbachev Factor* (Oxford, 1997), 24–129; Andrei Grachev, *Gorbachev's Gamble: Soviet Foreign Policy and the End of the Cold War* (Cambridge, 2008), 9–42.

5. Mikhail Gorbachev, *Zhizn'i reformy*, book 1, part 2, chapter 9 (Moscow, 1995); Valerii Boldin, *Krushenie p'edestala: Shtrikhi ko portretu M. S. Gorbacheva* (Moscow, 1995), 158.

6. Irina Lisnichenko, "Zastol'ia partiinoi élity," *Brestskii kur'er*, March 2013.

7. Grigori Medvedev, *The Truth About Chernobyl*, foreword by Andrei Sakharov (New York, 1991), 40–42.

8. Oleksandr Boliasnyi, "Pryskorennia tryvalistiu visim rokiv," *Kyïvs'ka pravda*, December 1, 1985; Aleksandr Boliasnyi, "Kogda iskliucheniia chasto

povtoriaiutsia, oni stanoviatsia normoi," *Vestnik* 7, no. 214 (March 30, 1999), www.vestnik.com/issues/1999/0330/koi/bolyasn.htm.

9. I. Kulykov and V. Shaniuk, "Vid nashoho novators'koho poshuku," *Kyïvs'ka pravda*, March 4, 1986; I. Kulykov, T. Lakhturova, and V. Strekal', "Plany partiï—plany narodu," *Kyïvs'ka pravda*, March 8, 1986; V. Losovyi, "Hrani kharakteru," *Prapor peremohy*, February 25, 1986; Oleksandr Boliasnyi, "Tsia budenna romantyka," *Kyïvs'ka pravda*, February 26, 1986.

10. *XXVII S"ezd Kommunisticheskoi partii Sovetskogo Soiuza, 25 fevralia—6 marta 1986: Stenograficheskii otchet* (Moscow, 1986), vol. 1, 3.

11. Viktor Loshak, "S"ezd burnykh aplodismentov," *Ogonek*, no. 7, February 2, 2016.

12. *XXVII S"ezd Kommunisticheskoi partii Sovetskogo Soiuza*, vol. 1, 23–121.

13. Ibid., 130–168.

14. "Na vershinakh nauki i vlasti: K stoletiiu Anatoliia Petrovicha Aleksandrova," *Priroda*, no. 2 (February 2003): 5–24.

15. *XXVII S"ezd Kommunisticheskoi partii Sovetskogo Soiuza*, vol. 1, 169–174.

16. V. P. Nasonov, "Slavskii Efim Pavlovich," Ministry sovetskoi epokhi, www.minister.su/article/1226.html; "Slavskii, E. P. Proshchanie s sablei," YouTube, published April 21, 2009, https://www.youtube.com/watch?v=KURb0EWtWLk&feature=related.

17. "Byvshii zamdirektora ChAĖS: My stali delat' takie AES iz-za Arkadiia Raikina," *Interfax*, April 23, 2016.

18. Igor Osipchuk, "Legendarnyi akademik Aleksandrov v iunosti byl belogvardeitsem," *Fakty*, February 4, 2014; Galina Akkerman, "Gorbachev: Chernobyl' sdelal menia drugim chelovekom," *Novaia gazeta*, March 2, 2006.

19. B. A Semenov, "Nuclear Power in the Soviet Union," *IAEA Bulletin* 25, no. 2 (1983): 47–59.

20. *XXVII S"ezd Kommunisticheskoi partii Sovetskogo Soiuza*, vol. 2, 29, 54, 139.

21. Ibid., 94–98.

22. Ibid., vol. 1, 141–142.

23. Boliasnyi, "Kogda iskliucheniia chasto povtoriaiutsia, oni stanoviatsia normoi"; O. Boliasnyi, "Dilovytist', realizm," *Kyïvs'ka pravda*, February 28, 1986.

CHAPTER 2: ROAD TO CHERNOBYL

1. I. Kulykov, T. Lakhturova, and V. Strekal', "Plany partiï—plany narodu," *Kyïvs'ka pravda*, March 8, 1986.

2. Mariia Vasil', "Byvshii director ChAĖS Viktor Briukhanov: 'Esli by nasli dlia menia rasstrel'nuiu stat'iu, to, dumaiu, menia rasstreliali by,'" *Fakty*, October 18, 2000.

3. Svetlana Samodelova, "Lichnaia katastrofa direktora Chernobylia," *Moskovskii komsomolets*, April 21, 2011; Vladimir Shunevich, "Byvshii direktor ChAËS Viktor Briukhanov: 'Kogda posle vzryva reaktora moia mama uznala,'" *Fakty*, December 1, 2010.

4. Shunevich, "Byvshii direktor ChAËS Viktor Briukhanov"; Grigori Medvedev, *The Truth About Chernobyl*, foreword by Andrei Sakharov (New York, 1991), 41–42; Oleksandr Boliasnyi, "Pryskorennia tryvalistiu visim rokiv," *Kyïvs'ka pravda*, December 1, 1985.

5. *Letopis' po Ipat'evskomu spisku: Polnoe sobranie russkikh letopisei*, vol. 2 (Moscow, 1998), cols. 676–677; Marina Heilmeyer, *Ancient Herbs* (Los Angeles, 2007), 15–18; Colin W. Wright, ed., *Artemisia* (London, 2002); Revelation 8:10–11; Lou Cannon, *President Reagan: The Role of a Lifetime* (New York, 1991), 860.

6. *Listy Aleksandra i Rozalii Lubomirskich* (Cracow, 1900); Alla Iaroshinskaia, *Chernobyl': Bol'shaia lozh'* (Moscow, 2011), Prologue.

7. *Słownik Geograficzny* (Warsaw, 1880), vol. 1, 752–754; L. Pokhilevich, *Skazaniia o naselennykh mestnostiakh Kievskoi gubernii* (Kyiv, 1864), 144–151; "Chernobyl'," *Ēlektronnaia evreiskaia ēntsiklopediia*, 1999, www.eleven.co.il /article/14672.

8. *Natsional'na knyha pam'iati zhertv Holodomoru: Kyïvs'ka oblast'* (Kyiv, 2008), 17, 1125–1131; "Knyha pam'iati zhertv Holodomoru," in *Chernobyl' i chernobyliane*, http://chernobylpeople.ucoz.ua/publ/istorija_chernobylja_i _rajona/golodomor/kniga_pamjati_zhertv_golodomora/31-1-0-97; *MAPA: Digital Atlas of Ukraine*, http://gis.huri.harvard.edu/images/flexviewers/huri_gis.

9. "Istoriia, Velikaia Otechestvennaia voina, Vospominaniia," in *Chernobyl' i chernobyliane*; "Okkupatsiia goroda Chernobyl'," Chernobyl', Pripiat', Chernobyl'skaia AES i zona otchuzhdeniia, http://chornobyl.in.ua/chernobil-war .html.

10. Petr Leshchenko, *Iz boia v boi* (Moscow, 1972). Cf. "Istoriia, Velikaia Otechestvennaia voina: Boi za Chernobyl'," in *Chernobyl' i chernobyliane*; "Okkupatsiia goroda Chernobyl'."

11. Vladimir Boreiko, *Istoriia okhrany prirody Ukrainy, X vek—1980*, 2nd ed. (Kyiv, 2001), chapter 9.

12. "Vybir maidanchyka," *Chornobyl's'ka AES*, http://chnpp.gov.ua/uk /history-of-the-chnpp/chnpp-construction/9-2010-09-08-09-57-419; Boreiko, *Istoriia okhrany prirody Ukrainy*, chapter 9; Alla Iaroshinskaia, *Chernobyl': 20 let spustia. Prestuplenie bez nakazaniia* (Moscow, 2006), 238; Petro Shelest, *Spravzhnii sud istorii shche poperedu: Spohady, shchodennyky, dokumenty*, comp. V. K. Baran, ed. Iurii Shapoval (Kyiv, 2003), 465–466; *Chernobyl'skaia atomnaia ēlektrostantsiia: Kul'turnoe i zhilishchno-bytovoe stroitel'stvo. General'nyi plan poselka* (Moscow, 1971), 10.

13. Grigorii Medvedev, *Chernobyl'skaia tetrad': Dokumental'naia povest'* (Kyiv, 1990), 28.

14. K. Myshliaiev, "Pervyi beton na Chernobyl'skoi atomnoi," *Pravda Ukrainy*, August 16, 1972; "Proektuvannia ta budivnytstvo," *Chornobyl's'ka AES*, http://chnpp.gov.ua/uk/history-of-the-chnpp/chnpp-construction/11-2010 -09-08-10-40-3911.

15. Viktor Briukhanov, "Enerhovelet pratsiuie na komunizm," *Radians'ka Ukraïna*, December 30, 1986.

16. Anatolii Diatlov, *Chernobyl': Kak èto bylo* (Moscow, 2003), chapter 4, http://lib.ru/MEMUARY/CHERNOBYL/dyatlow.txtl.

17. V. Lisovyi, "Hrani kharakteru," *Prapor peremohy*, February 25, 1986; "V gorodskom komitete Kompartii Ukrainy," *Tribuna ènergetika*, January 31, 1986.

18. Vladimir Dvorzhetskii, *Pripiat'—ètalon sovetskogo gradostroitel'stva* (Kyiv, 1985).

19. *Chernobyl'skaia atomnaia èlektrostantsiia*, 13.

20. Ivan Shchegolev, "Ètalonnyi sovetskii gorod: Vospominaniia pripiatchanina," *Ekologiia*, April 24, 2009, https://ria.ru/eco/20090424/169157074 .html; Artur Shigapov, *Chernobyl', Pripiat', dalee nigde* (Moscow, 2010), http: //royallib.com/read/shigapov_artur/chernobil_pripyat_dalee_nigde.html #20480.

21. "Interv'iu s Viktorom Briukhanovym, byvshim direktorom ChAÈS," *ChAÈS: Zona otchuzhdeniia*, http://chernobil.info/?p=5898.

22. Artur Shigapov, *Chernobyl', Pripiat', dalee nigde*; "Interv'iu s Viktorom Briukhanovym; Aleksandr Boliasnyi, "Kogda iskliucheniia chasto povtoriaiutsia, oni stanoviatsia normoi," *Vestnik* 7, no. 214 (March 30, 1999), www.vestnik .com/issues/1999/0330/koi/bolyasn.htm.

23. Vladimir Shunevich, "Byvshii direktor Cgernobyl'skoi atomnoi elektrostantsii Viktor Briukhanov: 'Noch'iu, proezzhaia mimo chetvertogo bloka uvidel, chto stroeniia nad reaktorom netu,'" *Fakty*, April 28, 2006, http: //fakty.ua/45760-byvshij-direktor-chernobylskoj-atomnoj-elektrostancii -viktor-bryuhanov-quot-nochyu-proezzhaya-mimo-chetvertogo-bloka-uvidel -chto-verhnego-stroeniya-nad-reaktorom-netu-quot.

24. I. Kulykov, T. Lakhturova, and V. Strekal', "Plany partiï—plany narodu," *Kyïvs'ka pravda*, March 8, 1986.

CHAPTER 3: POWER PLANT

1. "Obrashchenie kollektiva stroitelei i èkspluatatsionnikov Chernobyl'skoi AÈS," *Tribuna ènergetika*, March 21, 1986.

2. Vladimir Vosloshko, "Gorod, pogibshii v 16 let," *Soiuz Cgernobyl'*, January 24, 2002, www.souzchernobyl.org/?section=3&id=148.

3. E. Malinovskaia, "Est' 140 milliardov," *Tribuna ènergetika*, January 17, 1986; Proizvodstvennyi otdel, "Pochemu ne vypolnen plan 1985 goda po AÈS,"

ibid.; M. V. Tarnizhevsky, "Energy Consumption in the Residential and Public Services Sector," *Energy* 12, nos. 10–11 (October–November 1987): 1009–1012.

4. "Kizima Vasilii Trofimovich," Geroi strany, www.warheroes.ru/hero /hero.asp?Hero_id=16214; *MAPA: Digital Atlas of Ukraine*, http://harvard-cga .maps.arcgis.com/apps/webappviewer/index.html?id=d9d046abd7cd40a287 ef3222b7665cf3; Vosloshko, "Gorod, pogibshii v 16 let."

5. Artur Shigapov, *Chernobyl', Pripiat', dalee nigde.*

6. Iurii Shcherbak, *Chernobyl': Dokumental'noe povestvovanie* (Moscow, 1991), 31.

7. Boliasnyi, "Kogda iskliucheniia chasto povtoriaiutsia, oni stanoviatsia normoi," *Vestnik* 7, no. 214 (March 30, 1999), www.vestnik.com/issues/1999/0330 /koi/bolyasn.htm; Anatolii Tsybul's'kyi, "Kyïvs'ka pravda: Za 60 krokiv vid reaktora," Facebook page of newspaper *Kyivs'ka Pravda*, April 26, 2016, https://www .facebook.com/KiivskaPravda/posts/1257118560979876.

8. Tsybul's'kyi, "Kyïvs'ka pravda"; Natalia Filipchuk, "Vy stroite stantsiiu na prokliatom meste," *Golos Ukrainy*, September 26, 2007.

9. Filipchuk, "Vy stroite stantsiiu na prokliatom meste."

10. Minutes of meeting called by the deputy chairman of the Council of Ministers, V. E. Dymshits, April 1, 1980, Tsentral'nyi derzhavnyi arkhiv hromads'kykh ob'iednan' (TsDAHO), fond 1, op. 32, no. 2124, fols. 51–54; memo from Oleksii Tytarenko to Volodymyr Shcherbytsky, May 21, 1980, ibid., fols. 46–47.

11. Boliasnyi, "Kogda iskliucheniia chasto povtoriaiutsia."

12. L. Stanislavskaia, "Ne chastnoe delo: Distsiplina i kachestvo postavok," *Tribuna énergetika*, March 21, 1986.

13. "V gorodskom komitete Kompartii Ukrainy," *Tribuna énergetika*, January 31, 1986; Anatolii Diatlov, *Chernobyl': Kak éto bylo* (Moscow, 2003), chapter 4.

14. Diatlov, *Chernobyl': Kak éto bylo*, chapter 4.

15. Boris Komarov, "Kto ne boitsia atomnykh élektrostantsii," *Strana i mir* (Munich), no. 6 (1986): 50–59.

16. N. A. Dollezhal and Iu. I. Koriakin, "Iadernaia énergetika: Dostizheniia i problemy," *Kommunist* 14 (1979): 19–28.

17. A. Aleksandrov, "Nauchno-tekhnicheskii progress i atomnaia énergetika," *Problemy mira i sotsializma*, 6 (1979): 15–20.

18. Grigorii Medvedev, *Chernobyl'skaia tetrad': Dokumental'naia povest'* (Kyiv, 1990), 41.

19. "Spetsial'ne povidomlennia Upravlinnia Komitetu Derzhavnoi Bezpeky pry Radi Ministriv Ukrains'koi RSR [KDB URSR] po mistu Kyievu ta Kyivs'kii oblasti," August 17, 1976, *Z arkhiviv VChK—GPU-NKVD-KGB*, special issue, "Chornobyl's'ka trahediia v dokumentakh ta materialakh," vol. 16 (Kyiv, 2001), no. 2, 27–30.

20. Vladimir Voronov, "V predchustvii Chernobylia," *Sovershenno sekretno*, January 4, 2015.

21. Viktor Dmitriev, "Avariia 1982 g, na 1-m bloke ChAĖS," Prichiny Chernobyl'skoi avarii izvestny, November 30, 2013, http://accidont.ru/Accid82.html; "Povidomlennia Upravlinnia KDB URSR po mistu Kyievu," September 10, 1982, *Z arkhiviv*, no. 9, 44; "Povidomlennia Upravlinnia Komitetu Derzhavnoi Bezpeky URSR po mistu Kyievu," September 13, 1982, *Z arkhiviv*, no. 9, 45–46.

22. Maiia Rudenko, "Nuzhna li reabilitatsiia byvshemu direktoru ChAĖS?" *Vzgliad*, April 29, 2010.

23. "Spetsial'ne povidomlennia 6-ho viddilu Upravlinnia KDB URSR po mistu Kyievu," February 26, 1986, *Z arkhiviv*, no. 20, 64.

24. "Vsesoiuznoe soveshchanie," *Tribuna ėnergetika*, March 28, 1986.

25. David Marples, *Chernobyl and Nuclear Power in the USSR* (Edmonton, 1986), 117.

26. Medvedev, *Chernobyl'skaia tetrad'*, 15.

27. Ibid., 31; Grigori Medvedev, *The Truth About Chernobyl* (New York, 1991), 45–46; Nikolai Karpan, *Chernobyl': Mest' mirnogo atoma* (Moscow, 2006), 444.

28. Liubov Kovalevs'ka, "Ne pryvatna sprava," *Literaturna Ukraïna*, March 27, 1986.

29. Evgenii Ternei, "Zhivaia legenda mertvogo goroda," *Zerkalo nedeli*, April 28, 1995.

CHAPTER 4: FRIDAY NIGHT

1. "Vystuplenie tovarishcha Gorbacheva M. S. IX s"ezd Sotsialisticheskoi edinoi partii Germanii," *Pravda*, April 19, 1986; "V Politbiuro TsK KPSS," *Pravda*, April 25, 1986.

2. A. Esaulov, "Prazdnik truda," *Tribuna ėnergetika*, April 25, 1986; A. Petrusenko, "S polnoi otdachei," *Tribuna ėnergetika*, April 25, 1986; "Na uroven' masshtabnykh zadach: Ne plenume Pripiatskogo gorkoma Kompartii Ukrainy," *Tribuna ėnergetika*, April 18, 1986.

3. I. Nedel'skii, "Nerest ryby," *Tribuna ėnergetika*, April 25, 1986.

4. "U lisnykiv raionu," *Prapor peremohy*, April 26, 1986; "Khid sadinnia kartopli: Zvedennia," *Prapor peremohy*, April 26, 1986.

5. Iu. Vermenko and V. Kulyba, "Novi sorty kartopli dlia Kyïvs'koï oblasti," *Prapor peremohy*, April 26, 1986.

6. Svetlana Samodelova, "Lichnaia katastrofa direktora Chernobylia," *Moskovskii komsomolets*, April 21, 2011.

7. Stepan Mukha, chairman of the KGB of the Ukrainian SSR, to the Central Committee of the Communist Party of Ukraine, "Informatsionnoe soobshchenie," April 8, 1986, Archives of the Security Service of Ukraine (Archive SBU hereafter), fond 16, op. 1, no. 1113, 9.

8. "Na uroven' masshtabnykh zadach: Na plenume Pripiatskogo gorkoma Kompartii Ukrainy," *Tribuna energetika*, April 18, 1986; Nikolai Karpan, *Chernobyl': Mest' mirnogo atoma* (Moscow, 2006), 423–424.

9. R. I. Davletbaev, "Posledniaia smena," in *Chernobyl' desiat let spustia: Neizbezhnost' ili sluchainost'?* (Moscow, 1995), 367–368.

10. Vitalii Borets, "Kak gotovilsia vzryv Chernobylia," *Post Chornobyl'* 4, no. 28 (February 2006), www.postchernobyl.kiev.ua/vitalij-borec.

11. "Akt komissii po fizicheskomu pusku o zavershenii fizicheskogo puska reaktora RBMK-1000 IV bloka Chernobyl'skoi AĖS, 17 December 1983," Prichiny Chernobyl'skoi avarii izvestny, http://accidont.ru/phys_start.html; Viktor Dmitriev, "Kontsevoi effekt," Prichiny Chernobyl'skoi avarii izvestny, November 30, 2013, http://accidont.ru/PS_effect.html.

12. Borets, "Kak gotovilsia vzryv Chernobylia."

13. Grigori Medvedev, *Iadernyi zagar* (Moscow, 2002), 206; Ernest J. Sternglass, *Secret Fallouts: Low Level Radiation from Hiroshima to Three Mile Island* (New York, 1981), 120.

14. "Spetsial'ne povidomlennia 6-ho viddilu Upravlinnia Komitetu Derzhavnoi Bezpeky URSR po Kyievu," October 1984, *Z arkhiviv*, no. 17, 58–60.

15. Borets, "Kak gotovilsia vzryv Chernobylia."

16. Karpan, *Chernobyl': Mest' mirnogo atoma*, 326, 440.

17. "Pravoflanhovi p'iatyrichky," *Kyïvs'ka pravda*, December 29, 1985.

18. Igor' Kazachkov, in Iurii Shcherbak, *Chernobyl': Dokumental'noe povestvovanie* (Moscow, 1991), 366.

19. Ibid., 34–35.

20. "Spetsial'ne povidomlennia 6-ho viddilu Upravlinnia KDB URSR po Kyievu," February 4, 1986, *Z arkhiviv*, no. 19, 62–63.

21. Iurii Trehub, in Shcherbak, *Chernobyl'*, 38.

22. Ibid., 36–38; Karpan, *Chernobyl': Mest' mirnogo atoma*, 444.

CHAPTER 5: EXPLOSION

1. Anatolii Diatlov, *Chernobyl': Kak ėto bylo* (Moscow, 2003), chapter 4.

2. Vitalii Borets, "Kak gotovilsia vzryv Chernobylia," *Post Chornobyl'* 4, no. 28 (February 2006), www.postchernobyl.kiev.ua/vitalij-borec; Nikolai Karpan, *Chernobyl': Mest' mirnogo atoma* (Moscow, 2006), 440.

3. Recollections of V. V. Grishchenko, B. A. Orlov, and V. A. Kriat, in Diatlov, *Chernobyl': Kak ėto bylo*, appendix 8: "Kakim on parnem byl. Vospominaniia o A. S. Diatlove."

4. Iurii Trehub, in Iurii Shcherbak, *Chernobyl': Dokumental'noe povestvovanie* (Moscow, 1991), 38; Razim Davletbaev, in Grigori Medvedev, *Iadernyi zagar* (Moscow, 2002), 242.

5. Trehub, in Shcherbak, *Chernobyl'*, 38; Karpan, *Chernobyl': Mest' mirnogo atoma*, 330, 354.

6. Liubov Akimova, in Grigori Medvedev, *The Truth About Chernobyl* (New York, 1991), 148–149; "Toptunov, Leonid Fedorovich, 18.06.1960–14.05.1986," Slavutyts'ka zahal'no-osvitnia shkola, http://coolschool1.at.ua/index/kniga _pamjati_quot_zhivy_poka_pomnim_quot_posvjashhaetsja_tem_kto_pogib _v_chernobylskom_pekle_toptunov_l/0-417; Trehub, in Shcherbak, *Chernobyl'*, 39.

7. Trehub, in Shcherbak, *Chernobyl'*, 38–39.

8. R. I. Davletbaev, "Posledniaia smena," in *Chernobyl' desiat let spustia: Neizbezhnost' ili sluchainost'?* (Moscow, 1995), 381–382.

9. Trehub, in Shcherbak, *Chernobyl'*, 40–41; Karpan, *Chernobyl': Mest' mirnogo atoma*, 326–335, 350.

10. Diatlov, *Chernobyl': Kak èto bylo*, chapter 4; Karpan, *Chernobyl': Mest' mirnogo atoma*, 477, 478, 479.

11. Medvedev, *The Truth About Chernobyl*, 67–76; Karpan, *Chernobyl': Mest' mirnogo atoma*, 476.

12. Davletbaev, "Posledniaia smena," 370; Karpan, *Chernobyl': Mest' mirnogo atoma*, 336.

13. Karpan, *Chernobyl': Mest' mirnogo atoma*, 482; Diatlov, *Chernobyl': Kak èto bylo*, chapter 4.

14. Robert B. Cullen, Thomas M. DeFrank, and Steven Strasser, "Anatomy of Catastrophe: The Soviets Lift Lid on the Chernobyl Syndrome," *Newsweek*, September 1, 1986, 26–28; "Sequence of Events: Chernobyl Accident, Appendix," World Nuclear Association, November 2009, www.world-nuclear.org /information-library/safety-and-security/safety-of-plants/appendices/chernobyl -accident-appendix-1-sequence-of-events.aspx; "Xenon Poisoning," HyperPhysics, Department of Physics and Astronomy, Georgia State University, http://hyper physics.phy-astr.gsu.edu/hbase/nucene/xenon.html.

15. Davletbaev, "Posledniaia smena," 371; Trehub, in Shcherbak, *Chernobyl'*, 41–42.

16. Medvedev, *The Truth About Chernobyl*, 85–88.

CHAPTER 6: FIRE

1. "V Politbiuro TsK KPSS," *Pravda*, April 25, 1986, 1; *Izvestiia*, April 25, 1986, 1.

2. "Programma na nedeliu," TV Program in *Izvestiia*, April 19, 1986.

3. Halyna Kovtun, *Ia pysatymu tobi shchodnia: Povist' u lystakh* (Kyiv, 1989), 42.

4. Leonid Teliatnikov and Leonid Shavrei, in Iurii Shcherbak, *Chernobyl': Dokumental'noe povestvovanie* (Moscow, 1991), 49–50; Kovtun, *Ia pysatymu tobi shchodnia*, 52–54.

5. "Pozharnyi-Chernobylets Shavrei: My prosto vypolniali svoi dolg," *RIA Novosti Ukraina*, April 26, 2016, http://rian.com.ua/interview/20160426 /1009035845.html.

6. Ivan Shavrei, in Vladimir Gubarev, *Zarevo nad Pripiat'iu* (Moscow, 1987), 5; Andrei Chernenko, *Vladimir Pravik* (Moscow, 1988), 87; Shcherbak, *Chernobyl'*, 53.

7. Shavrei, in Shcherbak, *Chernobyl'*, 53–55.

8. Volodymyr Pryshchepa, in Gubarev, *Zarevo nad Pripiat'iu*, 5–6; Shavrei, in Shcherbak, *Chernobyl'*, 54.

9. Shavrei, in Shcherbak, *Chernobyl'*, 54; Grigori Medvedev, *The Truth About Chernobyl*, foreword by Andrei Sakharov (New York, 1991), 87.

10. Liudmyla Ihnatenko, in Svetlana Alexievich, *Voices from Chernobyl: The Oral History of a Nuclear Disaster* (New York, 2006), 5.

11. Hryhorii Khmel, in Shcherbak, *Chernobyl'*, 57–58.

12. Teliatnikov, in Shcherbak, *Chernobyl'*, 51.

13. Teliatnikov and Shavrei in Shcherbak, *Chernobyl'*, 50, 54; Kovtun, *Ia pysatymu tobi*, 52–54.

14. Teliatnikov, in Gubarev, *Zarevo nad Pripiat'iu*, 6–9; Kovtun, *Ia pysatymu tobi*, 62.

15. Stanislav Tokarev, "Byl' o pozharnykh Chernobylia," *Smena*, no. 1423 (September 1986); Valentyn Belokon, in Shcherbak, *Chernobyl'*, 62–63.

16. Belokon, in Shcherbak, *Chernobyl'*, 62–63.

17. Ibid., 63–64.

18. Gubarev, *Zarevo nad Pripiat'iu*, 7–9.

19. Anna Laba, "Pozharnyi-Chernobylets Shavrei: My prosto vypolniali svoi dolg," *RIA Novosti Ukraina*, April 26, 2016, http://rian.com.ua/interview /20160426/1009035845.html.

20. Tokarev, "Byl' o pozharnykh Chernobylia."

21. Khmel, in Shcherbak, *Chernobyl'*, 59; Tokarev, "Byl' o pozharnykh Chernobylia."

22. Liudmyla Ihnatenko, in Alexievich, *Voices from Chernobyl*, 6–7.

23. Kovtun, *Ia pysatymu tobi shchodnia*, 63–64.

CHAPTER 7: DENIAL

1. Mariia Vasil', "Byvshii direktor ChAĖS Briukhanov: 'Esli by nasli dlia menia rasstrel'nuiu stat'iu, to, dumaiu, menia rasstreliali by,'" *Fakty*, October 18, 2000; "Interv'iu s Viktorom Briukhanovym," *ChAES: Zona otchuzhdeniia*, http://chernobil.info/?p=5898; Svetlana Samodelova, "Lichnaia katastrofa direktora Chernobylia," *Moskovskii komsomolets*, April 21, 2011; V. Ia. Vozniak and S. N. Troitskii, *Chernobyl': Tak ėto bylo. Vzgliad iznutri* (Moscow, 1993), 163.

2. Sergei Parashin, in Iurii Shcherbak, *Chernobyl': Dokumental'noe povestvovanie* (Moscow, 1991), 76–77.

3. Sergei Babakov, "S pred"iavlennymi mne obvineniiami ne soglasen . . . ," *Zerkalo nedeli*, August 29, 1999; Briukhanov's court testimony, in Nikolai Karpan, *Chernobyl': Mest' mirnogo atoma* (Moscow, 2006), 419–420.

4. Rogozhkin's court testimony, in Karpan, *Chernobyl': Mest' mirnogo atoma*, 461–465; Parashin, in Shcherbak, *Chernobyl'*, 75–76.

5. Anatolii Diatlov, *Chernobyl': Kak èto bylo* (Moscow, 2003), chapter 5.

6. R. I. Davletbaev, "Posledniaia smena," in *Chernobyl' desiat let spustia: Neizbezhnost' ili sluchainost'?* (Moscow, 1995), 371.

7. Diatlov, *Chernobyl': Kak èto bylo*, chapter 5; Davletbaev, "Posledniaia smena," 372.

8. Iurii Trehub, in Shcherbak, *Chernobyl'*, 42–43.

9. Trehub, in Shcherbak, *Chernobyl'*, 43–44.

10. Diatlov, *Chernobyl': Kak èto bylo*, chapter 5; Diatlov's court testimony, in Karpan, *Chernobyl': Mest' mirnogo atoma*, 446–456; A. Iuvchenko's court testimony, in ibid., 479–480; Vozniak and Troitskii, *Chernobyl'*, 179.

11. Parashin, in Shcherbak, *Chernobyl'*, 76; Vozniak and Troitskii, *Chernobyl'*, 165, 179; Briukhanov's court testimony, in Karpan, *Chernobyl': Mest' mirnogo atoma*, 429; Diatlov, *Chernobyl': Kak èto bylo*, chapter 5.

12. Parashin, in Shcherbak, *Chernobyl'*, 78; Vladimir Chugunov's court testimony, in Karpan, *Chernobyl': Mest' mirnogo atoma*, 427; Viktor Smagin, in Grigori Medvedev, *The Truth About Chernobyl*, foreword by Andrei Sakharov (New York, 1991), 132.

13. Smagin, in Medvedev, *The Truth About Chernobyl*, 132–133.

14. Chugunov's court testimony, in Karpan, *Chernobyl': Mest' mirnogo atoma*, 427; Arkadii Uskov, in Shcherbak, *Chernobyl'*, 69–72; Vozniak and Troitskii, *Chernobyl'*, 181.

15. Smagin, in Medvedev, *The Truth About Chernobyl*, 130–131.

16. Ibid.; Uskov, in Shcherbak, *Chernobyl'*, 73–74; Parashin, in ibid., 77.

17. Vozniak and Troitskii, *Chernobyl'*, 150.

18. Sergei Babakov, "V nachale mne ne poveril dazhe syn," *Zerkalo nedeli*, April 23, 1999.

19. Parashin, in Shcherbak, *Chernobyl'*, 77; Babakov, "V nachale mne ne poveril dazhe syn"; Vozniak and Troitskii, *Chernobyl'*, 157.

20. Zhores Medvedev, *The Legacy of Chernobyl* (New York, 1990), 74–89; Viktor Haynes and Marko Bojcun, *The Chernobyl Disaster* (London, 1988), 32.

21. Babakov, "S pred"iavlennymi mne obvineniiami."

22. Vozniak and Troitskii, *Chernobyl'*, 35; Volodymyr Yavorivsky, Minutes of the Session of the Ukrainian Supreme Soviet, December 11, 1991, http://iportal .rada.gov.ua/meeting/stenogr/show/4642.html.

23. G. N. Petrov, in Medvedev, *The Truth About Chernobyl*, 88–89.

24. Ibid.

25. Liubov Kovalevskaia, in Shcherbak, *Chernobyl'*, 86–87.

26. Leonid Kham'ianov, *Moskva—Chernobyliu* (Moscow, 1988), excerpt in Karpan, *Chernobyl': Mest' mirnogo atoma*, appendix no. 1.

27. V. G. Smagin, in Medvedev, *The Truth About Chernobyl*, 172–173.

CHAPTER 8: HIGH COMMISSION

1. Galina Akkerman, "Gorbachev: Chernobyl' sdelal menia drugim chelove-kom," *Novaia gazeta*, March 2, 2006; cf. Mikhail Gorbachev, *Memoirs* (New York, 1996), 189.

2. Aleksandr Liashko, *Gruz pamiati: Vospominaniia*, vol. 3, *Na stupeniakh vlasti*, part 2 (Kyiv, 2001), 342–343; Elena Novoselova, "Nikolai Ryzhkov: Razdalsia zvonok pravitel'stvennoi sviazi—na Chernobyle avariia," *Rossiiskaia gazeta*, April 25, 2016.

3. "Srochnoe donesenie pervogo zamestitelia energetiki i elektrifikatsii SSSR A. N. Makukhina v TsK KPSS ob avarii na Chernobyl'skoi AES," April 26, 1986, in *Chernobyl': 26 aprelia 1986–dekabr' 1991. Dokumenty i materialy* (Minsk, 2006), 27.

4. Grigori Medvedev, *The Truth About Chernobyl*, foreword by Andrei Sakharov (New York, 1991), 128, 151–155; Valerii Legasov, "Ob avarii na Chernobyl'skoi AES," tape no. 1, Elektronnaluia biblioteka RoyalLib.Com, http://royallib.com/read/legasov_valeriy/ob_avarii_na_chernobilskoy_aes.html#0.

5. Mikhail Tsvirko, in Medvedev, *The Truth About Chernobyl*, 152.

6. Gennadii Shasharin, in Medvedev, *The Truth About Chernobyl*, 154–155, 157.

7. Medvedev, *The Truth About Chernobyl*, 157–158.

8. Legasov, "Ob avarii na Chernobyl'skoi AES," tape no. 1; Sergei Parashin, in Iurii Shcherbak, *Chernobyl': Dokumental'noe povestvovanie* (Moscow, 1991), 76–77.

9. Vladimir Shishkin, in Medvedev, *The Truth About Chernobyl*, 159–160.

10. Shishkin, in Medvedev, *The Truth About Chernobyl*, 162–165; Boris Prushinsky, in ibid., 165–166.

11. Prushinsky, in Medvedev, *The Truth About Chernobyl*, 165–166.

12. V. I. Andriianov and V. G. Chirskov, *Boris Shcherbina* (Moscow, 2009).

13. Novoselova, "Nikolai Ryzhkov."

14. Legasov, "Ob avarii na Chernobyl'skoi AES," tape no. 1.

15. Ibid.

16. Ibid.; Evgenii Ignatenko, in V. Ia Vozniak and S. N. Troitskii, *Chernobyl': Tak eto bylo. Vzgliad iznutri* (Moscow, 1993), 187.

17. Shasharin, in Medvedev, *The Truth About Chernobyl*, 166–167.

18. "Boris Evdokimovich Shcherbina," in *Chernobyl': Dolg i muzhestvo*, vol. 2 (Moscow, 2001); Colonel V. Filatov, in Medvedev, *The Truth About Chernobyl*, 179–180.

19. Leonid Kham'ianov, *Moskva—Chernobyliu* (Moscow, 1988); Armen Abagian, in Vozniak and Troitskii, *Chernobyl'*, 213.

20. Kham'ianov, *Moskva—Chernobyliu*.

21. Abagian, in Vozniak and Troitskii, *Chernobyl'*, 219–220; Kham'ianov, *Moskva—Chernobyliu*.

22. Legasov, "Ob avarii na Chernobyl'skoi AĖS," tape no. 1; Ivan Pliushch, Minutes of the Session of the Ukrainian Supreme Soviet, December 11, 1991, http://rada.gov.ua/meeting/stenogr/show/4642.html.

23. Pliushch, Minutes of the Session of the Ukrainian Supreme Soviet, December 11, 1991; Novoselova, "Nikolai Ryzhkov"; A. Perkovskaia, in Shcherbak, *Chernobyl'*, 92.

CHAPTER 9: EXODUS

1. Aleksandr Liashko, *Gruz pamiati: Vospominaniia*, vol. 3, *Na stupeniakh vlasti*, part 2 (Kyiv, 2001), 435.

2. *V masshtabe ėpokhi: Sovremenniki ob A. P. Liashko*, comp. V. I. Liashko (Kyiv, 2003).

3. Vasyl' Kucherenko and Vasyl' Durdynets', in *Chornobyl's'ka katastrofa v dokumentakh, faktakh ta doliakh liudei. MVS* (Kyiv, 2006), 83–84, 90; Dmitrii Kiianskii, "Pust' nash muzei budet edinstvennym i poslednim," *Zerkalo nedeli*, April 29, 2000.

4. Vladimir Shishkin, in Grigori Medvedev, *The Truth About Chernobyl*, foreword by Andrei Sakharov (New York, 1991), 162–163; *Chornobyl's'ka katastrofa v dokumentakh*, 91–93.

5. Durdynets', in *Chornobyl's'ka katastrofa v dokumentakh*, 83; Sergei Babakov, "V nachale mne ne poveril dazhe syn," *Zerkalo nedeli*, April 23, 1999; Oleksandr Liashko, quoted in report to the Ukrainian parliament by Volodymyr Yavorivsky, Minutes of the Session of the Ukrainian Supreme Soviet, December 11, 1991.

6. Lina Kushnir, "Valentyna Shevchenko: 'Provesty demonstratsiiu 1-ho travnia 1986-ho nakazaly z Moskvy,'" *Ukraïns'ka pravda*, April 25, 2011; Valentyna Shevchenko, quoted in report to the Ukrainian parliament by Volodymyr Yavorivsky, Minutes of the Session of the Ukrainian Supreme Soviet, December 11, 1991; Ivan Hladush to the Central Committee of the Communist Party of Ukraine, April 27, 1986, TsDAHO, fond 1, op. 25, no. 2996.

7. Liashko, *Gruz pamiati*, vol. 1, part 2, 347.

8. Ibid., 348; Oleksandr Liashko, quoted in report to the Ukrainian parliament by Volodymyr Yavorivsky, Minutes of the Session of the Ukrainian Supreme Soviet, December 11, 1991.

9. Liashko, quoted in report to the Ukrainian parliament by Volodymyr Yavorivsky, Minutes of the Session of the Ukrainian Supreme Soviet, December 11, 1991; "Povidomlennia Upravlinnia Kopmitetu derzhavnoi bezpeky URSR po Kyievu," April 26, 1986, in *Z arkhiviv*, no. 21, 65–66; "Povidomlennia KDB URSR to KDB SRSR," April 26, 1986, in ibid., no. 22.

10. Liashko, quoted in report to the Ukrainian parliament by Volodymyr Yavorivsky, Minutes of the Session of the Ukrainian Supreme Soviet, December 11, 1991; Volodymyr Lytvyn, *Polytychna arena Ukraïny: Diiovi osoby*

ta vykonavtsi (Kyiv, 1994), 178; *Chornobyl"s'ka katastrofa v dokumentakh*, 205; Memo: Ukrainian Ministry of Transport to the Central Committee in Kyiv, April 28, 1986, TsDAHO, Kyiv, fond 1, op. 25, no. 2996.

11. Shishkin, in Medvedev, *The Truth About Chernobyl*, 162–163; Liashko, *Gruz pamiati*, vol. 3, part 2, 352–355; Liashko, quoted in report to the Ukrainian parliament by Volodymyr Yavorivsky, Minutes of the Session of the Ukrainian Supreme Soviet, December 11, 1991.

12. Kovtun, *Ia pysatymu tobi shchodnia*, 64.

13. Leonid Shavrei, in Iurii Shcherbak, *Chernobyl': Dokumental'noe povestvovanie* (Moscow, 1991), 55–56.

14. Shavrei, in Shcherbak, *Chernobyl'*, 56.

15. Liudmyla Ihnatenko, in Svetlana Alexievich, *Voices from Chernobyl: The Oral History of a Nuclear Disaster* (New York, 2006), 6–7.

16. Viktor Smagin, in Medvedev, *The Truth About Chernobyl*, 169–173.

17. V. Ia. Vozniak and S. N. Troitskii, *Chernobyl': Tak èto bylo. Vzgliad iznutri* (Moscow, 1993), 207–208; Kate Brown, *Plutopia: Nuclear Families, Atomic Cities and the Great Soviet and American Plutonium Disasters* (New York, 2013), 172–176.

18. David L. Chandler, "Explained: Rad, Rem, Sieverts, Becquerels: A Guide to Terminology About Radiation Exposure," *MIT News*, March 28, 2011, http://news.mit.edu/2011/explained-radioactivity-0328.

19. "Acute Radiation Syndrome: A Fact Sheet for Clinicians," Centers for Disease Control and Prevention, https://emergency.cdc.gov/radiation/arsphysician factsheet.asp.

20. *Posledstviia oblucheniia dlia zdorov'ia cheloveka v rezul'tate Chernobyl'skoi avarii* (New York, 2012), 12.

21. Smagin, in Medvedev, *The Truth About Chernobyl*, 173.

22. Aleksandr Esaulov, in Shcherbak, *Chernobyl'*, 82–83.

23. Liudmyla Ihnatenko, in Alexievich, *Voices from Chernobyl*, 8–9.

24. Esaulov, in Shcherbak, *Chernobyl'*, 83–84; Vozniak and Troitskii, *Chernobyl'*, 207–208.

25. Shcherbak, *Chernobyl'*, 109–110.

26. Nadezhda Mel'nichenko, "Pripiat' 1986: Èvakuatsiia. Vospominaniia ochevidtsa," *Taimer*, April 26, 2013.

27. Liudmila Kharitonova and Volodymyr Voloshko, in Medvedev, *The Truth About Chernobyl*, 138, 141, 149.

28. Valerii Legasov, "Ob avarii na Chernobylskoi AES," tape no. 1, Elektronnaluia biblioteka RoyalLib.Com, http://royallib.com/read/legasov_valeriy/ob _avarii_na_chernobilskoy_aes.html#0; *Chornobyl's'ka katastrofa v dokumentakh*, 204–209.

29. Esaulov, in Shcherbak, *Chernobyl'*, 84–86.

30. "Sniato 26 aprelia 1986 g. v gorode Pripiat'," YouTube, April 26, 1986, published April 14, 2011, www.youtube.com/watch?v=XxGObvkLTg0; "Pripiat:

368   *Notes*

Ėvakuatsiia, April 27, 1986," YouTube, April 27, 1986, published April 25, 2011, www.youtube.com/watch?v=xAxCWNNyCpA.

31. Aneliia Perkovskaia, in Shcherbak, *Chernobyl'*, 90; Babakov, "V nachale mne ne poveril dazhe syn"; *Chornobyl's'ka katastrofa v dokumentakh*, 207.

32. "Sniato 26 aprelia 1986 g. v gorode Pripiat'," YouTube.

33. Liubov Kovalevskaia, in Shcherbak, *Chernobyl'*, 90; Liubov Kovalevskaia, "Preodolenie," in *Chernobyl': Dni ispytanii i pobed, Kniga svidetel'stv* (Kyiv, 1988), 77; "Sniato 26 aprelia 1986 g. v gorode Pripiat'," YouTube.

34. Elena Novoselova, "Nikolai Ryzhkov: Razdalsia zvonok pravitel'stvennoi sviazi—na Chernobyle avariia," *Rossiiskaia gazeta*, April 25, 2016.

35. Ivan Hladush to the Central Committee of the Communist Party of Ukraine, April 28, 1986, TsDAHO, Kyiv, fond 1, op. 25, no. 2996; Andrei Illei, "V trudnyi chas," in *Chernobyl': Dni ispytanii i pobed*, 121.

36. "Informatsiine povidomlennia KDB URSR do TsK KPU," April 28, 1986, *Z arkhiviv*, no. 23, 69–70.

CHAPTER 10: TAMING THE REACTOR

1. Gennady Shasharin, in Grigori Medvedev, *The Truth About Chernobyl*, foreword by Andrei Sakharov (New York, 1991), 192–193.

2. Valerii Legasov, "Ob avarii na Chernobyl'skoi AĖS," tape no. 1, Elektronnaluia biblioteka RoyalLib.Com, http://royallib.com/read/legasov_valeriy /ob_avarii_na_chernobilskoy_aes.html#0.

3. Aleksandr Liashko, *Gruz pamiati: Vospominaniia*, vol. 3, *Na stupeniakh vlasti*, part 2 (Kyiv, 2001), 354; A. Perkovskaia and Iu. Dobrenko, in Iurii Shcherbak, *Chernobyl': Dokumental'noe povestvovanie* (Moscow, 1991), 88–89; Anatoly Zayats, in Medvedev, *The Truth About Chernobyl*, 193; Valentyna Kovalenko, in V. Ia. Vozniak and S. N. Troitskii, *Chernobyl': Tak ėto bylo. Vzgliad iznutri* (Moscow, 1993), 235.

4. Liashko, *Gruz pamiati*, vol. 3, part 2, 356; Colonel Filatov and M. S. Tsvirko, in Medvedev, *The Truth About Chernobyl*, 194–195; Zhores Medvedev, *The Legacy of Chernobyl* (New York, 1990), 56; N. P. Baranovskaia, *Ispytanie Chernobylem* (Kyiv, 2016), 35.

5. Anastasiia Voskresenskaia, "Vertoletchik—likvidator Chernobyl'skoi avarii: 'My vstali v karusel' smerti,'" *Zashchishchat' Rossiu*, April 26, 2016, https:// defendingrussia.ru/a/vertoletchiklikvidator_avarii_na_chernobylskoj_aes-5793.

6. Medvedev, *The Truth About Chernobyl*, 194.

7. Legasov, "Ob avarii na Chernobyl'skoi AĖS," tape no. 1; Shasharin, in Medvedev, *The Truth About Chernobyl*, 201–202.

8. Legasov, "Ob avarii na Chernobyl'skoi AĖS," tape no. 1.

9. V. M. Fedulenko, "Koe-chto ne zabylos'," *Vklad kurchatovtsev v likvidatsiiu avarii na Chernobyl'skoi AĖS*, ed. V. A. Sidorenko (Moscow, 2012), 74–83.

10. Ibid.; Shasharin in Medvedev, *The Truth About Chernobyl*, 201–202.

11. "Iz rabochei zapisi zasedaniia Politbiuro TsK KPSS, April 28, 1986," in R. G. Pikhoia, *Sovetskii Soiuz: Istoriia vlasti, 1945–1991* (Novosibirsk, 2000), 429–431.

12. Medvedev, *The Truth About Chernobyl*, 194; Baranovskaia, *Ispytanie Chernobylem*, 35–36.

13. Shcherbak, *Chernobyl'*, 154.

14. Baranovskaia, *Ispytanie Chernobylem*, 31–32.

15. Report by Captain A. P. Stelmakh, deputy chief of the Prypiat police department, in *Chornobyl's'ka katastrofa v dokumentakh, faktakh ta doliakh liudei. MVS* (Kyiv, 2006), 425–426.

16. Fedulenko, "Koe-chto ne zabylos'."

17. Leonid Kham'ianov, *Moskva—Chernobyliu* (Moscow, 1988); *Chornobyl's'ka katastrofa v dokumentakh*, 277.

18. Fedulenko, "Koe-chto ne zabylos'."

19. Tatiana Marchulaite, in Vozniak and Troitskii, *Chernobyl*, 205; Aleksandr Esaulov, in Shcherbak, *Chernobyl'*, 233.

20. Kham'ianov, *Moskva—Chernobyliu*; *Chornobyl's'ka katastrofa v dokumentakh*, 277.

21. Lina Kushnir, "Valentyna Shevchenko: 'Provesty demonstratsiiu 1-ho travnia 1986-ho nakazaly z Moskvy,'" *Ukraïns'ka pravda*, April 25, 2011.

22. Liubov Kovalevskaia, in Shcherbak, *Chernobyl'*, 104.

23. Kushnir, "Valentyna Shevchenko."

CHAPTER 11: DEADLY SILENCE

1. Zhores A. Medvedev, *Nuclear Disaster in the Urals* (New York, 1980); Kate Brown, *Plutopia: Nuclear Families, Atomic Cities and the Great Soviet and American Plutonium Disasters* (New York, 2013), 231–246; V. A. Kostyuchenko and L. Yu. Krestinina, "Long-Term Irradiation Effects in the Population Evacuated from the East-Urals Radioactive Trace Area," *Science of the Total Environment* 142, nos. 1–2 (March 1994): 119–125.

2. *Chernobyl"skaia atomnaia ėlektrostantsiia: Kul'turnoe i zhilishchno-bytovoe stroitel'stvo. General'nyi plan poselka* (Moscow, 1971), 11.

3. "25 Years After Chernobyl, How Sweden Found Out," Radio Sweden—News in English, April 22, 2011, http://sverigesradio.se/sida/artikel.aspx ?programid=2054&artikel=4468603; Serge Schmemann, "Soviet Announces Nuclear Accident at Electric Plant," *New York Times*, April 29, 1986, A1.

4. "First Coverage of Chernobyl Disaster on Soviet TV, April 1986," YouTube, published April 29, 2011, https://www.youtube.com/watch?v =4PytcgdPuTI; Stephen Mulev, "The Chernobyl Nightmare Revisited," BBC News, April 18, 2006, http://news.bbc.co.uk/2/hi/europe/4918742.stm.

5. "Iz rabochei zapisi zasedaniia Politbiuro TsK KPSS, April 28, 1986," in R. G. Pikhoia, *Sovetskii Soiuz: Istoriia vlasti, 1945–1991* (Novosibirsk, 2000), 429–431.

6. Mikhail Gorbachev, *Memoirs* (New York, 1996), 189; Elena Novoselova, "Nikolai Ryzhkov: Razdalsia zvonok pravitel'stvennoi sviazi—na Chernobyle avariia," *Rossiiskaia gazeta*, April 25, 2016.

7. Wayne King and Warren Weaver Jr., "Briefing: Airline Business as Usual," *New York Times*, April 21, 1986; William J. Eaton, "PanAm and Aeroflot Resume Direct US-Soviet Air Service," *Los Angeles Times*, April 30, 1986.

8. "Festive Flight to Moscow Resumes US-Soviet Air Service," *New York Times*, April 30, 1986.

9. Schmemann, "Soviet Announces Nuclear Accident at Electric Plant."

10. "Statement by Principal Deputy Press Secretary Speakes on the Soviet Nuclear Reactor Accident at Chernobyl," May 3, 1986, Ronald Reagan Presidential Library and Museum, Public Papers of the President, www.reagan.utexas .edu/archives/speeches/1986/50386a.htm; "Implications of the Chernobyl Disaster," CIA Memo, April 29, 1986, www.foia.cia.gov/sites/default/files/document _conversions/17/19860429.pdf.

11. "Implications of the Chernobyl Disaster," April 29, 1986.

12. "Nuclear Disaster: A Spreading Cloud and an Aid Appeal; U.S. Offers to Help Soviet in Dealing with Accident," *New York Times*, April 30, 1986; Alex Brummer, "Reagan Offers U.S. Help," *Guardian*, April 25, 2005, www .theguardian.com/world/2005/apr/25/nuclear.uk.

13. "Statement by Principal Deputy Press Secretary Speakes on the Soviet Nuclear Reactor Accident at Chernobyl," May 1, 1986, Ronald Reagan Presidential Library and Museum, Public Papers of the President, www.reagan.utexas .edu/archives/speeches/1986/50186b.htm.

14. Luther Whitington, "Chernobyl Reactor Still Burning," United Press International, April 29, 1986, www.upi.com/Archives/1986/04/29/Chernobyl -reactor-still-burning/9981572611428; "Chernobyl Nuclear Power Plant Disaster Creates Radiation Scare," ABC News, April 30, 1986, http://abcnews.go.com /Archives/video/chernobyl-disaster-nuclear-reactor-fallout-1986-9844065.

15. Vladimir Fronin, "To vzlet, to posadka," in *Chernobyl': Dni ispytanii i pobed, Kniga svidetel'stv* (Kyiv, 1988), 125–129.

16. Schmemann, "Soviet Announces Nuclear Accident at Electric Plant"; Christopher Jarmas, "Nuclear War: How the United States and the Soviet Union Fought over Information in Chernobyl's Aftermath," *Vestnik*, August 31, 2015, www.sras.org/information_chernobyl_us_ussr.

17. "Ot Soveta ministrov SSSR," *Pravda*, April 30, 1986.

18. Stepan Mukha, head of the Ukrainian KGB, to the Central Committee in Kyiv, April 28, 1986, Archive SBU, fond 16, op. 1, no. 1113; Oles Honchar, *Shchodennyky (1984–1995)* (Kyiv, 2004), 90.

19. Stepan Mukha, head of the Ukrainian KGB, to Volodymyr Shcherbytsky, first secretary of the Central Committee of the Communist Party of Ukraine, April 29, 1986, Archive SBU, fond 16, op. 1, no. 1113.

20. Stepan Mukha, head of the Ukrainian KGB, to the Central Committee in Kyiv, April 28, 1986, Archive SBU, fond 16, op. 1, no. 1113.

21. Lina Kushnir, "Valentyna Shevchenko: 'Provesty demonstratsiiu 1-ho travnia 1986-ho nakazaly z Moskvy,'" *Ukraïns'ka pravda*, April 25, 2011.

22. KGB memo to the Central Committee of the Communist Party of Ukraine, April 28, 1986, TsDAHO, fond 1, op. 32, no. 2337; *Chornobyl's'ka katastrofa v dokumentakh, faktakh ta doliakh liudei. MVS* (Kyiv, 2006), 258; Aleksandr Kitral', "Gorbachev—Shcherbitskomu: Ne provedesh parad, sgnoiu!," *Komsomol'skaia pravda v Ukraine*, April 26, 2011; Novoselova, "Nikolai Ryzhkov."

23. "Ot Soveta ministrov SSSR," *Pravda*, May 1, 1986.

24. Alla Iaroshinskaia, *Chernobyl': Bol'shaia lozh'* (Moscow, 2011), 313.

25. Volodymyr Viatrovych, "'Cho eto oznachaet?' Abo borot'ba SRSR iz radiatsiieiu," in idem, *Istoriia z hryfom sekretno* (Kharkiv, 2014), 450–456. Cf. L. O. Dobrovol's'kyi, "Zakhody z likvidatsiï naslidkiv avariï na Chornobyl's'kii AES: Khronika podii," *Zhurnal z problem medytsyny pratsi*, no. 1 (2011): 7.

26. Kitral', "Gorbachev—Shcherbitskomu"; Elena Sheremeta, "Rada Shcherbitskaia: Posle Chernobylia Gorbachev skazal Vladimiru Vasil'evichu," *Fakty*, February 17, 2006.

27. Irina Lisnichenko, "Aleksandr Liashko: 'Kogda Iavorivskii chital svoi doklad, ia stoial u groba docheri,'" *Fakty*, April 27, 2001.

28. See photos in Kushnir, "Valentyna Shevchenko."

29. Natalia Petrivna, in Kseniia Khalturina, "Pervomai: Ot pervoi stachki 'za rabotu' do besplatnogo truda," *TopKyiv*, May 1, 2016, https://topkyiv.com /news/pervomaj-ot-pervoj-stachki-za-rabotu-do-besplatnogo-truda-chto -otmechaem-segodnya.

30. Natalia Petrivna, in Kseniia Khalturina, "Pervomai: Ot pervoi stachki"; Natalia Morozova, quoted in report to the Ukrainian parliament by Volodymyr Yavorivsky, Minutes of the Session of the Ukrainian Supreme Soviet, December 11, 1991, http://rada.gov.ua/meeting/stenogr/show/4642.html.

31. Honchar, *Shchodennyky*, 91; Heorhii Ral', quoted in report to the Ukrainian parliament by Volodymyr Yavorivsky, Minutes of the Session of the Ukrainian Supreme Soviet, December 11, 1991; Stepan Mukha, head of the Ukrainian KGB, to the Ukrainian Central Committee, Informatsionnoe soobshchenie, April 30, 1986, Archive SBU, fond 16, op. 1, no. 1113.

32. Galina Akkerman, "Gorbachev: Chernobyl' sdelal menia drugim chelovekom," *Novaia gazeta*, March 2, 2006.

CHAPTER 12: EXCLUSION ZONE

1. Evgenii Pasishnichenko, "My na RAFe s migalkami," *Rabochaia gazeta*, April 26, 2012.

2. Evgenii Chernykh, "Egor Ligachev: 'Stranno konechno, chto Gorbachev ne s"ezdil v Chernobyl','" *Komsomol'skaia pravda*, April 28, 2011; Aleksandr Liashko, *Gruz pamiati: Vospominaniia*, vol. 3, *Na stupeniakh vlasti*, part 2 (Kyiv, 2001), 358; Valerii Legasov, "Ob avarii na Chernobyl'skoi AĖS," tape no. 1, Elektronnaluia biblioteka RoyalLib.Com, http://royallib.com/read/legasov_valeriy /ob_avarii_na_chernobilskoy_aes.html#0.

3. Liashko, *Gruz pamiati*, vol. 3, part 2, 358; Elena Novoselova, "Nikolai Ryzhkov: Razdalsia zvonok pravitel'stvennoi sviazi—na Chernobyle avariia," *Rossiiskaia gazeta*, April 25, 2016.

4. Sergei Babakov, "S pred"iavlennymi mne obvineniiami ne soglasen . . . ," *Zerkalo nedeli*, August 29, 1999; Mariia Vasil', "Byvshii direktor ChAĖS Briukhanov: 'Esli by nasli dlia menia rasstrel'nuiu stat'iu, to, dumaiu, menia rasstreliali by,'" *Fakty*, October 18, 2000; Liashko, *Gruz pamiati*, vol. 3, part 2, 359.

5. Interview with Borys Kachura, in *Rozpad radians'koho Soiuzu: Usna istoriia nezalezhnoï Ukraïny*, http://oralhistory.org.ua/interview-ua/360.

6. Novoselova, "Nikolai Ryzhkov"; O. H. Rohozhyn, "Naslidky chornobyl's'koï katastrofy dlia zony vidchuzhennia ta sil Polissia," Informatsiinyi tsentr Polissia 2.0, November 2009, www.polissya.eu/2009/11/naslidki-chornobil skoi-katastrofi-zona.html.

7. Liashko, *Gruz pamiati*, vol. 3, part 2, 360.

8. Legasov, "Ob avarii na Chernobyl'skoi AĖES," tape no. 1.

9. Vasyl' Syn'ko, "Chornobyl's'kyi rubets'," *Sil's'ki visti*, April 26, 2013.

10. Elena Sheremeta, "Vitalii Masol: My tikhonechko gotovilis' k ėvakuatsii Kieva," *Fakty*, April 26, 2006; Syn'ko, "Chornobyl's'kyi rubets'."

11. Yurii Petrov, "Za parolem 'blyskavka': Spohady uchasnykiv likvidatsiï naslidkiv avariï na Chornobyl's'kii AES," in *Z arkhiviv* 16 (2001): 372–380; Pasishnichenko, "My na RAFe s migalkami."

12. KGB Memo to the Ukrainian Central Committee, April 28, 1986, TsDAHO, fond 1, op. 32, no. 2337; "Ot Soveta ministrov SSSR," *Pravda*, May 1, 1986; Anatolii Romanenko, Minister of Health of Ukraine, to the Ukrainian Central Committee, May 3 and 4, 1986, TsDAHO, fond 1, op. 25, no. 2996, fols. 11–12 and 17–18; "Materialy zasedanii operativnoi gruppy TsK Kompartii Ukrainy," May 3, 1986, TsDAHO, fond 1, op. 17, no. 385.

13. "Materialy zasedanii operativnoi gruppy TsK Kompartii Ukrainy," May 3, 1986, TsDAHO, fond 1, op. 17, no. 385; Maksym Drach, in Iurii Shcherbak, *Chernobyl': Dokumental'noe povestvovanie* (Moscow, 1991), 144–149.

14. *Prapor peremohy*, April 29, May 1, May 3, 1986.

15. "Materialy zasedanii operativnoi gruppy TsK Kompartii Ukrainy," no. 1, May 3, and no. 2, May 4, 1986, TsDAHO, fond 1, op. 17, no. 385.

16. "Materialy zasedanii operativnoi gruppy TsK Kompartii Ukrainy," May 4, 1986, TsDAHO, fond 1, op. 17, no. 385.

17. Zhores Medvedev, *The Legacy of Chernobyl* (New York, 1990), 57–59.

18. Syn'ko, "Chornobyl's'kyi rubets'."

19. Ibid.

20. Ibid.

21. Fr. Leonid, in Shcherbak, *Chernobyl'*, 97–100.

22. Syn'ko, "Chornobyl's'kyi rubets'."

23. Medvedev, *The Legacy of Chernobyl*, 59.

24. "Povidomlennia operhrup KDB SRSR ta KDB URSR," May 1, 1986, in *Z arkhiviv* 16, no. 24, 71–72; "Dovidka 6-ho upravlinnia KDB URSR," May 4, 1986, in *Z arkhiviv* 16, no. 25, 73–74; Syn'ko, "Chornobyl's'kyi rubets'."

CHAPTER 13: CHINA SYNDROME

1. Anatolii Cherniaev, "Gorbachev's Foreign Policy: The Concept," in *Turning Points in Ending the Cold War*, ed. Kiron G. Skinner (Stanford, CA, 2007), 128–129.

2. Alla Iaroshinskaia, *Chernobyl': Bol'shaia lozh'* (Moscow, 2011), 288.

3. Grigori Medvedev, *The Truth About Chernobyl*, foreword by Andrei Sakharov (New York, 1991), 203–204.

4. *The China Syndrome* (1979), DVD, Sony Pictures Home Entertainment, 2004.

5. Valerii Legasov, "Ob avarii na Chernobyl'skoi AĖS," tape no. 3, Elektronnaluia biblioteka RoyalLib.Com, http://royallib.com/read/legasov_valeriy /ob_avarii_na_chernobilskoy_aes.html#0; Protocol of meeting of the Politburo of the Central Committee of the Communist Party of Ukraine, May 8, 1986.

6. Legasov, "Ob avarii na Chernobyl'skoi AĖS," tape no. 3.

7. Zhores Medvedev, *The Legacy of Chernobyl* (New York, 1990), 58–59; Legasov, "Ob avarii na Chernobyl'skoi AĖS," tape no. 3.

8. Medvedev, *The Truth About Chernobyl*, 203.

9. Svetlana Samodelova, "Belye piatna Chernobylia," *Moskovskii komsomolets*, April 25, 2011; Stephen McGinty, "Lead Coffins and a Nation's Thanks for the Chernobyl Suicide Squad," *The Scotsman*, March 16, 2011; Legasov, "Ob avarii na Chernobyl'skoi AĖS," tape no. 3.

10. "Velikhov Evgenii Pavlovich," Geroi strany, www.warheroes.ru/hero /hero.asp?Hero_id=10689; "Legasov, Valerii Alekseevich," Geroi strany, www .warheroes.ru/hero/hero.asp?Hero_id=6709; Legasov, "Ob avarii na Chernobyl'skoi AĖS," tape no. 3; Medvedev, *The Truth About Chernobyl*, 223–224.

11. "Evgenii Velikhov o sebe v programme Liniia Zhizni," *Rossiia 1*, http: //tvkultura.ru/person/show/person_id/110366; Vladimir Naumov, "Interv'iu s Akademikom Evgeniem Velikhovym," *Vestnik*, October 23, 2001.

12. Iulii Andreev, "Neschast'e akademika Legasova," *Lebed, Nezavisimyi al'manakh,* October 2, 2005.

13. Medvedev, *The Legacy of Chernobyl,* 57–59.

14. Iurii Shcherbak, *Chernobyl': Dokumental'noe povestvovanie* (Moscow, 1991), 157.

15. Elena Sheremeta, "Vitalii Masol: My tikhonechko gotovilis' k évakuatsii Kieva," *Fakty,* April 26, 2006.

16. Interview with Borys Kachura, in *Rozpad Radians'koho Soiuzu: Usna istoriia nezalezhnoï Ukraïny,* http://oralhistory.org.ua/interview-ua/360.

17. "Dopovidna 6-ho upravlinnia KDB URSR," May 5, 1986, *Z arkhiviv,* no. 27, 76.

18. Valentyn Zgursky, mayor of Kyiv, to Volodymyr Shcherbytsky, first secretary of the Central Committee of the Communist Party of Ukraine, May 1986, TsDAHO, fond 1, op. 32, no. 2337, fol. 5; "Materialy zasedanii operativnoi gruppy TsK Kompartii Ukrainy," May 8, 1986, TsDAHO, fond 1, op. 17, no. 385, fol. 90.

19. Interview with Kachura, *Rozpad Radians'koho Soiuzu.*

20. Grigorii Kolpakov, "On bystro razbiralsia v étikh radiatsionnykh veshchakh," *Gazeta.ru,* January 23, 2014; Lina Kushnir, "Valentyna Shevchenko: 'Provesty demonstratsiiu 1-ho travnia 1986-ho nakazaly z Moskvy,'" *Ukraïns'ka pravda,* April 25, 2011; Aleksandr Liashko, *Gruz pamiati: Vospominaniia,* vol. 3, *Na stupeniakh vlasti,* part 2 (Kyiv, 2001), 372–375.

21. Medvedev, *The Legacy of Chernobyl,* 61.

22. Interview with Kachura, *Rozpad Radians'koho Soiuzu*; Kushnir, "Valentyna Shevchenko"; "Materialy zasedanii operativnoi gruppy TsK Kompartii Ukrainy," May 6, 1986, TsDAHO, fond 1, op. 17, no. 385, fol. 68.

23. Sheremeta, "Vitalii Masol"; Legasov, "Ob avarii na Chernobyl'skoi AÉS," tape no. 3.

24. Medvedev, *The Legacy of Chernobyl,* 61–62.

CHAPTER 14: COUNTING LIVES

1. Grigori Medvedev, *The Truth About Chernobyl,* foreword by Andrei Sakharov (New York, 1991), 223–224.

2. V. Gubarev and M. Odinets, "Gorod, more i reaktor," *Pravda,* May 8, 1986; interview with Borys Kachura, in *Rozpad Radians'koho Soiuzu: Usna istoriia nezalezhnoïi Ukraïny,* http://oralhistory.org.ua/interview-ua/360.

3. Gubarev and Odinets, "Gorod, more i reaktor."

4. A. P. Grabovskii, *Atomnyi avral* (Moscow, 2001), 129; A. Iu. Mitiunin, "Atomnyi shtrafbat: Natsional'nye osobennosti likvidatsii radiatsionnykh avarii v SSSR i Rossii," *Chernobyl', Pripiat', Chernobyl'skaia AÉs i zona otchuzhdeniia,* http://chornobyl.in.ua/atomniy-shtrafbat.html.

5. Mitiunin, "Atomnyi shtrafbat."

6. "Postanovlenie TsK KPSS i Soveta ministrov SSSR," May 29, 1986, no. 634-18, in *Sbornik informatsionno-normativnykh dokumentov po voprosam preodoleniia v Rossiiskoi Federatsii posledstvii Chernobyl'skoi katastrofy* (Moscow, 1993), parts 1, 2 (1986–1992), 21.

7. Valerii Legasov, "Ob avarii na Chernobyl'skoi AĖS," tape no. 3, Elektronnaluia biblioteka RoyalLib.Com, http://royallib.com/read/legasov_valeriy/ob_avarii_na_chernobilskoy_aes.html#0; N. D. Tarakanov, *Chernobyl'skie zapiski, ili razdum'ia o nravstvennosti* (Moscow, 1989), 136–172; Mitiunin, "Atomnyi shtrafbat."

8. Vitalii Skliarov, *Zavtra byl Chernobyl'* (Moscow, 1993), 169.

9. Sheremeta, "Vitalii Masol: My tikhonechko gotovilis' k ėvakuatsii Kieva"; Legasov, "Ob avarii na Chernobyl'skoi AĖS," tape no. 3; "Materialy zasedanii operativnoi gruppy TsK Kompartii Ukrainy," May 10, 1986, TsDAHO, fond 1, op. 17, no. 385, fol. 95.

10. Legasov, "Ob avarii na Chernobyl'skoi AĖS," tape no. 3.

11. Dmitrii Levin, "Chernobyl' glazami ochevidtsev spustia pochti chetvert' veka posle avarii," in *ChAĖS: Zona otchuzhdeniia*, http://chernobil.info/?p=5113; "CHAĖS: Likvidatsiia avarii," in *Chernobyl', Pripiat', Chernobyl'skaia AĖS*, http://chornobyl.in.ua/licvidacia-avarii.html.

12. Khem Salhanyk, in Iurii Shcherbak, *Chernobyl': Dokumental'noe povestvovanie* (Moscow, 1991), 202.

13. Liudmyla Ihnatenko, in Svetlana Alexievich, *Voices from Chernobyl: The Oral History of a Nuclear Disaster* (New York, 2006), 8–9.

14. Ibid., 10–12.

15. Arkadii Uskov, in Shcherbak, *Chernobyl'*, 129–132.

16. Ihnatenko, in Alexievich, *Voices from Chernobyl*, 13–21.

17. V. K. Ivanov, A. I. Gorski, M. A. Maksioutov, A. F. Tsyb, and G. N. Souchkevitch, "Mortality Among the Chernobyl Emergency Workers: Estimation of Radiation Risks (Preliminary Analysis)," *Health Physics* 81, no. 5 (November 2001): 514–521; M. Rahu, K. Rahu, A. Auvinen, M. Tekkel, A. Stengrevics, T. Hakulinen, J. D. Boice, and P. D. Inskip, "Cancer Risk Among Chernobyl Cleanup Workers in Estonia and Latvia, 1986–1998," *International Journal of Cancer* 119 (2006): 162–168.

CHAPTER 15: WAR OF WORDS

1. "Vystuplenie M. S. Gorbacheva po sovetskomu televideniiu," *Pravda*, May 15, 1986.

2. Luther Whitington, "Chernobyl Reactor Still Burning," United Press International, April 29, 1986, www.upi.com/Archives/1986/04/29/Chernobyl-reactor-still-burning/9981572611428; W. Scott Ingram, *The Chernobyl Nuclear Disaster* (New York, 2005), 56–59.

3. Ronald Reagan, "Radio Address to the Nation on the President's Trip

to Indonesia and Japan," May 4, 1986, www.reagan.utexas.edu/archives /speeches/1986/50486c.htm.

4. Jack Nelson, "Reagan Criticizes Disaster Secrecy," *Los Angeles Times*, May 4, 1986; David Reynolds, *Summits: Six Meetings That Shaped the Twentieth Century* (New York, 2007), 383–385.

5. "Statement on the Implications of the Chernobyl Nuclear Accident," May 5, 1986, 12 Summit, Ministry of Foreign Affairs of Japan, www.mofa.go.jp/policy /economy/summit/2000/past_summit/12/e12_d.html.

6. Stepan Mukha to Volodymyr Shcherbytsky, "Dokladnaia zapiska 'Ob operativnoi obstanovke v respublike v sviazi s avariei na Chernobyl'skoi AĖS," May 16, 1986, Archive SBU, fond 16, op. 1, no. 1113.

7. Nicholas Daniloff, *Of Spies and Spokesmen: My Life as a Cold War Correspondent* (Columbia, MO, 2008), 347–348; Volodymyr Kravets, minister of foreign affairs of Ukraine, to the Central Committee of the Communist Party of Ukraine, May 1, 1986, TsDAHO, fond 1, op. 25, no. 2996, fol. 14.

8. Daniloff, *Of Spies and Spokesmen*, 347–348; Stepan Mukha to the Central Committee of the Ukrainian Communist Party, May 11, 1986, Archive SBU, fond 16, op. 1, no. 1113; Stepan Mukha to the Central Committee of the Ukrainian Communist Party, May 12, 1986, ibid.; Stepan Mukha to Volodymyr Shcherbytsky, "Dokladnaia zapiska 'Ob operativnoi obstanovke v respublike v sviazi s avariei na Chernobyl'skoi AĖS," May 16, 1986, ibid.

9. Daniloff, *Of Spies and Spokesmen*, 343–344; Stepan Mukha to Volodymyr Shcherbytsky, "Spetsial'noe soobshchenie ob obstanovke sredi inostrantsev v sviazi z avariei na Chernobyl'skoi AĖS," April 30, 1986, Archive SBU, fond 16, op. 1, no. 1113.

10. Stepan Mukha to the Central Committee of the Ukrainian Communist Party, May 5, 1986, Archive SBU, fond 16, op. 1, no. 1113; Stepan Mukha to Volodymyr Shcherbytsky, "Dokladnaia zapiska 'Ob operativnoi obstanovke v respublike v sviazi s avariei na Chernobyl'skoi AĖS,'" May 16, 1986, ibid.

11. Valerii Legasov, "Ob avarii na Chernobyl'skoi AĖS," tape no. 3, Elektronnaluia biblioteka RoyalLib.Com, http://royallib.com/read/legasov_valeriy /ob_avarii_na_chernobilskoy_aes.html#0.

12. Anna Christensen, "The Area Around the Chernobyl Nuclear Plant Was Not Evacuated Until 36 Hours After a Fiery Explosion," United Press International, May 6, 1986, www.upi.com/Archives/1986/05/06/The-area-around -the-Chernobyl-nuclear-power-plant-was/1746515736000; Daniloff, *Of Spies and Spokesmen*, 344–345; Zhores Medvedev, *The Legacy of Chernobyl* (New York, 1990), 67–68.

13. "Soobshchenie TASS," *Pravda*, May 6, 1986; "News Summary," *New York Times*, May 6, 1986.

14. "Materialy zasedanii operativnoi gruppy TsK Kompartii Ukrainy," May 5 and 8, 1986, TsDAHO, fond 1, op. 17, no. 385; Philip Taubman, "Residents of Kiev Warned to Guard Against Radiation," *New York Times*, May 9, 1986.

15. Taubman, "Residents of Kiev Warned to Guard Against Radiation"; Aleksandr Liashko, *Gruz pamiati: Vospominaniia*, vol. 3, *Na stupeniakh vlasti*, part 2 (Kyiv, 2001), 357; Daniloff, *Of Spies and Spokesmen*, 343.

16. Robert G. Darst, *Smokestack Diplomacy: Cooperation and Conflict in East-West Environmental Politics* (Cambridge, MA, 2001), 149–152.

17. Evgenii Velikhov, "Ia na sanochkakh poedu v 35 god," in *Vklad kurchatovstev v likvidatsiiu posledstvii avarii na Chernobyl'skoi AÈS* (Moscow, 2012), 71–72; Alexander Nazaryan, "The Russian Massive Radar Site in the Chernobyl Exclusion Zone," *Newsweek*, April 18, 2014.

18. Walter Mayr, "Chernobyl's Aftermath: The Pompeii of the Nuclear Age. Part 3: A Dramatic Increase in Birth Defects," *Spiegel International*, April 17, 2006; "Press-konferentsiia v Moskve," *Pravda*, May 11, 1986; "Soviets Gaining Control at Chernobyl, Panel Says," *Los Angeles Times*, May 11, 1986; B. Dubrovin, "Blagorodnye tseli," *Pravda*, May 27, 1986; Medvedev, *The Legacy of Chernobyl*, 68.

19. Georgii Arbatov, "Bumerang," *Pravda*, May 9, 1986.

20. "Vystuplenie M. S. Gorbacheva po sovetskomu televideniiu," *Pravda*, May 15, 1986.

21. Pavel Palazchenko, *My Years with Gorbachev and Shevardnadze* (University Park, PA, 1997), 49.

22. John Murray, *The Russian Press from Brezhnev to Yeltsin: Behind the Paper Curtain* (Cheltenham, UK, 1994); Evgenii Velikhov, "Ia na sanochkakh poedu v 35 god," 71–72.

23. "Gorbachev Willing to Continue Talks," *Observer-Reporter*, May 16, 1986.

24. "Doctors Predict Chernobyl Death Toll Will Climb," *Observer-Reporter*, May 16, 1986; "Doctor Foresees More Chernobyl Deaths," *Standard Daily*, May 16, 1986.

25. "Bone Marrow Specialist Returns to Moscow," *Los Angeles Times*, May 25, 1986; William J. Eaton, "Gale Says Toll at Chernobyl Increases to 23," *Los Angeles Times*, May 30, 1986; David Marples, *The Social Impact of the Chernobyl Disaster* (Edmonton, 1988), 34–35; Anne C. Roark, "Chernobyl 'Hero': Dr. Gale—Medical Maverick," *Los Angeles Times*, May 5, 1988.

26. Jack Nelson, "Reagan Criticizes Disaster Secrecy: Soviets 'Owe World an Explanation' for Chernobyl Blast, President Says," *Los Angeles Times*, May 4, 1986.

27. Christopher Jarmas, "Nuclear War: How the United States and the Soviet Union Fought over Information in Chernobyl's Aftermath," *Vestnik*, August 31, 2015, www.sras.org/information_chernobyl_us_ussr.

28. Ibid.; Philippe J. Sands, ed., *Chernobyl: Law and Communication. Transboundary Nuclear Air Pollution* (Cambridge, 1988), xxxvii.

CHAPTER 16: SARCOPHAGUS

1. "Ukrytie dlia reaktora," Intenet muzei "U Chernobyl'skoi cherty," http: //museum.kraschern.ru/razdely-muzeya/uchastie-krasnoyartsev/ukrytie -dlya-reaktora.php.

2. "Ukrytie dlia reaktora"; V. Gubarev, "Sovremennye piramidy: Ukrytie dlia zemlian," *Literaturnaia gazeta*, December 12, 2001; Iulii Safonov, "Sistema Slavskogo," *Zerkalo nedeli*, April 19, 1996.

3. Iulii Safonov, "Chernobyl': Desiatyi god tragedii," *Zerkalo nedeli*, November 24, 1995.

4. "Povidomlennia OH KDB URSR ta KDB SRSR u misti Chornobyli," July 4, 1986, *Z arkhiviv*, no. 51, 118–119.

5. "Materialy zasedanii operativnoi gruppy TsK Kompartii Ukrainy," July 5, 1986, TsDAHO, fond 1, op. 17, no. 386, fol. 110; N. P. Baranovskaia, *Ispytanie Chernobylem* (Kyiv, 2016), 40; Safonov, "Sistema Slavskogo"; "Povidomlennia OH KDB URSR ta KDB SRSR u misti Chornovyli," July 25, 1986, *Z arkhiviv*, no. 55, 124–125.

6. Zhores Medvedev, *The Legacy of Chernobyl* (New York, 1990), 178.

7. Valerii Legasov, "Ob avarii na Chernobyl'skoi AĖS," tape no. 1, Elektronnaluia biblioteka RoyalLib.Com, http://royallib.com/read/legasov_valeriy /ob_avarii_na_chernobilskoy_aes.html#0; Valentyn Fedulenko, "22 goda Chernobyl'skoi katastrofe," *Pripyat.com*, http://pripyat.com/articles/22-goda -chernobylskoi-katastrofe-memuary-uchastnika-i-mnenie-eksperta-chast-1.html.

8. Fedulenko, "22 goda Chernobyl'skoi katastrofe"; Valentin Zhil'tsov, in Iurii Shcherbak, *Chernobyl': Dokumental'noe povestvovanie* (Moscow, 1991), 181–186.

9. "Iz rabochei zapisi zasedaniia Politbiuro TsK KPSS, April 28, 1986," in R. G. Pikhoia, *Sovetskii Soiuz: Istoriia vlasti, 1945–1991* (Novosibirsk, 2000), 434.

10. Sergei Babakov, "S pred"iavlennymi mne obvineniiami ne soglasen . . . ," *Zerkalo nedeli*, August 29, 1999; Vladimir Shunevich, "Byvshii director ChAĖS Viktor Briukhanov: 'Kogda posle vzryva reaktora moia mama uznala,'" *Fakty*, December 1, 2010.

11. Aleksandr Iakovlev, *Sumerki* (Moscow, 2003), 388.

12. Galina Akkerman, "Gorbachev: Chernobyl' sdelal menia drugim chelovekom," *Novaia gazeta*, March 2, 2006; "Chernobyl' do vostrebovaniia," *Rossiiskaia gazeta*, April 25, 2016.

13. Minutes of the Politburo Meeting of July 3, 1986, Fond Gorbacheva, www.gorby.ru/userfiles/protokoly_politbyuro.pdf.

14. Minutes of the Politburo meeting of July 3, 1986, Fond Gorbacheva; Alla Iaroshinskaia, *Chernobyl': 20 let spustia. Prestuplenie bez nakazaniia* (Moscow, 2006), 444–452.

15. Iaroshinskaia, *Chernobyl': 20 let spustia*, 444–452; Minutes of the Politburo meeting of July 3, 1986, Fond Gorbacheva; Nikolai Karpan, *Chernobyl': Mest' mirnogo atoma* (Moscow, 2006), 393–396.

16. "Tainy Chernobyl'skoi katastrofy," *Ukraina kriminal'naia*, April 27, 2015, http://cripo.com.ua/?sect_id=2&aid=192439.

17. Minutes of the Politburo meeting of July 3, 1986, Fond Gorbacheva.

18. Minutes of the Politburo meeting of July 3, 1986, Fond Gorbacheva.

19. Walter Patterson, "Chernobyl: The Official Story," *Bulletin of the Atomic Scientists* 42, no. 9 (November 1986): 34–36; Stuart Diamond, "Experts in Vienna Outline New Plan for A-Plant Safety," *New York Times*, August 30, 1986.

20. "Informatsiia ob avarii na Chernobyl'skoi AĖS i ee posledstviiakh, podgotovlennaia dlia MAGATE," Institut atomnoi energii imeni I. V. Kurchatova, http://magate-1.narod.ru/4.html.

21. "Tainy Chernobyl'skoi katastrofy," *Ukraina kriminal'naia*, April 27, 2015.

22. Politburo meeting, October 2, 1986; Minutes of the Politburo meeting of July 3, 1986, Fond Gorbacheva.

23. Minutes of the Politburo meeting of July 3, 1986, Fond Gorbacheva.

24. "Ukrytie dlia reaktora," Internet muzei "U Chernobyl'skoi cherty"; Gubarev, "Sovremennye piramidy."

25. Artem Troitskii, *Ėnergetika strany i liudi iz vlasti: Vospominaniia, khronika, razmyshleniia* (Moscow, 2013), 155; Safonov, "Sistema Slavskogo."

26. Taras Shevchenko, "A Cherry Orchard by the House," translated by Boris Dralyuk and Roman Koropeckyj, *Ukrainian Literature* 4 (2004), http://sites.utoronto.ca/elul/Ukr_Lit.

27. "Slavskii, E. P. Proshchanie s sablei," documentary film, YouTube, published September 17, 2013, www.youtube.com/watch?v=bFGxtpRshHI; Vitalii Skliarov, *Zavtra byl Chernobyl'* (Moscow, 1993), 6–11.

CHAPTER 17: CRIME AND PUNISHMENT

1. Iulii Andreev, "Neschast'e akademika Legasova," *Lebed, Nezavisimyi al'manakh*, October 2, 2005; Mariia Vasil', "Familiiu akademika Legasova," *Fakty*, April 28, 2001.

2. Andreev, "Neschast'e akademika Legasova."

3. V. Legasov, V. Demin, and Ia. Shevelev, "Nuzhno li znat' meru v obespechenii bezopasnosti?" *Ėnergiia i ėkologiia* 4 (1984): 9–17.

4. Vasil', "Familiiu akademika Legasova."

5. Valerii Legasov, "Iz segodnia v zavtra: Mysli vslukh," *Chernobyl' i bezopasnost'* (St. Petersburg, 1998), 146; Vasil', "Familiiu akademika Legasova."

6. Valerii Legasov, "Ob avarii na Chernobyl'skoi AĖS," tape no. 1, Elektronnaluia biblioteka RoyalLib.Com, http://royallib.com/read/legasov_valeriy/ob_avarii_na_chernobilskoy_aes.html#0; Andreev, "Neschast'e akademika Legasova."

7. James Reason, *Managing the Risks of Organizational Accidents* (London, 1997), 15.

8. Vasil', "Familiiu akademika Legasova."

9. Elena Shmaraeva, "Radioaktivnyi protsess: 30 let nazad obviniaemykh po delu avarii na Chernobyl'skoi AĖS sudili priamo v zone otchuzhdeniia," Mediazona, *Deutsche Welle*, April 26, 2016, https://zona.media/article/2016/26/04 /chernobyl.

10. Nikolai Karpan, *Chernobyl': Mest' mirnogo atoma* (Moscow, 2006), 416–418.

11. V. Ia. Vozniak, *Ot Tiumeni do Chernobylia (zapiski Chernobyl'skogo ministra)* (Moscow, 2016), 130; Sergei Babakov, "S pred"iavlennymi mne obvineniiami ne soglasen . . . ," *Zerkalo nedeli*, August 29, 1999.

12. Svetlana Samodelova, "Lichnaia katastrofa direktora Chernobylia," *Moskovskii komsomolets*, April 21, 2011.

13. Samodelova, "Lichnaia katastrofa direktora Chernobylia."

14. Shmaraeva, "Radioaktivnyi protsess."

15. Babakov, "S pred"iavlennymi mne obvineniiami ne soglasen"; Anatolii Diatlov, *Chernobyl': Kak ėto bylo* (Moscow, 2003), chapter 9.

16. Karpan, *Chernobyl': Mest' mirnogo atoma*, 419; Shmaraeva, "Radioaktivnyi protsess"; "Interv'iu s Viktorom Briukhanovym," *ChAĖS: Zona otchuzhdeniia*, http://chernobil.info/?p=5898; Vladimir Shunevich, "Viktor Briukhanov: Iz partii menia iskliuchili priamo na zasedanii Politbiuro TsK KPSS," *Fakty*, July 7, 2012.

17. Karpan, *Chernobyl': Mest' mirnogo atoma*, 433.

18. Ibid., 444–457; Diatlov, *Chernobyl': Kak ėto bylo*, chapter 9.

19. Diatlov, *Chernobyl': Kak ėto bylo*, chapter 10; "Interv'iu s Viktorom Briukhanovym"; *Biulleten' Verkhovnogo suda SSSR* (Moscow, 1987), 20; Karpan, *Chernobyl': Mest' mirnogo atoma*, 499–508.

20. Babakov, "S pred"iavlennymi mne obvineniiami ne soglasen"; Vladimir Shunevich, "Byvshii direktor ChAĖS Viktor Briukhanov: 'Kogda posle vzryva reaktora moia mama uznala,'" *Fakty*, December 1, 2010; Mariia Vasil', "Byvshii direktor ChAĖS Viktor Briukhanov: 'Esli by nasli dlia menia rasstrel'nuiu stat'iu, to, dumaiu, menia rasstreliali by,'" *Fakty*, October 18, 2000.

21. Legasov, "Iz segodnia—v zavtra," *Pravda*, October 5, 1987.

22. Vasil', "Familiiu akademika Legasova."

CHAPTER 18: WRITERS' BLOCK

1. Yurii Mushketyk, first secretary of the Ukrainian Writers' Union, to the Central Committee of the Communist Party of Ukraine, January 20, 1988, TsDAHO, fond 1, op. 32, no. 2455, fols. 3–4.

2. Stepan Mukha, head of the Ukrainian KGB, to Volodymyr Shcherbytsky, first secretary of the Central Committee of the Communist Party of Ukraine, "Ob operativnoi obstanovke v sviazi s avariei na Chernobyl'skoi AĖS," June 2, 1986, 4, 5. Archive SBU, fond 16, op. 1, no. 1113.

3. Stepan Mukha, head of the Ukrainian KGB, to Volodymyr Shcherbytsky,

first secretary of the Central Committee of the Communist Party of Ukraine, "O prebyvanii korrespondentov SShA," November 20, 1986, 1, 3, Archive SBU, fond 16, op. 1, no. 1114; Mike Edwards, photographs by Steve Raymer, "Ukraine," *National Geographic* 171, no. 5 (May 1987): 595–631; Mike Edwards, photographs by Steve Raymer, "Chernobyl—One Year After," *National Geographic* 171, no. 5 (May 1987): 632–653.

4. "Ales' Adamovich predskazal strashnye posledstviia Chernobylia i spas Belarus' ot iadernykh boegolovok," *TUT.BY*, April 26, 2007, http://news.tut.by /society/86832.html.

5. Oles' Honchar, *Sobor* (Kyiv, 1968), 14–15; Roman Solchanyk, "Introduction," in *Ukraine: From Chernobyl to Sovereignty* (New York, 1992), xiii.

6. Honchar, *Shchodennyky (1984–1995)* (Kyiv, 2004), 99, 107; M. P. Vozna, "Ekolohichni motyvy v 'Shchodennykakh' Olesia Honchara," *Tekhnolohiï i tekhnika drukarstva*, no. 3 (2006): 136–145.

7. *Mystetstvo Ukraïny: Bibliohrafichnyi dovidnyk*, ed. A. V. Kudryts'kyi (Kyiv, 1997), 357; Oleksandr Levada, *Zdrastui Prypiat'* (Kyiv, 1974), 69.

8. Levada, *Zdrastui Prypiat'*, 56.

9. Ivan Drach, *Korin' i krona* (Kyiv, 1976), 27–31.

10. Ivan Drach, "Chornobyl's'ka madonna," *Vitchyzna*, no. 1 (1988): 46–62; Larissa M. L. Zaleska-Onyshkevych, "Echoes of Glasnost: Chornobyl in Soviet Ukrainian Literature," in *Echoes of Glasnost in Soviet Ukraine*, ed. Romana Bahry (North York, Ontario, 1989), 151–170.

11. Volodymyr Lytvyn, *Politychna arena Ukraïny: Diiovi osoby ta vykonavtsi* (Kyiv, 1994), 110–111; Ivan Drach, *Polityka: Statti, dopovidi, vystupy, interv'iu* (Kyiv, 1997), 334.

12. Borys Oliinyk's address, in *Vsesoiuznaia konferentsiia Kommunisticheskoi partii Sovetskogo Soiuza, 28 iiunia–1 iiulia 1988 g. Stenograficheskii otchet v dvukh tomakh* (Moscow, 1988), vol. 2, 31.

13. Stepan Mukha, head of the Ukrainian KGB, to the Ukrainian Central Committee, "Informatsionnoe soobshchenie za 17 iiulia 1986 g.," p. 3, Archive SBU, fond 16, op. 1, no. 1114; Nikolai Galushko, head of the Ukrainian KGB, to the Ukrainian Central Committee, "Informatsionnoe soobshchenie za 29 iiunia 1987 g.," p. 2, Archive SBU, fond 16, op. 1, no. 1117.

14. Iurii Shcherbak, *Chernobyl: A Documentary Story*, foreword by David R. Marples (Edmonton, 1989); Lytvyn, *Politychna arena Ukraïny*, 182.

15. Stepan Mukha, head of the Ukrainian KGB, to Volodymyr Shcherbytsky, first secretary of the Central Committee of the Communist Party of Ukraine, "Informatsionnoe soobshchenie," April 16, 1987; "Obrashchenie v TsK KPSS, Prezidium Verkhovnoho Soveta SSR, Ministerstvo atomnoi énergetiki SSSR i gazetu 'Pravda,'" Archive SBU, fond 16, op. 1, no. 1116.

16. N. Galushko, I. Gladush, and P. Osipenko to Volodymyr Shcherbytsky, "O gotoviashcheisia aktivistami t.n. 'ukrainskogo kul'turologicheskoho kluba' antiobshchestvennoi aktsii," April 25, 1988, Archive SBU, fond 16, op. 1, no.

1119; Oles Shevchenko's statement at the December 11, 1991, session of the Supreme Soviet of Ukraine, www.rada.gov.ua/zakon/skl1/BUL14/111291_46.htm.

17. David R. Marples, *Ukraine Under Perestroika: Ecology, Economics and the Workers' Revolt* (New York, 1991), 137–141.

18. Nikolai Galushko, head of the Ukrainian KGB, to the Ukrainian Central Committee, "O sostoiavshemsia v g. Kieve mitinge po problemam ėkologii," November 14, 1988, Archive SBU, fond 16, op. 1, no. 1120.

19. Marples, *Ukraine Under Perestroika*, 141–142; V. V. Ovsiienko, "Makar, Ivan Ivanovych," in *Dysydents'kyi rukh Ukraïny: Virtual'nyi muzei*, http://archive .khpg.org/index.php?id=1184058826; Ihor Mel'nyk, "Pershyi mitynh u L'vovi: Spohady ochevydtsia," *Zbruch*, June 13, 2013.

20. Nikolai Galushko, head of the Ukrainian KGB, to Volodymyr Shcherbytsky, first secretary of the Ukrainian Central Committee, "O sozdanii initsiativnoi gruppy v podderzhku perestroiki v Soiuze pisatelei Ukrainy," November, 24, 1988, Archive SBU, fond 16, op. 1, no. 1120; Drach, *Polityka*, 334.

21. Volodymyr Shcherbytsky to the Central Committee in Moscow, January 7, 1989, TsDAHO, fond 1, op. 32, no. 2671, fols. 1–3.

CHAPTER 19: NUCLEAR REVOLT

1. "Vspominaia Chernobyl'skuiu katastrofu," *NewsInPhoto*, March 19, 2011, http://newsinphoto.ru/texnologii/vspominaya-chernobylskuyu-katastrofu.

2. Nikolai Galushko, head of the Ukrainian KGB, to Volodymyr Shcherbytsky, first secretary of the Ukrainian Central Committee, "O nekotorykh problemakh likvidatsii posledstvii avarii na Chernobyl'skoi AĖS," December 6, 1988, Archive SBU, fond 16, op. 1, no. 1120.

3. Archie Brown, *The Gorbachev Factor* (Oxford, 1997); Chris Miller, *The Struggle to Save the Soviet Economy: Mikhail Gorbachev and the Collapse of the USSR* (Chapel Hill, NC, 2016).

4. Vakhtang Kipiani and Vladimir Fedorin, "Shcherbitskii skazal—kakoi durak pridumal slovo 'perestroika'?" *Ukraïns'ka pravda*, September 11, 2011.

5. Nikolai Galushko, head of the Ukrainian KGB, to Volodymyr Shcherbytsky, first secretary of the Ukrainian Central Committee, "Ob otklikakh na vstrechu general'nogo sekretaria TsK KPSS s gruppoi pisatelei," February 27, 1989, Archive SBU, fond 16, op. 1, no. 1122; Oleksii Haran', *Vid stvorennia Rukhu do bahatopartiinosti* (Kyiv, 1992).

6. "Prohrama narodnoho Rukhu Ukraïny za perebudovu," *Literaturna Ukraïna*, no. 7 (February 16, 1989); "Program of the Popular Movement for Restructuring of Ukraine," in *Toward an Intellectual History of Ukraine: An Anthology of Ukrainian Thought from 1710 to 1995*, eds. Ralph Lindheim and George S. N. Luckyj (Toronto, 1996), 353–354.

7. Paul Josephson, Nicolai Dronin, Ruben Mnatsakanian, Aleh Cherp, Dmitry Efremenko, and Vladislav Larin, *The Environmental History of Russia* (Cambridge, 2013), 274–284.

8. David R. Marples, *Ukraine Under Perestroika: Ecology, Economics and the Workers' Revolt* (New York, 1991), 155.

9. Alla Iaroshinskaia, *Bosikom po bitomu steklu: Vospominaniia, dnevniki, dokumenty*, vol. 1 (Zhytomyr, 2010); Vakhtang Kipiani, "Yaroshyns'ka, shcho ty robysh u Narodyts'komu raioni?" *Ukraïns'ka pravda*, April 29, 2006.

10. Alla Iaroshinskaia, *Chernobyl': Bol'shaia lozh'* (Moscow, 2011), 1–40. Cf. Alla A. Yaroshinskaya, *Chernobyl: Crime Without Punishment* (New Brunswick, NJ, 2011), 1–23.

11. Vitalii Karpenko's statement at the December 11, 1991, session of the Ukrainian parliament, www.rada.gov.ua/zakon/skl1/BUL14/111291_46.htm; Volodymyr Yavorivsky's report to the Ukrainian parliament, December 11, 1991, in ibid.; Yaroshinskaya, *Chernobyl: Crime Without Punishment*, 46–47.

12. Yaroshinskaya, *Chernobyl: Crime Without Punishment*, 25.

13. Ales' Adamovich, "Chestnoe slovo, bol'she ne vzorvetsia, ili mnenie nespetstialista—otzyvy spetsialistov," *Novyi mir*, no. 9 (1988): 164–179; "Ales' Adamovich predskazal strashnye posledstviia Chernobylia i spas Belarus' ot iadernykh boegolovok," *TUT.BY*, April 26, 2007, http://news.tut.by/society/86832.html.

14. Marples, *Ukraine Under Perestroika*, 50–52.

15. Yaroshinskaya, *Chernobyl: Crime Without Punishment*, 32–45; Yaroshinskaia, *Bosikom po bitomu steklu*, vol. 2, 7–55; Nikolai Galushko, head of the Ukrainian KGB, to Volodymyr Shcherbytsky, first secretary of the Ukrainian Central Committee, "Ob obstanovke v Narodicheskom raione Zhitomirskoi oblasti," June 16, 1990, Archive SBU, fond 16, op. 1, no. 1123.

16. David Marples, *Belarus: From Soviet Rule to Nuclear Catastrophe* (London, 1996), 121–122.

17. Jane I. Dawson, *Eco-Nationalism: Anti-Nuclear Activism and National Identity in Russia, Lithuania, and Ukraine* (Durham, NC, 1996), 59–60.

18. Volodymyr Lytvyn, *Politychna arena Ukraïny: Diiovi osoby ta vykonavtsi* (Kyiv, 1994), 201–208.

19. Nikolai Galushko, head of the Ukrainian KGB, to Volodymyr Ivashko, first secretary of the Ukrainian Central Committee, "O protsessakh, sviazannykh so stroitel'stvom i ekspluatatsiei AĖS v respublike," April 29, 1990, Archive SBU, fond 16, op. 1, no. 1125; Nikolai Galushko, head of the Ukrainian KGB, to Vitalii Masol, head of the Council of Ministers of Ukraine, "O neblagopoluchnoi obstanovke skladyvaiushcheisia vokrug Khmel'nitskoi AĖS," May 11, 1990, in ibid., no. 1126.

20. H. I. Honcharuk, *Narodnyi rukh Ukraïny: Istoriia* (Odesa, 1997); D. Efremenko, "Eco-nationalism and the Crisis of Soviet Empire (1986–1991)," *Irish Slavonic Studies* 24 (2012): 17–20.

CHAPTER 20: INDEPENDENT ATOM

1. "1991 rik: Pershi dni usvidomlennia. Fotoreportazh," *UkrInform*, August 23, 2013, www.ukrinform.ua/rubric-other_news/1536034-1991_rik_pershi _dni_usvidomlennya_fotoreportag_1856500.html.

2. Serhii Plokhy, *The Last Empire: The Final Days of the Soviet Union* (New York, 2014), 73–151.

3. "Verkhovna rada Ukraïny, Stenohrama plenarnoho zasidannia," August 24, 1991, http://iportal.rada.gov.ua/meeting/stenogr/show/4595.html.

4. "Chomu akt proholoshennia nezalezhnosti Ukraïny zachytav komunist," *Hromads'ke radio*, August 25, 2016, https://hromadskeradio.org/programs/kyiv -donbas/chomu-akt-progoloshennya-nezalezhnosti-ukrayiny-zachytav-komunist; author's interview with Leonid Kravchuk, November 21, 2016, Harvard University.

5. Volodymyr Iavorivs'kyi, *Mariia z polynom u kintsi stolittia* (Kyiv, 1988) (Journal publication, 1987); Oles' Honchar, *Lysty* (Kyiv, 2008), 301.

6. Nikolai Galushko, head of the Ukrainian KGB, to Volodymyr Yavorivsky, head of the Chernobyl Commission of the Ukrainian parliament, "O nekotorykh problemakh likvidatsii posledstvii avarii na Chernobyl'skoi AĖS," May 23, 1991, 5, Archive SBU, fond 16, op. 1, no. 1129; Volodymyr Iavorivs'kyi, "Usi my zhertvy i vynuvattsi katastrofy," *Oikumena*, no. 2 (1991); Volodymyr Iavorivs'kyi, "Pravda Chornobylia: kolo pershe," *Oikumena*, no. 5 (1991); Volodymyr Iavorivs'kyi, "Khto zapalyv zoriu Polyn?" *Nauka i suspil'stvo*, no. 9 (1991).

7. "Briukhanov—menia privezli k mestu predpolagaemogo stroitel'stva," *Pripiat.com*, http://pripyat.com/people-and-fates/bryukhanov-menya-privezli -k-mestu-predpolagaemogo-stroitelstva-les-pole-i-snegu-po-.

8. Vladimir Shunevich, "Viktor Briukhanov: Iz partii menia iskliuchili priamo na zasedanii Politbiuro TsK KPSS," *Fakty*, July 7, 2012; Anatolii Diatlov, *Chernobyl': Kak ėto bylo* (Moscow, 2003), chapter 5.

9. *INSAG-7: Chernobyl'skaia avariia. Dopolnenie k INSAG-1. Doklad mezhdunarodnoi konsul'tativnoi gruppy po iadernoi bezopasnosti* (Vienna, 1993).

10. N. P. Baranovskaia, *Ispytanie Chernobylem* (Kyiv, 2016), 221–222. Aleksandr Liashko, *Gruz pamiati: Vospominaniia*, vol. 3, *Na stupeniakh vlasti*, part 2 (Kyiv, 2001), 436, 439–440.

11. "Soglashenie o sozdanii Sodruzhestva nezavisimykh gosudarstv," in *Rspad SSSR: Dokumenty i fakty (1986–1992)*, vol. 1, ed. Sergei Shakhrai (Moscow, 2009), 1028–1031; Plokhy, *The Last Empire*, 295–387.

12. "Verkhovna rada Ukraïny, Stenohrama plenarnoho zasidannia," December 11, 1991, http://static.rada.gov.ua/zakon/skl1/BUL14/111291_46.htm.

13. Liashko, *Gruz pamiati*, vol. 3, part 2, 442–454.

14. "Ugolovnoe delo protiv rukovoditelei Ukrainy: Chernobyl'skaia avariia. Chast' 4," *Khroniki i kommentarii*; April 21, 2011, https://operkor.wordpress .com/2011/04/21; Alla Iaroshinskaia, *Chernobyl' 20 let spustia. Prestuplenie bez*

nakazaniia (Moscow, 2006), 464–492; Natalia Baranovs'ka, "Arkhivni dzherela vyvchennia Chornobyl's'koï katastrofy," *Arkhivy Ukrainy*, nos. 1–6 (2006): 170–184.

15. Pekka Sutella, "The Underachiever: The Ukrainian Economy Since 1991," Carnegie Endowment for International Peace, March 9, 2012, http://carnegieendowment.org/2012/03/09/underachiever-ukraine-s-economy-since -1991#.

16. Volodymyr Iavorivs'kyi, "Same z kniahyni Ol'hy ia b pochynav istoriiu Ukraïny," *Vechirnii Kyiv*, May 2, 2016; Baranovskaia, *Ispytanie Chernobylem*, 185–192; Adriana Petryna, "Chernobyl's Survivors: Paralyzed by Fatalism or Overlooked by Science?" *Bulletin of the Atomic Scientists* (2011); Adriana Petryna, *Life Exposed: Biological Citizens After Chernobyl* (Princeton, NJ, 2002), 4, 23–25.

17. V. P. Udovychenko, "Ukraina—svit—Chornobyl: Problemy i perspektyvy," in *Naukovi ta tekhnichni aspekty mizhnarodnoho spivrobitnytstva v Chornobyli*, vol. 3 (Kyiv, 2001), 664–665.

CHAPTER 21: GLOBAL SHELTER

1. Yuri Shcherbak, *The Strategic Role of Ukraine: Diplomatic Addresses and Lectures* (Cambridge, MA, 1998); Serhii Plokhy, *The Last Empire: The Final Days of the Soviet Union* (New York, 2014), 175–179.

2. Bohdan Harasymiw, *Post-Communist Ukraine* (Edmonton, 2002), Taras Kuzio, *Ukraine: State and Nation Building* (London, 1998); Kataryna Wolczuk, *The Molding of Ukraine: The Constitutional Politics of State Formation* (Budapest, 2001); Serhii Plokhy, *The Gates of Europe: A History of Ukraine* (New York, 2015), 323–336.

3. Iurii Kostenko, *Istoriia iadernoho rozzbroiennia Ukraïny* (Kyiv, 2015), 369–399; Steven Greenhouse, "Ukraine Votes to Become a Nuclear-Free Country," *New York Times*, November 17, 1994; Khristina Lew, "Ukraine's President Arrives for State Visit in the U.S.: U.S. Promises Additional $200 Million in Assistance," *Ukrainian Weekly*, November 27, 1994, 1; "For the Record: President Clinton's Remarks Welcoming President Kuchma," in ibid., 3; "Ukraine, Nuclear Weapons and Security Assurances at a Glance," Arms Control Association, https://www.armscontrol.org/factsheets/Ukraine-Nuclear-Weapons.

4. "Budapest Memorandum on Security Assurances," Council on Foreign Relations, December 5, 1994, www.cfr.org/nonproliferation-arms-control-and -disarmament/budapest-memorandums-security-assurances-1994/p32484; "Ukrainian Parliament Appeals to the Budapest Memorandum Signatories," *Interfax Ukraine*, February 28, 2014, http://en.interfax.com.ua/news/general /193360.html; Editorial Board, "Condemnation Isn't Enough for Russian Actions in Crimea," *Washington Post*, February 28, 2014; Thomas D. Grant, "The Budapest Memorandum and Beyond: Have the Western Parties Breached a Legal Obligation?" *European Journal of International Law*, February 18, 2015,

www.ejiltalk.org/the-budapest-memorandum-and-beyond-have-the-western
-parties-breached-a-legal-obligation.

5. "Joint Summit Statement by President Clinton and President of Ukraine Leonid D. Kuchma," White House, Office of the Press Secretary, November 22, 1994, http://fas.org/spp/starwars/offdocs/j941122.htm.

6. Robert G. Darst, *Smokestack Diplomacy: Cooperation and Conflict in East-West Environmental Politics* (Cambridge, MA, 2001), 164–167.

7. Ibid., 177; "Ukraine: Chernobyl Plant Could Be Closed Down," Associated Press, April 13, 1995, www.aparchive.com/metadata/youtube/f4d94438a 4ca1ea9d078a2472ea6612e; Marta Kolomayets, "Ukraine to Shut Down Chornobyl by 2000," *Ukrainian Weekly*, April 16, 1995, 1, 4.

8. Adriana Petryna, *Life Exposed: Biological Citizens After Chernobyl* (Princeton, NJ, 2002), 92–93.

9. "Halifax G-7 Summit Communiqué," June 16, 1995, www.g8.utoronto .ca/summit/1995halifax/communique/index.html; "Chernobyl Closure Agreed, But Who Foots the Bill?" *Moscow Times*, April 15, 1995; "Talks Open on Pulling Plug on Plant," Reuters, November 2, 1955.

10. "Memorandum of Understanding Between the Governments of the G-7 Countries and the Commission of the European Communities and the Government of Ukraine on the Closure of the Chernobyl Nuclear Plant," University of South Carolina Research Computing Facility, http://www-bcf.usc .edu/~meshkati/G7.html.

11. Darst, *Smokestack Diplomacy*, 179–180.

12. Svetlana Samodelova, "Lichnaia katastrofa direktora Chernobylia," *Moskovskii komsomolets*, April 21, 2011.

13. Darst, *Smokestack Diplomacy*, 181–183; "Nuclear Power in Ukraine," World Nuclear Association, October 2016, www.world-nuclear.org/information -library/country-profiles/countries-t-z/ukraine.aspx; "EBRD Approves K2R4 Loan—Campaign Continues," *Nuclear Monitor* 540 (December 15, 2000), www.wiseinternational.org/nuclear-monitor/540/ebrd-approves-k2r4 -loan-campaign-continues.

14. "Chernobyl'skaia AÉS: Desiat' let posle chasa 'Ch,'" Gorod.cn.ua, December 20, 2010, www.gorod.cn.ua/news/gorod-i-region/22424-chernobylskaja-aes -desjat-let-posle-chasa-ch.html.

15. David Marples, "Nuclear Power Development in Ukraine: Déjà Vu?" *New Eastern Europe*, November 14, 2016, www.neweasterneurope.eu /articles-and-commentary/2186-nuclear-power-development-in-ukraine-deja-vu.

16. Paul Josephson, Nicolai Dronin, Ruben Mnatsakanian, Aleh Cherp, Dmitry Efremenko, and Vladislav Larin, *An Environmental History of Russia* (Cambridge, 2013), 267; "Health Effects of the Chernobyl Accident: An Overview," World Health Organization, April 2006, www.who.int/ionizing_radiation /chernobyl/backgrounder/en; Adriana Petryna, "Chernobyl's Survivors: Paralyzed by Fatalism or Overlooked by Science?" *Bulletin of the Atomic Scientists*

(2011); Keith Baverstock and Dillwyn Williams, "The Chernobyl Accident 20 Years On: An Assessment of the Health Consequences and the International Response," *Environmental Health Perspectives* 114 (September 2016): 1312–1317; Marples, "Nuclear Power Development in Ukraine: Déjà Vu?"

17. David Marples, *Belarus: From Soviet Rule to Nuclear Catastrophe* (London, 1996), 46–52; Josephson et al., *An Environmental History of Russia*, 263–266.

18. Darst, *Smokestack Diplomacy*, 179; "NOVARKA and Chernobyl Project Management Unit Confirm Cost and Time Schedule for Chernobyl New Safe Confinement," European Bank for Reconstruction and Development, April 8, 2011, http://archive.li/w2pVU; "Chernobyl Confinement Reaches Final Stage, But Funds Need Boost," *World Nuclear News*, March 17, 2015; "Chernobyl Donor Conference Raises Extra $200 Million for New Safe Confinement Project," *Russia Today*, April 30, 2015, https://www.rt.com/news/254329-chernobyl-sarcophagus-project-funding.

19. Darst, *Smokestack Diplomacy*, 135.

20. Marples, "Nuclear Power Development in Ukraine: Déjà Vu?"

21. Hartmut Lehmann and Jonathan Wadsworth, "Chernobyl: The Long-Term Health and Economic Consequences," Centre Piece Summer, 2011, http://cep.lse.ac.uk/pubs/download/cp342.pdf; Marc Lallanilla, "Chernobyl: Facts About the Nuclear Disaster," LiveScience, September 25, 2013.

22. Alexis Madrigal, "Chernobyl Exclusion Zone Radioactive Longer Than Expected," *Wired*, December 15, 2009, https://www.wired.com/2009/12/chernobyl-soil; Serhii Plokhy, "Chornobyl: A Tombstone of the Reckless Empire," Harvard Ukrainian Research Institute, April 21, 2016, www.huri.harvard.edu/news/news-from-huri/248-chornobyl-tombstone-of-reckless-empire.html.

EPILOGUE

1. Alexander J. Motyl, "Decommunizing Ukraine," *Foreign Affairs*, April 28, 2015.

2. David Lochbaum, Edwin Lyman, Susan Q. Stranahan, and the Union of Concerned Scientists, *Fukushima: The Story of a Nuclear Disaster* (New York, 2014).

3. James Conca, "Bill Gates Marking Progress on Next Generation of Nuclear Power—in China," *Forbes*, October 2, 2015.

INDEX

Susan Wilson

SERHII PLOKHY is the Mykhailo Hrushevsky Professor of Ukrainian History and director of the Ukrainian Research Institute at Harvard University. The award-winning author of *Lost Kingdom*, *The Gates of Europe*, and *The Last Empire*, among many other works, Plokhy lives in Arlington, Massachusetts.